ALGAE: NUTRITION, POLLUTION CONTROL AND ENERGY SOURCES

ALGAE: NUTRITION, POLLUTION CONTROL AND ENERGY SOURCES

KRISTIAN N. HAGEN
EDITOR

Nova Science Publishers, Inc.
New York

For permission to use material from this book please contact us:
Telephone 631-231-7269; Fax 631-231-8175
Web Site: http://www.novapublishers.com

LIBRARY OF CONGRESS CATALOGING-IN-PUBLICATION DATA

Algae : nutrition, pollution, control, and energy sources / Kristian N. Hagen (editor).
p. cm.
ISBN 978-1-60692-008-4 (hardcover)
1. Algae. 2. Algae--Biotechnology. 3. Algae--Control. 4. Algae as food. I. Hagen, Kristian N.
QK564.A44 2008
579.8'16--dc22

2008032285

Published by Nova Science Publishers, Inc. ✣ New York

CONTENTS

PREFACE

Algae are photosynthetic organisms that occur in most habitats, ranging from marine and freshwater to desert sands and from hot boiling springs to snow and ice. They vary from small, single-celled forms to complex multicellular forms, such as the giant kelps of the eastern Pacific that grow to more than 60 meters in length and form dense marine forests. Algae are found in the fossil record dating back to approximately 3 billion years in the Precambrian. They exhibit a wide range of reproductive strategies, from simple, asexual cell division to complex forms of sexual reproduction.

Algae are important as primary producers of organic matter at the base of the food chain. They also provide oxygen for other aquatic life. Algae may contribute to mass mortality of other organisms, in cases of algal blooms, but they also contribute to economic well- being in the form of food, medicine and other products. In tropical regions, coralline algae can be as important as corals in the formation of reefs.

Seaweeds are larger algae that live in the marine (salt or brackish water) environment. Kelps are large brown seaweeds in the genera Pelagophycus, Laminaria, Macrocystis, etc. In the Pacific, individual kelp plants may reach 65 meters in length.

Marine algae, as primary producers, are ecologically important, and economically have been used as food and medicines for centuries. Today, various species of marine algae provide not only food but also produce extracts such as agar, carrageenans, and alginates. These extracts are used in numerous food, pharmaceutical, cosmetic, and industrial applications.

This new book presents the latest research from around the world in this increasingly important field.

In the early history of the Earth, atmospheric oxygen was built up by photosynthetic microalgae. These prokaryotic and eukaryotic microorganisms use sunlight energy to drive the synthesis of organic molecules from CO_2 and water, and liberate oxygen as a by-product. The remarkable adaptation capacity of unicellular algae to produce oxygen in different extreme environments can used in various fields of constructive human endeavor. Numerous innovative solutions have been proposed to utilize photosynthetic microorganisms in pollution control to treat oxygenated wastewater, to regenerate the atmosphere in the enclosed weightlessness environment in space and in medicine to develop an extracorporeal bioartificial lung and oxygenated implanted bioartificial pancreas.

Chapter 1 focuses mainly on the various methodologies employed recently in the development of microalga-based photosynthetic oxygen generators for such different areas as pollution control, space life support and medicine. Progress in these fields require a multidisciplinary approach to explore and optimize the interaction between components of the biological system and different technological processes. The authors discuss their research findings related to the application of microalgae for oxygen delivery to implantable bioartificial pancreas, emphasizing the importance of the close cooperation of multidisciplinary teams engaged in the development of such bio-hybrid medical devices. Special attention is paid to the advantages of microalgae, compared with other technologies for oxygenation, to serve as a safe, cheap and smart tool for oxygen supply.

One of the proposed mechanisms for metal tolerance in bacteria and algae is the sequestration of metal cations by long-chain polymers of inorganic polyphosphate. The highly metal-tolerant alga *Chlamydomonas acidophila* KT-1 (KT-1) accumulates within the cell large amounts of polyphosphate that are degraded in response to cadmium (Cd) stress. Chapter 2 investigates the mechanism of action for this important phosphate-based metal-detoxifying function of KT-1. After shifting KT-1 cells from a Cd-free to a Cd-plus medium, high molecular weight forms of polyphosphate were degraded and very low molecular weight polyphosphate disappeared from the cells. Coincidently, the total intracellular phosphate decreased to only 10% of control levels, and subsequent phosphate uptake from the medium was markedly enhanced. Phosphate limitation causes a similar effect, but the authors suspect that Cd stress and phosphate deprivation may have different molecular mechanisms in KT-1 cells. Upon shifting from a Cd-plus to a Cd-free medium, intracellular Cd decreased by 43%, total intracellular phosphate increased nine fold. The authors propose a two-step mechanism involving the chelation and extracellular transport of Cd to explain the important metal detoxifying effects of polyphosphate in KT-1.

The biosorption of Pb^{2+} and Cd^{2+} by the organic components (exopolymers, nonliving cells, living cells, exopolysaccharides and extracellular proteins of the exopolymers) of the dominant microorganisms obtained from the natural biofilms of the South Lake of Changchun, China and effects of biosorption time, temperature and initial pH value were investigated in Chapter 3.

The results showed that Pb^{2+} and Cd^{2+} biosorption process by the exopolymers, the nonliving cells and the living cells was divided into fast phase and slow phase, biosorption equilibrium time was 360, 100 and 100 min respectively; the maximum biosorption amounts were all obtained at pH=6; For the exopolymers, the biosorption amounts increased with the temperature increasing in the range of 15-30°C and then decreased, the effect of temperature on the biosorption ability of the nonliving cells and the living cells was negligible; both Langmuir and Freundlich isotherms could describe the course of the biosorption of Pb^{2+} and Cd^{2+}. The biosorption ability of the exopolymers is much higher than that of the nonliving cells or the living cells.

The results of the biosorption of Pb^{2+} and Cd^{2+} by the exopolysaccharides and extracellular proteins showed that both Langmuir and Freundlich isotherms could also be used to describe the thermodynamics adsorption processes. Temperature and pH value could also affect the adsorption processes. The biosorption ability of the exopolysaccharides is higher than that of the extracellular proteins.

Several green algal groups are known to tolerate transient desiccation in their vegetative state. This appears to be a substantial advantage when spreading to new habitats. Only recently the physiology of algae form arid habitats or man-made surfaces has been investigated extensively. The phylogenetic relationships of desiccation tolerant green algae have been determined at a molecular level, and substantial progress has been made in understanding avoidance or damage repair strategies. In addition to physiological adaptations, structural conditions like thickened cell walls covered with or containing substances that reduce evaporation and prevent desiccation are an important strategy in these organisms. Very limited information is available on the structure and ultrastructure of desiccated vegetative algal samples. This however appears to be important, when attempting to understand how a quick recovery from desiccation stress is achieved. In Chapter 4 transmission electron microscopic data will demonstrate possible effects of desiccation in algal samples. Moreover it is demonstrated that in desiccation tolerant algae organelles like the chloroplast have the capacity to remain structurally integer during desiccation. These are new results that contribute to our understanding of desiccation tolerance in green algae.

As presented in Chapter 5, algae are present in most kinds of environments. They are adapted to saline and freshwater even to wetland. As they mostly live in an environment connected with water where concentration of essential nutrients is usually low, they had to develop mechanisms to concentrate nutrients in their body. For that reason they can accumulate several inorganic ions up to high levels. This capability can be used in the processes of pollutant removal. Metal ions represent an important group of hazardous contaminants in industrial wastewater, and algae can be effectively used to remove them. One of the possibilities represents bioaccumulation when metals are accumulated inside the cells as the result of metabolic activity. The other possibility is to use the ability of algae to adsorb metals onto their cell surface in the process called biosorption. In some situations the enzymatic apparatus of algae can cause the precipitation of metals such as gold or silver present in the environment. But the use of algae for the degradation of pollutants is much wider. They are able to decrease concentration of nutrients, especially nitrogen in wastewater. They assimilate NH_4^+ ions produced during the biodegradation of organic compounds and decrease their concentration more than twice compared with the biodegradation without algae. Their role in organic removal is also well known. They were applied to the dye concentration reduction in wastewater, olive oil mill wastewater treatment, water from paper industry treatment, etc. Because of their high sensitivity to high concentrations of pollutants, they are also used in toxicity tests or as sensitive bioindicators of ecological changes. It is obvious that algae have great potential in pollution control and in the processes applied to decrease the level of pollution in the environment.

Overgrowth of cyanobacteria has caused great serious water environment problems and threatens the drinking water supply worldwide. Many methods have been researched and practiced for its control. Chapter 6 investigates thoroughly the ultrasonic control of cyanobacteria growth in water, the ultrasonic enhancement of algal cell removal via coagulation, and the degradation of algal toxins by sonication. The paper also examines the important operational parameters and aqueous matrix factors.

The results showed that proper ultrasonic irradiation could effectively control the growth of some cyanobacteria species. Gas vacuole in the algal cells can be easily destroyed during ultrasonic irradiation by acting as the 'nuclei' for acoustic cavitation and collapse during the "bubble crush" period, which results in the cyanobacteria settlement and growth inhibition. Sonication can also instantly decrease the antenna complexes like chlorophyll *a* and phycocyanins. Direct damage on cell surface or even cell fracturing was also observed. But the dominant mechanism is the acoustic cavitation of gas vacuole. The same mechanism also explains the enhancement of coagulation by sonication. As a result, cyanobacteria with gas vacuoles can be easily controlled using ultrasonic waves, but algae without gas vacuoles are virtually immune to sonication.

On the other hand, the reason for ultrasonic degradation of algal toxins is because of the chemical changes caused by sonication, especially the formation of hydroxyl free radicals in water. Ultrasonic frequency, intensity, and duration play important roles in the algal growth control and ultrasonic-coagulation removal, and ultrasonic toxin degradation.

As explained in Chapter 7, the CO_2 is the substrate for the carboxylation reaction of the Rubisco that is the key enzyme implied in production of organic carbon from CO_2 by the primary producers. In the present ocean, the CO_2 concentration is not enough to saturate the Rubisco carboxylation rates. However, an increase of the CO_2 concentration by 50% with respect to the current value is expected by the year 2100, due to the acidification caused by the emission of CO_2 from the fossil fuel burning. Therefore, it could be expected that the acidification will alter the photosynthesis rates in the ocean. This hypothesis is examined by revising the results of laboratory experiments in which the effects of changes in CO_2 concentration have been researched. Most of alga taxonomic groups feature photosynthesis rates almost non-sensitive to short-term changes in the CO_2 concentration due to they have developed mechanisms for using HCO_3^- for photosynthesis (whose concentration is one magnitude order higher than CO_2) which permit to increase the CO_2 concentration around Rubisco (the so-called carbon concentrating mechanisms, CCM). Limitation of the photosynthesis rates by the CO_2 concentration (other resources being non-limited) in air-equilibrated seawater has been only described in a few alga species, including coccolithophorids (a phytoplankton group that episodically produces blooms in vast areas of the ocean) and some species of benthic brown and red algae. These latter species are shade plants whose growth at their natural habitat is primary limited by the light intensity. In spite of the photosynthesis saturation at the actual CO_2 concentration for many species, the growth at high CO_2 induces an acclimation response that consists of down-regulation of the CCMs. Furthermore, effects of the high CO_2 on photosynthetic apparatus, growth rates and carbohydrates and protein contents have been described, although this response is un-uniform among the alga species examined. This finding suggests that acidification could induce changes in the taxonomic composition of the communities. However, the few experiments of high CO_2 performed with natural assemblages of phytoplankton do not demonstrate substantial changes in the community structure or the succession patterns during the development of alga blooms.

Sulfated polysaccharides, in which at least some of the hydroxyl groups of the sugar residues are substituted by sulfate groups, are universal and characteristic ingredients of algae. So far, various types of polysaccharides have been isolated from algae and studied their chemical structures. Sulfated galactans including carrageenans and agaroids and sulfated fucans (fucoidans) are well known sulfated polysaccharides isolated from red algae and brown algae, respectively. Various types of sulfated polysaccharides have been isolated from green and blue-green algae. On the other hand, sulfated polysaccharides are well known to possess multiple biological activities such as anticoagulant, antiviral, antitumor, antioxidant and immunomodulating effects. Therefore, algal sulfated polysaccharides are promising candidates for developing novel pharmaceuticals. In addition, the third function of food has been attracted a great deal of attention to maintain human health, recently, therefore, it is possible to develop functional foods based on the biological potencies of sulfated polysaccharides. In Chapter 8, the authors will summarize the chemical characteristics and biological activities of sulfated polysaccharides from algae.

As presented in Chapter 9, marine algae traditionally serve as food and medicine in East-Asian countries as well as in Celtic areas, such as Scotland, Ireland, and Brittany. In recent years the use of seaweeds in cuisine and as health foods has been adopted in many European and North American countries. An increasing number of studies support anti-cancer properties of seaweed. Furthermore, it has been found to be a very nutritious food source containing large amounts of antioxidants, vitamins, and trace elements. For instance, seaweeds are known to comprise high quantities of iodine. Iodine concentrations vary widely among seaweed species, with geographic and seasonal variations as well as post-harvest storage conditions as further contributing factors.

In humans, iodine plays a crucial role in thyroid hormone synthesis. Thus, iodine excess following the consumption of seaweeds or seaweed-based products may affect thyroid gland function. In countries where marine algae are traditionally consumed as food, regular seaweed intake appears to be frequently associated with thyroid enlargement, hypofunction, and Hashimoto's thyroiditis, a chronic inflammatory autoimmune disease of the thyroid gland. In contrast, in patients with underlying thyroid disease excessive consumption of seaweed or seaweed-containing dietary supplements may cause thyroid hyperfunction.

The multicellular organisms the authors shall treat in Chapter 10 range from unicellular *Gonium* to *Volvox* having thousands of cells. By introducing a seminal notion of complexity hierarchy, the authors have constructed a simple classification of extracellular matrices in the framework of formal language theory. Then, this classification has enabled us to understand the necessity of various forms of multicells.

The authors have also constructed another model for generalized Anabaena, which is a genus of filamentous cyanobacteria or blue-green algae, using a Lindenmayer system (an L-system). An L-system is a parallel rewriting system that was introduced originally to model the development of multicellular organisms. The aim of this model is the derivation of exact algebraic equations between the proliferation and transition rates for high cell-type diversity. For simplicity, the authors have studied 'generalized Anabaena' having three cell types: A, B and C on the assumption of $A \Rightarrow B \Rightarrow C$ cell lineage. In this model, the high cell-type diversity corresponds to the condition that A, B and C cells are well mingled. This model has revealed that the condition for high cell-type diversity can be described as an elliptic curve under a certain condition that is deeply related to 'Fermat's last theorem.'

Cells and organelles are surrounded by at least one membrane that controls the exchange of energy and matter between the cytoplasm and its environment. Ion transport through membranes is an essential process for life. For instance, algae nutrition, osmotic and hydrostatic pressure regulation and cell signaling are typical cellular functions that are directly controlled by the transport of ions through the plasma membrane. The quantitative description of membrane transport through algae (and plants in general) is usually restricted to the uniform steady state case. However, spatial and temporal dynamics arising from the nonlinear properties of ion transport have recently been revealed to be of prime importance for cell signaling and developmental axis emergence. In Chapter 11 the authors will review the mechanisms of ion transport (active and passive transport) in marine algae and their implication in cell physiology, morphology and homeostasis. The basic principles of ion transport will be explain and the authors will show how the nonlinear coupling between different ion transport systems such as ATPases, channels and co-transports can give rise to self-organized spatiotemporal events such as action potentials or stationary patterns of transcellular currents observed in marine algae.

In Chapter 12, the authors introduce a new method for the detection of D-amino acids and present the results of this analysis of D-amino acids in marine macro- and micro-algae.

Chapter 13 demonstrates that by operating at a low current density, 2.5 A/m^2, electrocoagulation/flotation (ECF) performed superior to the chemical coagulation (CC) for the removal of algae. For the similar metal ion doses compared, the ECF can remove 11-12% more *Chlorophyll-a* than CC method. A colloidal titration study shows that the charge of flocs generated by the ECF was more positive than that by CC over the full doses studied and then the charge effect contributes to the superior performance of the ECF. In addition to this, a direct reduction of algal numbers on the surface of anodes is speculated and the flotation of aggregated flocs by hydrogen bubbles, resulting from the ECF process, should contribute to the overall algal removal.

The implications of diet on health sustainability have assumed a major importance, supported by considerable epidemiological evidences, and is well recognized by the scientific community and general public, on developed countries. Microalgae are able to enhance the nutritional content of conventional food and feed preparation and hence to positively affect humans and animal health due to their original chemical composition, namely high protein content, with balanced amino acids pattern, carotenoids, fatty acids, vitamins, polysaccharides, sterols, phycobilins and other biologically active compounds, more efficiently than traditional crops. The aim of Chapter 14 is to review the most important features of microalgae in animal and human nutrition, particularly in the development of novel design-foods rich in carotenoids and polyunsaturated fatty acids with antioxidant effect and other beneficial health properties.

RESEARCH AND REVIEW STUDIES

In: Algae: Nutrition, Pollution Control and Energy Sources ISBN: 978-1-60692-008-4
Editor: Kristian N. Hagen, pp. 3-11

Chapter 1

MICROALGAE AS PHOTOSYNTHETIC OXYGEN GENERATORS FOR POLLUTION CONTROL, LIFE SUPPORT SYSTEMS AND MEDICINE

Konstantin Bloch* and Pnina Vardi
Felsenstein Medical Research Center, Sackler Faculty of Medicine,
Tel-Aviv University, Israel

Abstract

In the early history of the Earth, atmospheric oxygen was built up by photosynthetic microalgae. These prokaryctic and eukaryotic microorganisms use sunlight energy to drive the synthesis of organic molecules from CO_2 and water, and liberate oxygen as a by-product. The remarkable adaptation capacity of unicellular algae to produce oxygen in different extreme environments can used in various fields of constructive human endeavor. Numerous innovative solutions have been proposed to utilize photosynthetic microorganisms in pollution control to treat oxygenated wastewater, to regenerate the atmosphere in the enclosed weightlessness environment in space and in medicine to develop an extracorporeal bioartificial lung and oxygenated implanted bioartificial pancreas.

This paper focuses mainly on the various methodologies employed recently in the development of microalga-based photosynthetic oxygen generators for such different areas as pollution control, space life support and medicine. Progress in these fields require a multidisciplinary approach to explore and optimize the interaction between components of the biological system and different technological processes. The authors discuss their research findings related to the application of microalgae for oxygen delivery to implantable bioartificial pancreas, emphasizing the importance of the close cooperation of multidisciplinary teams engaged in the development of such bio-hybrid medical devices. Special attention is paid to the advantages of microalgae, compared with other technologies for oxygenation, to serve as a safe, cheap and smart tool for oxygen supply.

---

* E-mail address: kbloch@post.tau.ac.il. Tel: +972-3-9376280. Fax: +972-3-9211478. Corresponding author: Dr. Konstantin Bloch, Diabetes and Obesity Research Laboratory, Felsenstein Medical Research Center, Sackler Faculty of Medicine, Tel-Aviv University, 49100 Petah Tikva, Israel

Introduction

The number of algal species has been estimated at between one and ten million, most of which are microalgae (*Metting, 1996*). The photosynthetic microalgae occur in most habitats, ranging from freshwater to salt-saturated water of Dead Sea and from hot boiling springs to Arctic ice. Common applications of microalgae for human requirements utilize algal organic compounds synthesized during photosynthesis from carbon dioxide and water, using light energy. Microalgae have been used for food and medicine in different civilizations since ancient times. The modern applications of algae or algal-derived chemical compounds in the food industry, pollution control, medicine and energy supply are well known and have been reviewed comprehensively (*Cannell, 1990; Borowitzka, 1999, Richmond, 2000*). Much less is known about the biotechnological application of microalgae as a photosynthetic oxygen generator. The oxygen produced by plant cells is the byproduct of the photosynthetic reaction occurring in the early stage of life evolution on the Earth. In this paper we focus on various aspects of algal biotechnology utilizing oxygen generated by unicellular organisms during photosynthesis, for medicine, pollution control and life support in space.

Photosynthetic Oxygen and the Evolution of Life

At first, the Earth had no oxygen in the atmosphere and initial oxygen enrichment was due to photosynthetic microorganisms such as cyanobacteria or blue green algae, that could live anaerobically and aerobically (*Kasting, 2001; Kerr, 2005*). Based on the evidence from organic biomarkers in well-preserved sedimentary rocks, these first photosynthetic microalgae are thought to have emerged about 2.7 billion years ago (*Brocks et al., 1999*). Comparative analysis of ribosomal RNA from cyanobacteria and portions of the DNA inside chloroplasts suggest that all eukaryotes, including algae and higher plants, derive their photosynthetic capabilities from cyanobacteria through endosymbiosis (*Reyes-Prieto et al., 2006*).

For several million years since the first cyanobacteria colonized the oceans and land, the earth's atmosphere contained almost no oxygen. Iron and sulfur compounds mopped up almost all of the free oxygen immediately. When the oxygen absorbing compounds in the oceans and rocks were exhausted, the atmospheric concentration of oxygen increased rapidly. This dramatic rise of oxygen in earth's atmosphere was described as the Great Oxidation Event, believed to occur about 2.5 billion years ago (*Farquhar et al., 2007; Anbar et al., 2007)* This rise in atmospheric oxygen was followed by the evolutionary emergence of organisms with oxygen-based metabolism.

The use of Photosynthetic Oxygenation for Bioremediation Processes

During evolution heterotrophic microorganisms developed the capacity to utilize organic compounds, toxic for other organisms, as the energy source for their metabolism. This capacity is used today for bioremediation, intended as pollution control technology based on utilization of microorganisms to break down the organic pollutants in different environments. Typically, bioremediation of a contaminated site works either by anaerobic or by aerobic biodegradation. Anaerobic biodegradation is the breakdown in anoxic environment of organic contaminants by microorganisms. The aerobic process takes place in the presence of oxygen . Generally, aerobic microorganisms use oxygen as an electron acceptor breaks down organic chemicals, producing carbon dioxide and water. Many toxic compounds are more easier to degrade aerobically than anaerobically.

One of the most serious limitation in aerobic biodegradation of waste materials in water is insufficient oxygenation, due to low solubility of oxygen in water. The oxygen deficiency results in microbial growth inhibition and the decreased rate of toxin degradation by heterotrophic microorganisms such as bacteria and fungi. In order to solve the problem of limited oxygen supply, different technological solutions have been suggested including mechanical aeration and air bubbling. Indeed, all these technologies provide an increased level of dissolved oxygen for aerobic biodegradation. However, they are expensive and can provoke contamination of the atmosphere due to enhanced evaporation of hazardous pollutants (*Brandi et al., 2000; Hamoda, 2006*). Another approach to enhance oxygen supply for waste water treatment is the use of symbiotic interactions between photosynthetic microalgae and heterotrophic bacteria. The pioneer in the field of the industrial application of algal bacterial association for wastewater treatment was William J. Oswald. About five decades ago he described the main principles of photosynthetic oxygen supply to the heterotrophic bacteria for the efficient aerobic biodegradation of pollutants (*Golueke et al, 1957; Oswald, 1962*). For the past years the concept of algal bacterial consortium has been intensively studied by various research teams (*Safonova et al., 1999, 2004; Cohen, 2002; Borde et al., 2003; Muñoz et al., 2004; Muñoz & Guieysse, 2006*). The design of such microbial symbiosis is based on the algal cells' capacity to use light energy for the synthesis of organic molecules from CO_2 and water, and liberate the oxygen required by aerobic bacteria to breakdown hazardous organic pollutants. In turn, microalgae use CO_2 released during bacterial respiration for photosynthesis.

Prokaryotic and eukaryotic microalgae have undergone remarkable environmental adaptation and are able to liberate oxygen in different extreme conditions of low and high pH, temperature and salinity (*Rothschild & Mancinelli, 2001*). The photosynthetic potential of microalgae can be utilized for oxygenation in various technologies requiring aerobic degradation of organic compounds by bacteria. However, the main obstacle to this concept is the fact that unicellular algae are much more sensitive to toxic pollutants (*Chen and Lin, 2006*) and their multiplication rate is less intensive compared to bacteria (*Aiba, 1982*). In addition, algae and bacteria may require different pH and temperature for cultivation. Thus, special attention must be given in selecting optimal partners for such consortia, otherwise it is likely that the symbiotic algal bacterial balance will be unstable during their co-cultivation (*Muñoz & Guieysse, 2006*; *Arranz et al., 2007*).

Typically, in a closed system, the mixture of microalgae and bacteria is placed in a stirred reservoir of a photobioreactor where microorganisms can be co-cultured as free cells (*Muñoz et al., 2004*) or as cells immobilized into various solid carriers preventing them from being washed off during industrial wastewater treatment (*Safonova et al., 2004; Mallick, 2002; de-Bashan et al., 2002*). The disadvantage of cell immobilization in solid or hydrogel matrices may be the poor mass transfer characteristic of chemicals and gases (*Moreno-Garrido, 2007*). Another approach to improve the oxygen supply for aerobic biodegradation of toxins by microbial communities, is implementation of phototrophic biofilms. The biofilms contain photoautotrophs, chemoautotrophs and heterotrophs, attached to a solid support and used for effective aerobic wastewater treatment (*Nicolella et al., 2000; Wolf et al., 2007*). In addition to oxygen supply to bacterial partners, microalgae might remove xenobiotics from polluted environments. Algal biomass produced during photosynthetic oxygenation can further be used for agriculture, biodiesel production and pharmaceutical applications, thus reducing cost of wastewater treatment (*Borowitzka, 1999; Richmond, 2000; Spolaore et al, 2006; Chisti, 2007*). Although photosynthetic microorganisms provide efficient, cheap and safe tools to enhance oxygen supply for the aerobic biodegradation of waste materials, further efforts towards optimization of algal bacterial consortia are needed for the industrial application of this technology.

Photosynthetic Oxygenation for Atmosphere Regeneration in an Enclosed Space Ecosystem

A life support system must enable the survival of humanbeings in long-term space exploration missions. The major tasks of such a system are to provide food production, water purification and waste management, oxygen regeneration and CO_2 utilization (*Eckart , 1995*). This system can be based on two different concepts: physico–chemical and biological. An example of first one is atmosphere regeneration by electrochemical oxygen generation . Such an approach was used by NASA for development of Environmental Control and Life Support System (ECLSS) in the international space station (*Tatara and Roman, 1998*). However, during the long-term extraterrestrial conditions this physico-chemical life support system may provide and maintain suitable life conditions for only a limited period of time *(Drysdale et al. 2003).*

Since the first Russian satellite carrying a dog abroad was launched into space nearly 50 years ago, various technologies utilizing photosynthetic microorganisms for life support in enclosed space environment have been suggested (*Myers, 1964; Mori at al., 1989; Brechignac at al., 1992; Hendrickx et al., 2006, Lehto et al., 2006*). Using an interdisciplinary approach the European Space Agency in collaboration with several independent organizations developed Micro Ecological Life Support System Alternative (MELiSSA) with the use of microorganisms living in interconnected controllable bioreactors (*Hendrickx et al., 2006*). An integral part of this system is microalgal photobioreactor serving as a producer of oxygen and edible biomass containing carbohydrates, protein, fat, vitamins and minerals. The photosynthetic cyanobacteria *Arthrospira* (Spirulina) was selected for the MELiSSA ecosystem as the most effective oxygen producer and essential food supplement. These blue-green algae can grow in warm water up to 70°C and pH 10.0 keeping the culture free of other microbial contaminations (*Khan et al., 2005*). In a Russian life support Hybrid Biosphere Systems (BIOS), photosynthetic eukaryotic unicellular *Chlorella* algae were utilized as a potential tool for atmosphere regeneration, food supplementation and waste product treatments in enclosed space ecosystem (*Salisbury et al., 1997; Gitelson et al., 1976*).

The main problem related to the implementation of a microbial life support system as a reliable bioregenerative tool, is the long-term stability of the ecosystem. At present, influences of different physical, chemical and biological factors on the stability of the microbial community cultivated in closed space environment, are still not fully understood. For example, it was suggested that the destabilizing factors in the MELiSSA loop may be the external physical conditions such as microgravity, space radiation, electromagnetism and vibration. Genetic changes and a stress response during long-term culture is to be expected and might have an undesirable impact on stability of microbial ecosystem (*Hendrickx et al., 2006*). The optimal combination of micralgae-based life support system and a physico-chemical regenerative system could possibly present an attractive solution for oxygen regeneration, food production and waste management in future long-distance space exploration missions.

Photosynthetic Oxygen Generator for Artificial Organs

There are at least two potential applications of photosynthetic oxygenation for artificial organs: the development of an artificial lung and a bioartificial pancreas. Feasibility studies for a photosynthetic artificial lung and a photosynthetic oxygen generator for a bioartificial pancreas have recently been published (*Basu et al., 1994, 1997; Bloch et al, 2006 a,b*). Although the authors showed proof of concept for both these technologies in vitro, many problems related to the optimization of biological, chemical and physical parameters must be solved before the application of such medical devices for in vivo trials.

In the case of the artificial lung, the authors suggested using the thermophilic photosynthetic microalga *Chlorella pyrenoidosa* that can remove carbon dioxide and generate a sufficient amount of oxygen for extracorporeal life support for pulmonary respiratory failure in children and adults. Such a photosynthetic–based device is more portable and may be a potential alternative for the currently used mechanical devices to ventilate the lung. Optimization of parameters (e.g. alga growth, light intensity, substrate concentration, and pH changes) affecting photosynthetic oxygenation in an enclosed system enable the demonstration of the ability of this system to supply O_2 and remove CO_2 at rate of 0.55 mmoles/L/hr, that is close to 1.0 mmoles/L/hr required for physiologic applications. To achieve sufficient oxygenation, microalga strains with more efficient O_2 production and suppressed photoinhibition capacity must be selected or genetically engineerted. Special attention should be paid to potentially toxic alga strains. An important issue is also optimization of gas-exchange parameters between patient's blood and alga-based photobioreactor (*Basu et al., 1994, 1997*).

In contrast to the artificial lung which is relatively large device used as an extracorporeal life support system, the bioartificial pancreas (BAP) is a small medical device that can be implanted subcutaneously or intra-peritoneally to diabetic patients . Implantable BAP contains insulin-producing pancreatic islets protected from immune rejection. Selective islet immune protection is achieved by their encapsulation in semi-permeable matrixes, such as alginate, or implantation of islets in solid devices surrounded by diffusion membranes. However, immunoisolation leads to total loss of the vasculature, resulting in the substantially decreased oxygen level. As pancreatic insulin producing beta cells are extremely sensitive to oxygen deficiency, the limitations in oxygen diffusion leads to the establishment of severe hypoxic conditions, with a decrease in stimulated insulin secretion, and subsequent beta cell death. This process is believed to be the major obstacle to a successful cure for diabetes by the implantation of bio-artificial pancreas (*Colton, 1995*). Recently, we described a new technology for microalga-based, photosynthetic oxygen supply to encapsulated islets, in which a thermophylic strain of the unicellular alga *Chlorella sorokiniana* was used as a natural photosynthetic oxygen generator. To test the feasibility of algal-based oxygen generation supporting insulin secretion of pancreatic islets in an anoxic environment, alginate capsules containing both pancreatic islets and unicellular algae were perifused with oxygen-free medium. The system was adapted for the determination of insulin secretion, oxygen consumption and photosynthetic efficiency. We found that alga-based photosynthetic oxygen production could compensate the oxygen consumption of islets and provide optimal insulin secretion from encapsulated islets, perifused with oxygen-free medium (*Bloch et al., 2006a*). However, the close proximity of unicellular algae and islets in the same compartment for a long-term period could lead to some undesirable effects on the functional activity of co-immobilized cells, such as algal overgrowth, a switch from photosynthetic to heterotrophic nutrition of the algae, the accumulation of toxic metabolites, and competition for nutrients between plant and mammalian cells. To overcome such difficulties we studied a different BAP system containing two compartments: one for oxygen producing algal cells and another for insulin secreting cells (*Bloch et al., 2006b*). These two compartment were separated by semi-permeable matrix. In this new model, the algae were capable of producing sufficient amounts of oxygen to restore the insulin response of the islets located in a separate compartment. However, in this system, the number of alga cells needed to support the functional activity of one islet was significantly higher compared to the single compartment design. Further optimization of photosynthetic oxygen production to compensate islet respiration demand should be made for future clinical application of this technology. Such optimization studies include the selection of highly effective photosynthetic algal strains adapted to body environments and the development of a biocompatible gas-permeable matrix to separate islet and alga compartments in the BAP. Progress in the fields of light transmission and microelectronics would enable the miniaturization of such systems with new light emitting diodes as an efficient energy supply for future intra-body illuminations of photosynthetic microorganisms immobilized in a photobioreactor.

Conclusion

The potential applications of microalgae as photosynthetic oxygen generators for aerobic biodegradation of organic pollutants in enclosed photobiorectors, life support systems for long-term space missions and artificial organs for patient care have been successfully proved in feasibility studies. However, none of these technologies have yet been implemented for extensive commercial use. To practically apply photosynthetic oxygenation as a safe, cheap and smart tool for different technologies, special attention should be paid to the selection/engineering alga strains with an appropriate level of photosynthesis in extreme environments; the optimization of microalgal mass production and harvesting; the improvement of mass transfer characteristic of gas permeable materials and the better understanding of microalgal behavior as a symbiotic partner of microbial consortium.

Acknowledgement

This work was supported by a grant from the Research Foundation of Tel Aviv University. We thank Sara Dominitz for editorial help.

References

Aiba, S. Growth kinetics of photosynthetic microorganisms, *Adv. Biochem. Eng.*, 1982, 23, 85–156.

Anbar, AD; Duan, Y; Lyons, TW; Arnold, GL; Kendall, B; Creaser, RA; Kaufman, AJ; Gordon, GW; Scott, C; Garvin, J; Buick, R. A whiff of oxygen before the great oxidation event? *Science*. 2007, 317,1903-1906.

Arranz, A; Bordel, S; Villaverde, S; Zamarreño, JM; Guieysse, B; Muñoz, R.
Modeling photosynthetically oxygenated biodegradation processes using artificial neural networks. *J. Hazard Mater.*, 2007, Nov 17; [Epub ahead of print]

Basu, S; Salley, SO; Whittlesey, GC; Klein, MD. Feasibility studies for a photosynthetic artificial lung. Optimization of parameters affecting photosynthesis. *ASAIO J.*, 1994, 40, 743-746.

Basu-Dutt, S; Fandino, MR; Salley, SO; Thompson, IM; Whittlesey, GC; Klein, MD. Feasibility of a photosynthetic artificial lung. *ASAIO J.*, 1997, 43, 279-83.

Bloch, K; Papismedov, E; Yavriyants, K; Vorobeychik, M; Beer, S; Vardi, P. Photosynthetic oxygen generator for bio-artificial pancreas. *Tissue Engineering*, 2006 a, 12, 337-344.

Bloch, K; Papismedov, E; Yavriyants, K; Vorobeychik, M; Beer, S; Vardi, P. Immobilised microalgal cells as an oxygen supply system for encapsulated pancreatic islets: A feasibility study. *Artificial Organs*, 2006 b, 30, 715-718.

Borde, X; Guieysse, B; Delgado, O; Munoz, R; Hatti-Kaul, R; Nugier-Chauvin, C; Patin, H; Mattiasson, B. Synergistic relationships in algal–bacterial microcosms for the treatment of aromatic pollutants, *Bioresourc. Technol*, 2003, 86, 293–300.

Borowitzka, MA. Commercial production of microalgae: ponds, tanks, tubes and fermenters, *J. Biotechnol.*, 1999, 70, 313–321.

Brandi, G; Sisti, M; Amagliani, G. Evaluation of the environmental impact of microbial aerosols generated by wastewater treatment plants utilizing different aeration systems. *J. Appl. Microbiol.* 2000, 88, 845–852.

Brechignac, F; Schiller, P. Pilot CELSS based on a maltose-excreting Chlorella: concept and overview on the technological developments. *Adv. Space Res.* 1992, 12, 33-36.

Brocks, JJ; Logan, GA; Buick, R; Summons, RE. Archean molecular fossils and the early rise of eukaryotes. *Science*, 1999, 285, 1033-1036.

Cannell, RJ. Algal biotechnology. *Appl. Biochem. Biotechnol.*, 1990, 26, 85-105.

Chen, C.Y. & Lin, J.H. Toxicity of chlorophenols to *Pseudokirchneriella subcapitata* under air-tight test environment, *Chemosphere*, 2006, 62, 503–509.

Chisti, Y. Biodiesel from microalgae. *Biotechnol Adv.*, 2007, 25, 294-306.

Cohen, Y. Bioremediation of oil by marine microbial mats. *Int Microbiol.* 2002, 5, 189-93.

Colton, CK. Implantable biohybrid artificial organs. *Cell Transplant.*, 1995, 4, 415-436.

de-Bashan, LE; Moreno, M; Hernandez, JP; Bashan, Y. Removal of ammonium and phosphorus ions from synthetic wastewater by the microalgae Chlorella vulgaris coimmobilized in alginate beads with the microalgae growth-promoting bacterium Azospirillum brasilense. *Water Res*. 2002, 36, 2941-2948.

Drysdale, AE; Ewert, MK; Hanford, AJ. Life support approaches for Mars missions. *Adv. Space Res*., 2003, 31, 51-61.

Eckart, P. Life Support and Biospherics-a handbook. *Life Support Biosph. Sci*., 1995,2, 103-106.

Farquhar, J; Peters, M; Johnston, DT; Strauss, H; Masterson, A; Wiechert, U; Kaufman, AJ. Isotopic evidence for Mesoarchaean anoxia and changing atmospheric sulphur chemistry. *Nature*, 2007, 449, 706-709.

Gitelson, II; Terskov, IA; Kovrov, BG; Sidko, FY; Lisovsky, GM; Okladnikov, YN; Belyanin, VN; Trubachov, IN; Rerberg, MS. Life support system with autonomous control employing plant photosynthesis. *Acta Astronaut*., 1976, 3, 633-650.

Golueke, CG; Oswald, WJ; Gotaas, HB. Anaerobic Digestion Of Algae. *Appl Microbiol.* 1957, 5, 47-55.

Hamoda, MF. Air pollutants emissions from waste treatment and disposal facilities, *J. Environ. Sci. Health Part A—Toxic/Hazard. Subst. Environ. Eng.* 2006, 41, 77–85.

Hendrickx, L; De Wever, H; Hermans, V; Mastroleo, F; Morin, N; Wilmotte, A; Janssen, P; Mergeay, M. Microbial ecology of the closed artificial ecosystem MELiSSA (Micro-Ecological Life Support System Alternative): reinventing and compartmentalizing the Earth's food and oxygen regeneration system for long-haul space exploration missions. *Res Microbiol*., 2006, 157, 77-86.

Kasting, JF. Earth history. The rise of atmospheric oxygen. *Science*. 2001, 293, 819-820.

Kerr, RA. Earth science. The story of O2. *Science*. 2005, 308, 1730-1732.

Khan, Z; Bhadouria, P; Bisen, PS. Nutritional and therapeutic potential of Spirulina. *Curr Pharm Biotechnol.* 2005, 6, 373-379.

Lehto, KM; Lehto, HJ; Kanervo, EA. Suitability of different photosynthetic organisms for an extraterrestrial biological life support system. *Res Microbiol*., 2006, 157, 69-76.

Mallick, N. Biotechnological potential of immobilized algae for wastewater N, P and metal removal: a review. *Biometals*, 2002, 15, 377-390.

Metting, FB. Biodiversity and application of microalgae. *Journal of Industrial Microbiology and Biotechnology,* 1996, 17, 477-489.

Moreno-Garrido, I. Microalgae immobilization: Current techniques and uses. *Bioresour Technol.,* 2007, Jul 6; [Epub ahead of print]

Mori, K; Ohya, H; Matsumoto, K; Furuune, H; Isozaki, K; Siekmeier, P. Design for a bioreactor with sunlight supply and operations systems for use in the space environment. *Adv Space Res*., 1989, 9, 161-168.

Muñoz, R; Guieysse, B. Algal-bacterial processes for the treatment of hazardous contaminants: a review. *Water Res*., 2006, 40, 2799-2815.

Muñoz, R; Köllner, C; Guieysse, B; Mattiasson, B. Photosynthetically oxygenated salicylate biodegradation in a continuous stirred tank photobioreactor, *Biotechnol. Bioeng,*. 2004, 87, 797–803.

Myers, J. Use of algae for support of the human in space. *Life Sci Space Res.* 1964, 2:323-36.

Nicolella, C; van Loosdrecht, MC; Heijnen JJ. Wastewater treatment with particulate biofilm reactors. *J Biotechnol*., 2000, 80,1-33.

Oswald, WJ. The coming industry of controlled photosynthesis. *Am J Public Health Nations Health,* 1962, 52, 235-242.

Reyes-Prieto, A; Hackett, JD; Soares, MB; Bonaldo, MF;Bhattacharya, D. Cyanobacterial contribution to algal nuclear genomes is primarily limited to plastid functions. *Curr Biol.*, 2006,16, 2320-2325.

Richmond, A. Microalgal biotechnology at the turn of the millennium: a personal view, *J. Appl. Phycol.*, 2000, 12, 441–451.

Rothschild, LJ; Mancinelli, RL. Life in extreme environments. *Nature,* 2001, 409, 1092-1101.

Safonova, E; Dmitrieva, I.A; Kvitko, K.V. The interaction of algae with alcanotrophic bacteria in black oil decomposition, *Resourc. Conserv. Recycl.* 1999, 27, 193–201.

Safonova, E; Kvitko, K.V; Iankevitch, M.I; Surgko, L.F; Afri I.A; Reisser, W. Biotreatment of industrial wastewater by selected algal–bacterial consortia, *Eng. Life Sc.* 2004, 4, 347–353.

Salisbury, FB; Gitelson, JI; Lisovsky, GM. Bios-3: Siberian experiments in bioregenerative life support. *Bioscience.* 1997, 47, 575-585.

Spolaore, P; Joannis-Cassan, C; Duran, E; Isambert, A. Commercial applications of microalgae. *J Biosci Bioeng.* 2006, 101, 87-96.

Tatara, JD; Roman, MC. An overview of ISS ECLSS life testing at NASA, MSFC. *Life Support Biosph Sci.* 1998, 5, 13-21.

Wolf ,G; Picioreanu, C; van Loosdrecht, MC. Kinetic modeling of phototrophic biofilms: the PHOBIA model. *Biotechnol Bioeng.* 2007, 97, 1064-1079.

In: Algae: Nutrition, Pollution Control and Energy Sources
Editor: Kristian N. Hagen, pp. 13-21
ISBN: 978-1-60692-008-4

Chapter 2

POLYPHOSPHATE CONTRIBUTES TO CD TOLERANCE IN *CHLAMYDOMONAS ACIDOPHILA* KT-1

Kahoko Nishikawa[1,2], Noriko Tominaga[1], Tadashi Uchino[3], Ayumi Oikawa[1] and Hiroshi Tokunaga[4]

[1] Institute of Environmental Science for Human Life, Ochanomizu University, Japan
[2] Department of Traumatology and Critical Care Medicine, National Defense Medical College, Japan
[3] Division of Environmental Chemistry, National Institute of Health Sciences, Japan
[4] Pharmaceuticals and Medical Devices Agency Office of Compliance and Standards, Japan

Abstract

One of the proposed mechanisms for metal tolerance in bacteria and algae is the sequestration of metal cations by long-chain polymers of inorganic polyphosphate. The highly metal-tolerant alga *Chlamydomonas acidophila* KT-1 (KT-1) accumulates within the cell large amounts of polyphosphate that are degraded in response to cadmium (Cd) stress. We investigated the mechanism of action for this important phosphate-based metal-detoxifying function of KT-1. After shifting KT-1 cells from a Cd-free to a Cd-plus medium, high molecular weight forms of polyphosphate were degraded and very low molecular weight polyphosphate disappeared from the cells. Coincidently, the total intracellular phosphate decreased to only 10% of control levels, and subsequent phosphate uptake from the medium was markedly enhanced. Phosphate limitation causes a similar effect, but we suspect that Cd stress and phosphate deprivation may have different molecular mechanisms in KT-1 cells. Upon shifting from a Cd-plus to a Cd-free medium, intracellular Cd decreased by 43%, total intracellular phosphate increased nine fold. We propose a two-step mechanism involving the chelation and extracellular transport of Cd to explain the important metal detoxifying effects of polyphosphate in KT-1.

1. Introduction

Cadmium (Cd) is generally a very toxic heavy metal that can inhibit cell growth and kill cells. Some algae have evolved functions to tolerate high levels of Cd in the environment, including: cell wall binding; chelation with phytochelatins (PCs) and other substrates; vacuolar compartmentalisation; and/or cell excretion of Cd. *Chlamydomonas reinhardtii* uses cell wall binding, but *C. reinhardtii* cells still accumulate and tolerate Cd, despite cell wall binding; thus, other metal-tolerance functions aside from cell wall binding have been proposed to function concurrently (Macfie et al., 2000).

The acidophilic alga *Chlamydomonas acidophila* KT-1 (KT-1) accumulates and tolerates higher intracellular levels of heavy metals, including Cd, than do other algae (Nishikawa, 2001). Chelation by PCs is one function (Grill, 1989) that has been confirmed in KT-1 tolerance to Cd stress (Nishikawa, 2006a). However, we suspect that additional metal-tolerance functions operate in KT-1 because the amount of PC is not sufficient to account for the increased accumulation of tolerable concentrations of Cd in cells. Furthermore, the identification of both Cd and phosphate in KT-1 vacuoles by transmission electron microscopy (TEM) and energy-dispersive X-ray (EXD) analysis following Cd stress suggests that vacuolar compartmentalisation of Cd also functions in KT-1 (Nishikawa, 2003). Consequently, we investigated the possible contribution of a phosphate-based function to explain Cd tolerance in KT-1. We hypothesised that phosphate can detoxify Cd by cytoplasmic complexation and vacuolar compartmentalisation. The sequestration of metal cations with long-chain polymers of inorganic polyphosphate has been proposed as a mechanism for metal tolerance due to the binding of various metal ions to polyphosphate polymers (Kornberg, 1999). Studies of phosphate metabolism using ^{31}P-NMR *in vivo* demonstrated that Cd induces the degradation of high molecular weight forms of polyphosphate and increases intracellular concentrations of inorganic phosphate (Pi) in KT-1 cells (Nishikawa, 2003).

A mechanism for heavy metal tolerance that is associated with the degradation of polyphosphate has been especially studied in bacteria in recent decades. Van veen and co-workers have shown that *Acinetobacter johnsonii* can reversibly transport metal-phosphates (van Veen, 1994a; van Veen, 1994b). The recombinant study of Keasling and Hupf (1996) indicated that not only a large quantity of intracellular polyphosphate, but also the ability to synthesise and degrade polyphosphate, are important functions for heavy metal tolerance in *E. coli*. In addition, they proposed that the metal phosphates are transported out of the cell by the inorganic phosphate transport (PIT) system (Keasling, 1997).

We analysed phosphate metabolism in KT-1 cells, with a particular focus on polyphosphate degradation as a possible function to explain Cd tolerance. We quantified the kinetics of degradation for high molecular weight polyphosphate, uptake of Pi, and secretion of phosphate that was possibly complexed with Cd, by KT-1 cells.

2. Material and Methods

2.1. Strain and Growth Conditions

Chlamydomonas acidophila KT-1 (KT-1) was isolated from an acidic lake, Lake Katanuma in Miyagi prefecture, Japan (Nishikawa et al., 2001). KT-1 was cultured at 20°C under a light intensity of 150 μmol m^{-2} s^{-1} provided byn white fluorescent lamps for plants (National Co., Japan) with a 12 h/12 h light/dark cycle in modified Sager-Granick medium (basal medium; 1.22 mM $MgSO_4.7H_2O$, 3.75 mM NH_4NO_3, 0.57 mM K_2HPO_4, 0.73 mM KH_2PO_4, 0.36 mM $CaCl_2$, 37 μM $FeCl_3.6H_2O$, 16.2 μM H_3BO_4, 3.5 μM $ZnSO_4.H_2O$, 2.02 μM $MnCl_4.H_2O$, 0.24 μM $CuSO_4$ $5H_2O$, 0.84 μM $CoCl_2.6H_2O$, 0.83 μM $Na_2MoO.4H_2O$; Sager, 1953). To maintain each optimum pH, 50 mM succinic acid was added to basal medium for KT-1 (adjusted to pH 4.0). Pi-deficient medium (P medium) contained 1.25 mM of Pi (2% basal medium). Cadmium treatment was carried out using basal medium or P medium containing 20 μM $CdSO_4.8/3H_2O$, and cells were incubated for 3 days in the standard conditions described above.

2.2. Extraction and Observation of Polyphosphate

To select a suitable polyphosphate extraction method, three polyphosphate extraction methods were evaluated: the acetone-glassmilk method (Ault-Riche, 1998), acid extraction (Pick, 1990), and hot water extraction (Kuroda, 2002). According to the results (data not shown), low molecular weight polyphosphate was best extracted by acid extraction, and high molecular weight polyphosphate was best extracted by the acetone-glassmilk method. To observe the degradation of high molecular weight polyphosphate, 10 μl of acetone-glassmilk extract was applied to 20% polyacrylamide gel or 2% agarose gel in TAE buffer containing 0.04 M Tris base, 0.02 M glacial acetic acid, and 1 mM EDTA (pH 8.0) and electrophoresed at 20 mA for 1.5 h (Powerstation 1000 VP, Atto). Polyphosphate glass Type 25 and 95 (Sigma) were used as polyphosphate standards. They were not strictly purified and showed a rather broad band, but still functioned sufficiently as control indicators. After electrophoresis, gels were fixed in 10% acetone in methanol for 60 min with shaking. Polyphosphate was stained with 0.05% toluidine blue, 2.5% methanol, and 0.5% glycerol and then destained with a solution of 25% methanol, 5% glycerol, and 5% acetate.

2.3. The Uptake of [^{32}P] Phosphate in KT-1

KT-1 cells were precultured in either P medium or basal medium for 2 weeks prior to the phosphate uptake measurements. The uptake measurements were performed in basal medium containing 1 μCi/ml [^{32}P] supplemented with or without 20 μM Cd. Uptake experiments were initiated by injecting precultured cells (approximately 1 x 10^6 cells/ml) into the incubation medium and incubating for 3 days in the standard conditions described above. The tracer content of the cells was determined by sampling 1 ml of cells and measuring radioactivity by liquid scintillation counting (Hitachi).

2.4. Intracellular Phosphate and Cd Accumulation

All reagents were of analytical grade. The method followed was that of Roychowdhury (2002). A stock solution of cadmium at a concentration of 1000 mg/L (Cica-Teagent, Cho-ku, Tokyo, Japan) was used to prepare standard solutions for ICP-MS. The sample digestions were carried out with concentrated nitric acid and high-purity hydrogen peroxide (30–35.5%). A microwave digestion system (MARS 5; CEM Corporation, Matthews, North Carolina 28106, USA) with a rotor for 14 Teflon HP-500 digestion vessels was used for sample digestion. An inductively coupled plasma mass spectrophotometer (ICP-MS; Hewlett-Packard 4500) was used as a chromatographic detector. The ICP-MS system was equipped with a Shimazu LC-10ADVP liquid chromatography solvent delivery pump, a Shimadzu DGU-12A degasser, and double-pass Scott-type spray chamber (water cooled, 2; Orion, RKS-1500V-C). Off-line data from the ICP-MS were processed using special chromatographic software (HP ChemStation).

3. Results

3.1. Degradation of High Molecular Weight Polyphosphate and Disappearance of Low Molecular Weight Polyphosphate

The treatment of KT-1 cells with 20 μM Cd induced the time-dependent degradation of high molecular weight polyphosphate (Figure 1). Modest degradation occurred at day 1 (lane C),

Figure 1. High molecular weight polyphosphate by acetone-extraction were degraded by cadmium. Standard polyphosphate 95 (Sigma); (A), Acetone-extraction from no treatment cells (control); (B), Acetone-extraction from cells with cells treated with 20 μM Cd for 1 day; (C), Acetone-extraction from cells treated with 20 μM Cd for 2 days; (D), Acetone-extraction from 20 μM Cd treatment for 4 days; (E).

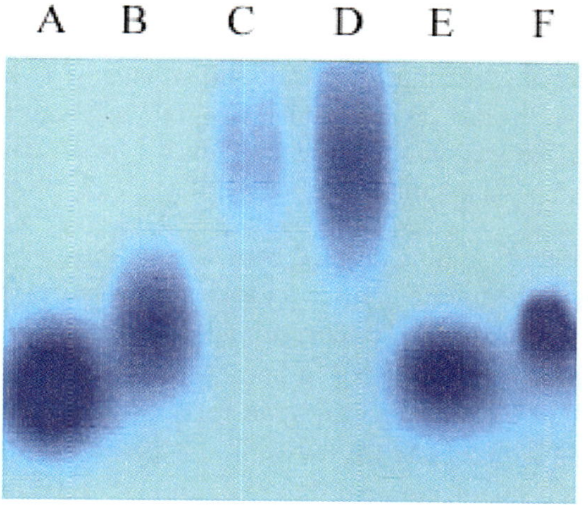

Figure 2. Effect of cadmium on high and low weight molecular polyphosphate. Standard polyphosphate 95 (Sigma); (A), standard polyphosphate 25 (Sigma); (B), acetone-extraction (high weight) from non-treatment cells; (C), acetone-extraction (high weight) from cells treated with 20 μM Cd for 3 days; (D), acid-extraction (low weight) from non-treatment cells; (E), acid-extraction (low weight) from cells treated with 20 μM Cd for 3 days.

but by days 2 and 4, lower molecular weight polyphosphate degradation products appeared (lanes D and E). The total polyphosphate was determined from the combined acid extract (low molecular weight) and acetone extract (high molecular weight) applied on the same gel (Figure 2). It was necessary to use 2% agarose to resolve the broad range of polyphosphate sizes on the same gel. The high molecular weight polyphosphate from cells treated with Cd (Figure 2, lane D) showed the lower molecular weight polyphosphate band consistent with the results in Figure 1. In contrast, in the low molecular weight acid extract (Figure 2, lane F), the lowest band disappeared and the upper band increased, so a portion of the higher band in lane F was considered to come from the degradation of high molecular weight polyphosphate induced by Cd stress. Accordingly, Cd treatment induced both the degradation of high molecular weight polyphosphate and the disappearance of some of the lowest molecular weight polyphosphate, possibly including orthophosphate.

3.2. Intracellular Phosphate and Cd Accumulation in KT-1

The relationship between phosphate levels and Cd accumulation in cells was analysed using ICP-MS (Table 1). The intracellular phosphate concentration was regulated by the extracellular phosphate concentration in KT-1 cells (Nishikawa, 2006) so we used both P medium and basal medium to control the levels of intracellular phosphate. The intracellular phosphate in basal medium was 100.5 ± 0.48 μg/mg dry weight (wt), and this level decreased dramatically to 14.79 ± 1.06 μg/mg dry wt (~85% decrease) when cells were incubated for > 2 weeks in P medium. The intracellular phosphate level of cells treated with 20 μM Cd for 3 days in basal medium was 8.47 ± 0.47 μg/mg dry wt (~90% decrease). Surprisingly, Cd exposure in addition to Pi deprivation produced an additional decrease in intracellular phosphate to only 4.78 ± 0.87 μg/mg dry wt (~95% decrease). Thus, Cd stress induced phosphate efflux both with and without phosphate deprivation. The maximum Cd accumulation was 1.26 ± 0.11 μg Cd/mg dry wt when cells were treated with 20 μM Cd. When cells that were treated with 20 μM Cd were shifted to Cd-free basal medium for 1 day, Cd accumulation decreased from 1.26 ± 0.11 to 0.74 ± 0.07 μg Cd/mg dry wt (~41% decrease in Cd). Simultaneously, intracellular phosphate recovered to 77.01 ± 4.04 μg/mg dry wt (~77% of control levels in basal medium) after only 1 day.

Table 1. Change of total phosphate and Cd contents by phosphate and Cd stress in *C. acidophila* KT-1. Total phosphate and Cd levels were measured by ICP-MS (Hewlett-Packard 4500). Non-treatment 3d; cells were incubated in basal medium for 3 days, Basal with Cd 3d; cells were incubated in basal medium containing 20 μM Cd for 3 days, Basal with Cd 3d to Basal; cells were transferred to basal medium without Cd for 1 day after cells treated with 20 μM Cd for 3 days, P-deficient 3d; cells were incubated in P-deficient medium (1/52 phosphate of basal medium) for 3 days, P-deficient with Cd 3d; cells were incubated in P-deficient medium containing 20 μM Cd for 3 days

	P (mg/mg dry weight)	**Cd (mg/mg dry weight)**
Basal 3d	100.50 ± 0.48	n.d.
Basal with Cd 3d	8.47 ± 0.47*1	1.26 ± 0.11*1
Basal with Cd 3d to Basal	77.01 ± 4.04*2	0.74 ± 0.07*2
P-deficient 3d	14.79 ± 1.06*1*3	n.d.
P-deficient with Cd 3d	4.78 ± 0.87*1	1.24 ± 0.05*1

1 showed statistically significant differences between non-treatment cells and treatment groups ($P < 0.05$ as calculated by one-way ANOVA).

*2 was statistically significant differences with Basal with Cd 3d,

*3 was with P-deficient 3d. n.d.; An amount of Cd was below detection limit.

3.3. Uptake of Phosphate from the Extracellular Environment Induced by Cd Stress and Pi Deficiency

The uptake of [^{32}P] was activated by both Cd stress and Pi limitation (Figure 3). Cd treatment produced an 76% increase in the uptake of extracellular phosphate over that of untreated KT-1 cells in basal medium. The pre-incubation of KT-1 cells (no Cd) in P medium (Pi-limited cells) produced almost 400% increase in extracellular phosphate uptake. Cells exposed to Cd in P medium (double stress) exhibited 10 time more increase in the uptake of extracellular phosphate than that of control KT-1 cells in basal medium. Thus, both Cd and phosphate stress activate the uptake of extracellular phosphate.

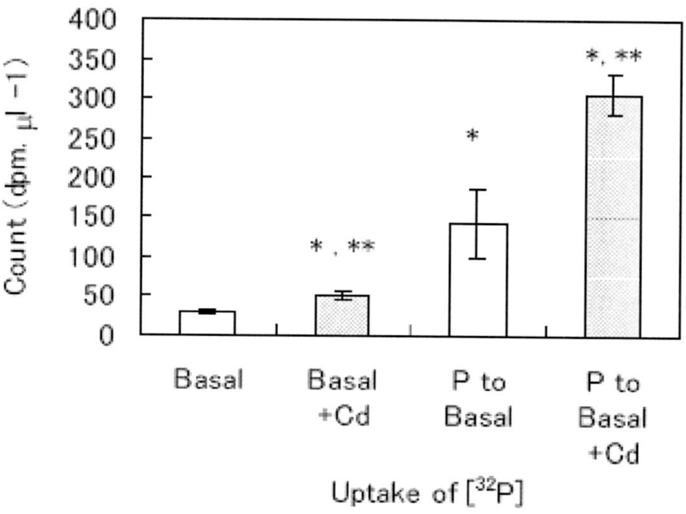

Figure 3. Uptake of [^{32}P] in *C. acidophila* KT-1. Cells were cultured for 3 days with [^{32} P] (1 mCi/l, a 5 Ci/mol inorganic phosphate) in basal medium and/or P medium supplemented with or without 20 µM Cd supplement.

4. Discussion

The presence of large quantities of intracellular polyphosphate and the ability to synthesise and degrade polyphosphate are important functions for heavy metal tolerance in *E. coli* (Keasling, 1996). The degradation of intracellular polyphosphate induced by heavy metals has also been reported in *Acidithiobacillus ferrooxidans* (Alvarez, 2004). Because the initial detoxifying function of polyphosphate is most likely the chelation of metal cations by orthophosphate (Pi) from polyphosphate, it is essential that a phosphatase such as exopolyphosphatase (PPX) catalyse the removal of the terminal phosphate from polyphosphate. We observed the Cd-induced degradation of polyphosphate, decreased intracellular phosphate, and increased uptake of extracellular phosphate in KT-1 cells. Phosphate limitation induces the degradation of polyphosphate and decreases intracellular phosphate (Nishikawa, 2006b). We demonstrated that Cd enhances these responses, even in cells that were previously deprived of phosphate, suggesting that Cd stress and phosphate limitation function through separate pathways with distinct molecular mechanisms. Additionally, we showed that the intracellular phosphate concentration does not correspond with Cd accumulation, and cells that were switched from a Cd-containing to Cd-free medium actively released Cd into the medium.

Keasling (1997) proposed a two-step bacterial metal detoxification model for which the first step involved the degradation of polyphosphate and the chelation of Cd, possibly by the formation of a Cd-Pi complex, and the second step involved the excretion of the chelated Cd complex from the cells. A key aspect of the bacterial model was the provision that the intracellular metal cation concentration would regulate the activity of PPX, which in turn would degrade polyphosphate to generate Pi so that metal cations could be co-transported from the cell with Pi through the Pit system characterised in *E. coli*.

The alga *Chlamydomonas reinhardtii* was reported to use the phosphate transporter to incorporate arsenate from the environment (Kobayashi, 2003). This report suggested to us the possibility that *Chlamydomonas spp*. could use the phosphate transporter for both the influx and efflux of heavy metals. With regard to *Chlamydomonas acidophila* KT-1, further characterisation of the phosphate transport system will help to clarify this concept.

Polyphosphate is an intriguing compound that can function as a phosphorus (Pi) reservoir, a substitute for ATP in sugar metabolism (Bonting, 1991; Phillips, 1993), and a factor in regulatory responses to stress and nutritional deficiencies (Rao, 1996; Kuroda, 2001). Recently, Werner et al. (2007) proposed that cell wall polyphosphates might be partially responsible for sequestering heavy metals, and the polyphosphate chain size was shown to affect the degree of heavy metal tolerance.

In summary, our results demonstrate that polyphosphate can contribute to heavy metal tolerance in *Chlamydomonas acidophila* KT-1 by chelating Cd through the formation of a Cd-phosphate complex that can ultimately be excreted from the cell. We confirmed the presence of three of the four metal tolerance functions in KT-1 cells: chelation, vacuolar compartmentalisation, and excretion. We propose that polyphosphate/phosphate might contribute to metal tolerance in *Chlamydomonas acidophila* KT-1 as both a chelator and co-transporter to sequester and excrete Cd from cells, respectively.

References

Alvarez, S. and Jerez, C. A. (2004). Copper ions stimulate polyphosphate degradation and phosphate efflux in Acidithiobacillus ferrooxidans. *Appl Environ Microbiol.* **70**, 5177-82.

Ault-Riche, D., Fraley, C. D., Tzeng, C. M. and Kornberg, A. (1998). Novel assay reveals multiple pathways regulating stress-induced accumulations of inorganic polyphosphate in Escherichia coli. *J Bacteriol.* **180**, 1841-7.

Bonting, C. F., Kortstee, G. J. and Zehnder, A. J. (1991). Properties of polyphosphate: AMP phosphotransferase of Acinetobacter strain 210A. *J Bacteriol.* **173**, 6484-8.

Grill, E., Loffler, S., Winnacker, E. L. and Zenk, M. H. (1989). Phytochelatins, the heavy-metal-binding peptides of plants, are synthesized from glutathione by a specific gamma-glutamylcysteine dipeptidyl transpeptidase (phytochelatin synthase). *Proc Natl Acad Sci U S A.* **86**, 6838-6842.

Keasling, J. D. (1997). Regulation of intracellular toxic metals and other cations by hydrolysis of polyphosphate. *Ann N Y Acad Sci.* **829**, 242-9.

Keasling, J. D. and Hupf, G. A. (1996). Genetic manipulation of polyphosphate metabolism affects cadmium tolerance in Escherichia coli. *Appl Environ Microbiol.* **62**, 743-6.

Kobayashi, I., Fujiwara, S., Shimogawara, K., Kaise, T., Usuda, H. and Tsuzuki, M. (2003). Insertional mutagenesis in a homologue of a Pi transporter gene confers arsenate resistance on chlamydomonas. *Plant Cell Physiol.* **44**, 597-606.

Kornberg, A. (1999). Inorganic polyphosphate: a molecule of many functions. *Prog Mol Subcell Biol.* **23**, 1-18.

Kuroda, A., Nomura, K., Ohtomo, R., Kato, J., Ikeda, T., Takiguchi, N., Ohtake, H. and Kornberg, A. (2001). Role of inorganic polyphosphate in promoting ribosomal protein degradation by the Lon protease in E. coli. *Science,* **293**, 705-8.

Kuroda, A., Takiguchi, N., Gotanda, T., Nomura, K., Kato, J., Ikeda, T. and Ohtake, H. (2002). A simple method to release polyphosphate from activated sludge for phosphorus reuse and recycling. *Biotechnol Bioeng.* **78**, 333-8.

Macfie, S. M. and Welbourn, P. M. (2000). The cell wall as a barrier to uptake of metal ions in the unicellular green alga Chlamydomonas reinhardtii (Chlorophyceae). *Arch Environ Contam Toxicol.* **39**, 413-9.

Nishikawa, K. and Tominaga, N. (2001). Isolation, growth, ultrastructure, and metal tolerance of the green alga, Chlamydomonas acidophila (Chlorophyta). *Biosci Biotechnol Biochem.* **65**, 2650-6.

Nishikawa, K., Yamakoshi, Y., Uemura, I. and Tominaga, N. (2003). Ultrastructural changes in Chlamydomonas acidophila (Chlorophyta) induced by heavy metals and polyphosphate metabolism. *FEMS Microbiology Ecology,* **44**, 253-259.

Nishikawa, K., Onodera, A. and Tominaga, N. (2006a). Phytochelatins do not correlate with the level of Cd accumulation in Chlamydomonas spp. *Chemosphere,* **63**, 1553-9.

Nishikawa, K., Machida, H., Yamakoshi, Y., Ohtomo, R., Saito, K., Saito, M. and Tominaga, N. (2006b). Polyphosphate metabolism in an acidophilic alga Chlamydomonas acidophila KT-1 (Chlorophyta) under phosphate stress. *Plant Science,* **170**, 307-313.

Phillips, N. F., Horn, P. J. and Wood, H. G. (1993). The polyphosphate- and ATP-dependent glucokinase from Propionibacterium shermanii: both activities are catalyzed by the same protein. *Arch Biochem Biophys.* **300**, 309-19.

Pick, U., Bental, M., Chitlaru, E. and Weiss, M. (1990). Polyphosphate-hydrolysis--a protective mechanism against alkaline stress? *FEBS Lett.* **274**, 15-8.

Rao, N. N. and Kornberg, A. (1996). Inorganic polyphosphate supports resistance and survival of stationary-phase Escherichia coli. *J Bacteriol.* **178**, 1394-400.

Roychowdhury, T., Uchino, T., Tokunaga, H. and Ando, M. (2002). Arsenic and other heavy metals in soils from an arsenic-affected area of West Bengal, India. *Chemosphere,* **49**, 605-18.

Sager, R. and Granick, S. (1953). Nutritional studies with Chlamydomonas reinhardi. *Ann N Y Acad Sci.* **56**, 831-8.

van Veen, H. W., Abee, T., Kortstee, G. J., Konings, W. N. and Zehnder, A. J. (1994a). Substrate specificity of the two phosphate transport systems of Acinetobacter johnsonii 210A in relation to phosphate speciation in its aquatic environment. *J Biol Chem.* **269**, 16212-6.

van Veen, H. W., Abee, T., Kortstee, G. J., Pereira, H., Konings, W. N. and Zehnder, A. J. (1994b). Generation of a proton motive force by the excretion of metal-phosphate in the polyphosphate-accumulating Acinetobacter johnsonii strain 210A. *J Biol Chem.* **269**, 29509-14.

Werner, T. P., Amrhein, N. and Freimoser, F. M. (2007). Inorganic polyphosphate occurs in the cell wall of Chlamydomonas reinhardtii and accumulates during cytokinesis. *BMC Plant Biol.* **7**, 51.

In: Algae: Nutrition, Pollution Control and Energy Sources
Editor: Kristian N. Hagen, pp. 23-39
ISBN: 978-1-60692-008-4

Chapter 3

STUDY ON LEAD AND CADMIUM ADSORPTION BY THE ORGANIC COMPONENTS OF NATURAL BIOFILMS

Chunli Kang, Deming Dong[1], Ping Guo, Xiuyi Hua, Fei Peng, Chunyan Su, Jing Guo and Yuxia Zhao

College of Environment and Resources, Jilin University, Changchun, 130012, P. R. China

Abstract

The biosorption of Pb^{2+} and Cd^{2+} by the organic components (exopolymers, nonliving cells, living cells, exopolysaccharides and extracellular proteins of the exopolymers) of the dominant microorganisms obtained from the natural biofilms of the South Lake of Changchun, China and effects of biosorption time, temperature and initial pH value were investigated.

The results showed that Pb^{2+} and Cd^{2+} biosorption process by the exopolymers, the nonliving cells and the living cells was divided into fast phase and slow phase, biosorption equilibrium time was 360, 100 and 100 min respectively; the maximum biosorption amounts were all obtained at pH=6; For the exopolymers, the biosorption amounts increased with the temperature increasing in the range of 15-30°C and then decreased, the effect of temperature on the biosorption ability of the nonliving cells and the living cells was negligible; both Langmuir and Freundlich isotherms could describe the course of the biosorption of Pb^{2+} and Cd^{2+}. The biosorption ability of the exopolymers is much higher than that of the nonliving cells or the living cells.

The results of the biosorption of Pb^{2+} and Cd^{2+} by the exopolysaccharides and extracellular proteins showed that both Langmuir and Freundlich isotherms could also be used to describe the thermodynamics adsorption processes. Temperature and pH value could also affect the adsorption processes. the biosorption ability of the exopolysaccharides is higher than that of the extracellular proteins.

Keywords: lead, cadmium, adsorption, natural biofilms

---

[1] E-mail address: dmdong@mail.jlu.edu.cn (Corresponding author.)

1. Introduction

Biofilms play an important role in natural aquatic environment and wastewater treatment [Akiyoshi O et al., 1996, Dong De-ming et al. 2000, Motoyuki S 1997].In recent years people pay more and more attention to the research on the adsorption of heavy metals by the natural biofilms and most people think the inorganic components such as iron and manganese oxides are much more important in the adsorption process [Dong De-ming et al. 2000]. A few research is reported about the adsorption of heavy metals by the organic components of natural biofilms although the organic components is the necessary part of the natural biofilms and is composed of 80-90% of the natural biofilms by dry weight [Zhang Xiao-qi et al., 1998, Liu Yan-zhi et al., 2005, Andreas J et al., 1998]. But this kind of work is very helpful to study the transport and transformation of heavy metals in water and reveal the adsorption mechanism of the biofilms. It is known that organic components of the biofilms mainly include exopolymers, nonliving cells and living cells and exopolymers are mainly composed of exopolysaccharides and extracellular proteins [Zhang Xiao-qi et al., 1998].

In this chapter we investigated the biosorption of Pb^{2+} and Cd^{2+} by the organic components (exopolymers, nonliving cells, living cells, exopolysaccharides and extracellular proteins of the exopolymers) of the dominant microorganisms obtained from the natural biofilms of the South Lake of Changchun, China and effects of biosorption time, temperature and initial pH value.

2. Experimental Section

2.1. Preparation of the Organic Components of Natural Biofilms

In this experiment, biofilms were cultivated on slides in the natural water of the South Lake of Changchun, China [Lu Wen-xi et al. 2002]. Dominant microorganisms of the cultivated natural biofilms were obtained by the dilution plate method. The culture solution containing the dominant microorganisms was cultivated in an oscillation incubator for 3~4 d (28~30 °C) and then centrifuged for 20 min (17 000 r/min). The matter on the bottom of the centrifugal tube was living cells and after freeze-dried it became nonliving cells. The supernatant was filtrated for 2~3 times and then freeze-dried to get the exopolymer samples. The extracellular proteins were deposited from the filtrated supernatant when ammonium sulfate was added. The exopolysaccharides were deposited from the deproteinized supernatant with 90% ethanol [Flemming H-C, 1995]. The extracellular proteins and the exopolysaccharides were freeze-dried for use as adsorbent.

2.2. Experimental Methods

In the biosorption experiments, the initial concentrations of Cd^{2+} and Pb^{2+} were 4.45 and 19.3 μM respectively and the concentrations in equilibrium solutions were measured by GFAAS or FAAS. All the data are the mean of two paralleled experimental results.

2.2.1. Biosorption by the Exopolymers

Took the 20mL exopolymer solution (63mg/L) into a dialysis bag and put the bag in the 200mL solution containing Cd^{2+} or Pb^{2+} and stirred (100r/min) by a magnetic stirrer untill the adsorption equilibrium and sampled the equilibrium solutions in or out of the bag and measured the concentrations. The biosorption amounts of the exopolymers were calculated by the differences of the concentrations in or out of the bag.

2.2.2. Biosorption by the Living or the Nonliving Cells

1 mg freeze-dried nonliving cells or 0.25 mL suspension solution (dry weight 1 mg) of living cells were respectively put into 5 mL solution containing Cd^{2+} or Pb^{2+}.The solutions were oscillated until the adsorption equilibrium, then sampled and measured the concentrations of Cd^{2+} or Pb^{2+}.

2.2.3. Biosorption by the Exopolysaccharides and the Extracellular Proteins

Took the 20mL exopolysaccharide solution (60mg/L) or the 20mL extracellular protein solution (20mg/L) into a dialysis bag and put the bag in the 200mL solution containing Cd^{2+} or Pb^{2+} and stirred (100r/min) by a magnetic stirrer until the adsorption equilibrium and sampled the equilibrium solutions in or out of the bag and measured the concentrations. The biosorption amounts of the the exopolysaccharides and the extracellular proteins were calculated by the differences of the concentrations in or out of the bag.

3. Results and Discussion

3.1 $Cd^{2+}$and Pb^{2+} Biosorption by the Exopolymers, the Living or the Nonliving Cells

3.1.1. Effect of Biosorption Time on the Adsorption by the Exopolymers, the Living or the Nonliving Cells

Pb^{2+} and Cd^{2+} biosoption dynamic process by the exopolymers, nonliving cells and living cells was divided into fast phase and slow phase (Fig1-1, 1-2).As for exopolymers, 50% of the maximum biosorption amount was gotten within 1h and the adsorption equilibrium reached within 360min. But for the living or nonliving cells, over 90% of the maximum biosorption amount was obtained within 30 min and the adsorption equilibrium reached within 100min. Many reasons were able to result in the rate differences such as structure of the bio-adsorbent, special characters of functional groups, surface binding sites as well as surface area and others.

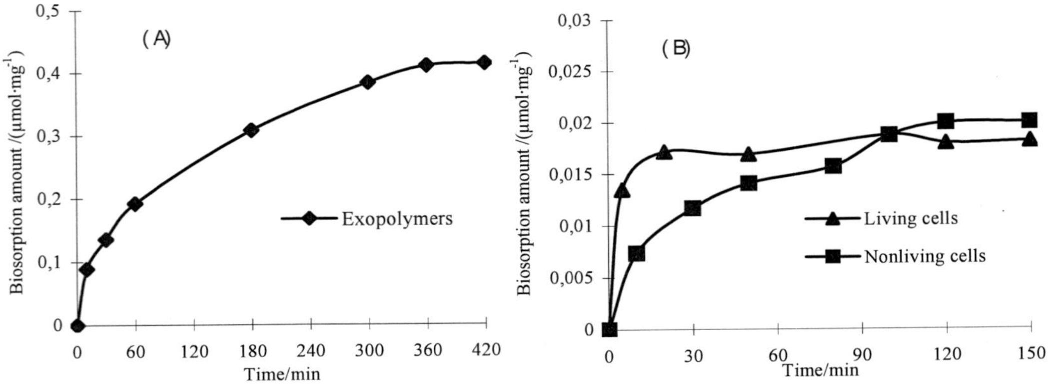

Figure 1.1. Effect of biosorption time on Cd^{2+} adsorption by the exopolymers, living or nonliving cells (25°C, pH=6).

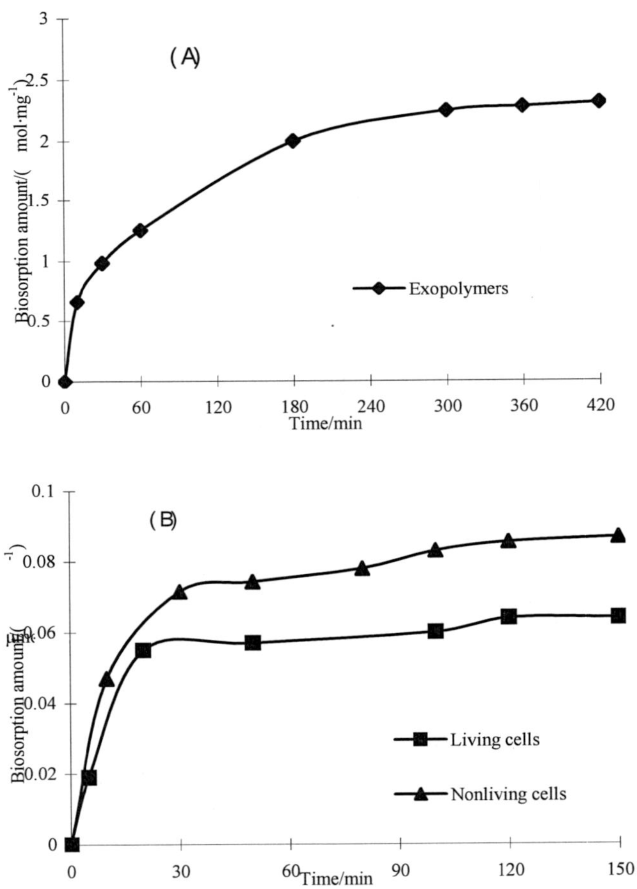

Figure 1.2. Effect of biosorption time on Pb^{2+} adsorption by the exopolymers, the living or the nonliving cells (25°C, pH=6).

3.1.2. Effect of pH Value on the Adsorption by the Exopolymers, Living or Nonliving Cells

pH value is one of the most important factors affecting the biosorption [Volesky B et al., 1993; Liehr S K 1995]. In this experiment, the maximum biosorption amounts by the three organic components were all obtained at pH=6 (Fig.2-1, 2-2) .This was because under the pH<6 acidic conditions large amounts of hydrogen ions competed the surface binding sites with Cd^{2+} or $Pb^{2+}$and under the pH > 6 alkaline conditions Cd^{2+} and Pb^{2+} easily form hydroxyl complex compounds which binding forces with the organic components were weaker than that of the Cd^{2+} or Pb^{2+}[Kang Chun-li et al., 2005a, 2005b]. So both the acidic and the alkaline conditions prohibited the biosorption. But the prohibition of Pb^{2+} biosorption was not obvious at pH=6-8.

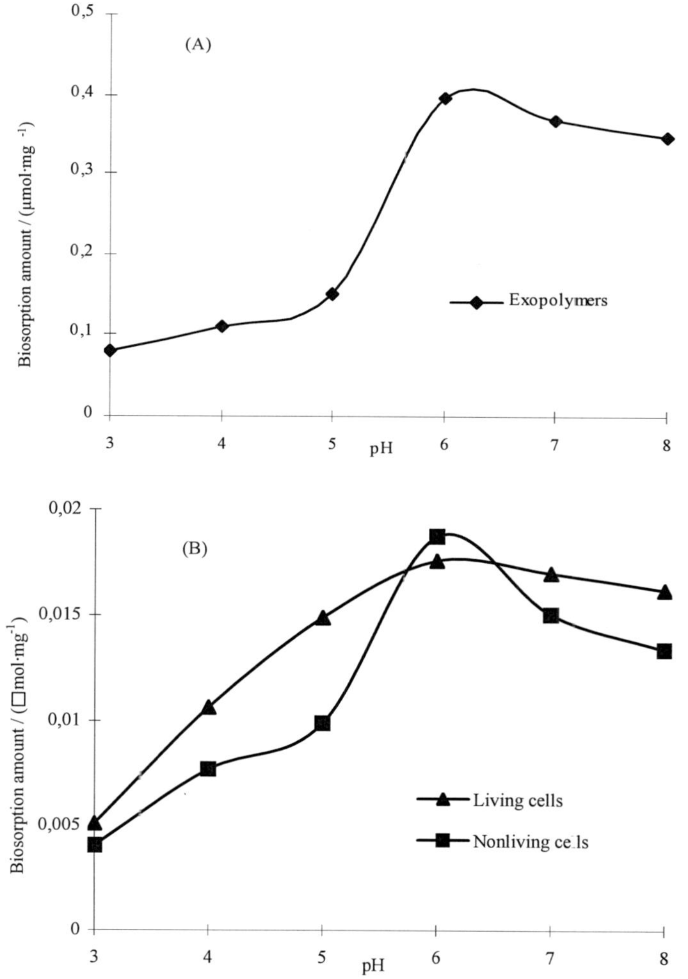

Figure 2.1. Effect of pH value on Cd^{2+} adsorption by the exopolymers, the living or the nonliving cells (25°C).

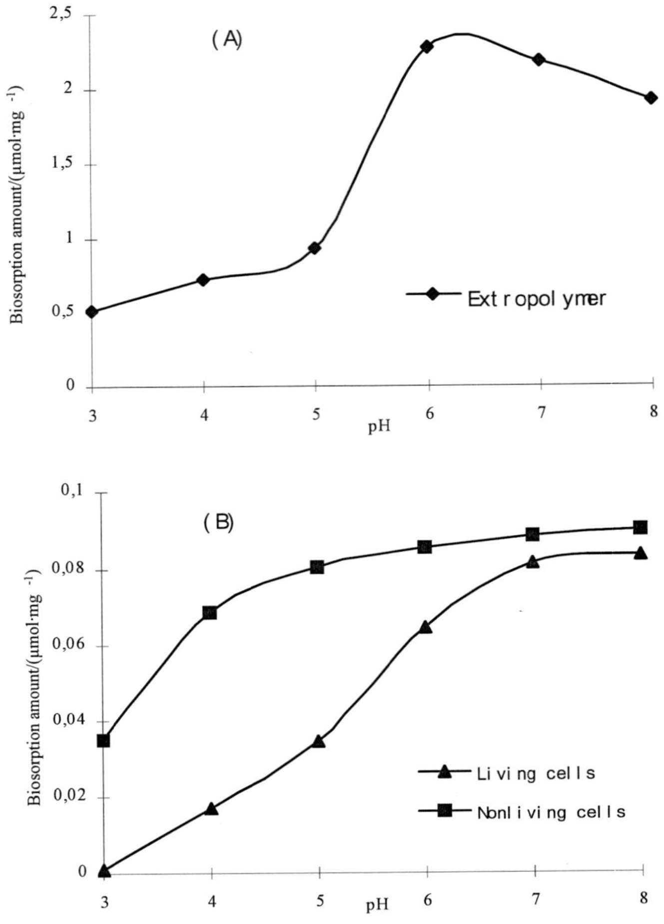

Figure 2.2. Effect of pH value on Pb^{2+} adsorption by the exopolymers, the living or the nonliving cells (25°C).

3.1.3. Effect of Temperature on the Adsorption by the Exopolymers, the Living or the Nonliving Cells

Figure 3.1 and Figure 3.2 showed the effect of temperature on the adsorption. For exopolymers, the biosorption amounts increased with the temperature increasing in the range of 15-30 °C and then decreased. The reason was that raising temperature was benefit for chemical adsorption reaction but the too much higher temperature would affect the bioactivity of exopolymers and make the biosorption amounts decreased, as for nonliving cells and living cells, the effect of temperature on the adsorption was negligible in the range of 15-35 °C.

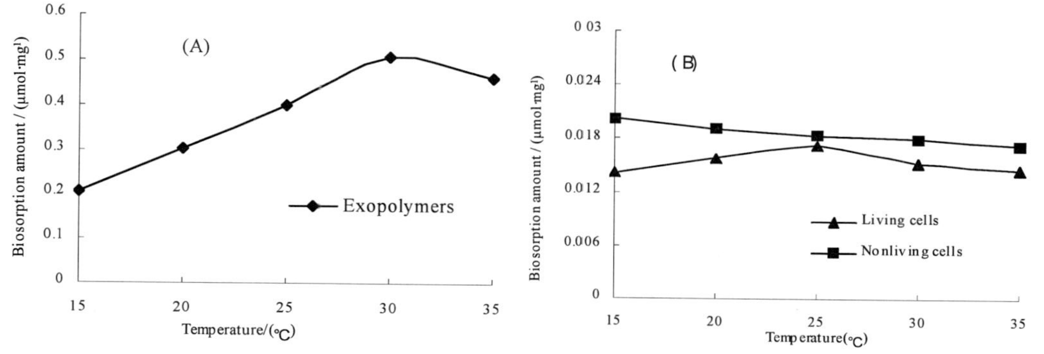

Figure 3.1. Effect of temperature on Cd^{2+} adsorption by the exopolymers, the living or the nonliving cells (pH=6).

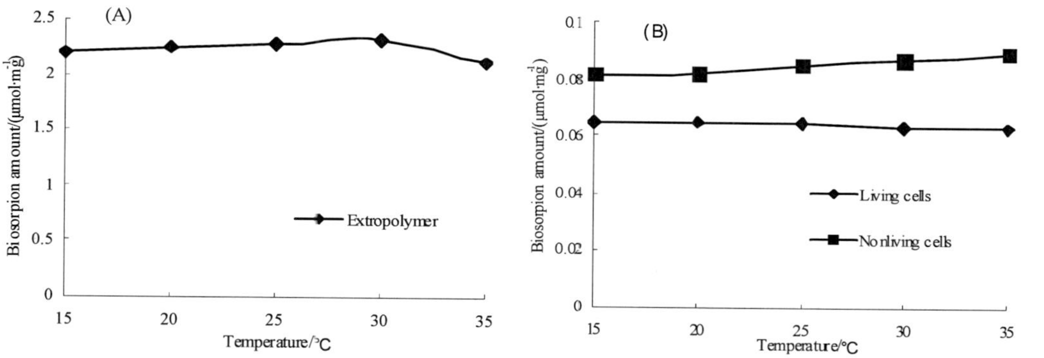

Figure 3.2. Effect of temperature on Pb^{2+} adsorption by the exopolymers, the living or the nonliving cells (pH=6).

3.1.4. Cd^{2+} and Pb^{2+}Adsorption Isotherms by the Exopolymers, Living or Nonliving Cells

The adsorption isotherm curves of Cd^{2+}, Cd^{2+} with Pb^{2+} present, Pb^{2+} or Pb^{2+} with Cd^{2+} present were showed in Fig.4-1 and Fig.4-2 respectively. Both Langmuir and Freundlich isotherms were applied to estimate the adsorption parameters (Table 1-1, 1-2, 2-1, 2-2) which reflected the adsorption ability of adsorbent, the stability of the metal ions after adsorbed, the effect of the concentrations of the metal ions and the affinity between the adsorbents and the metal ions [Ridyan S et al., 1999]. The K_d, N and K listed in the table showed that the biosorption ability of the exopolymers was much higher than that of the nonliving cells or living cells.

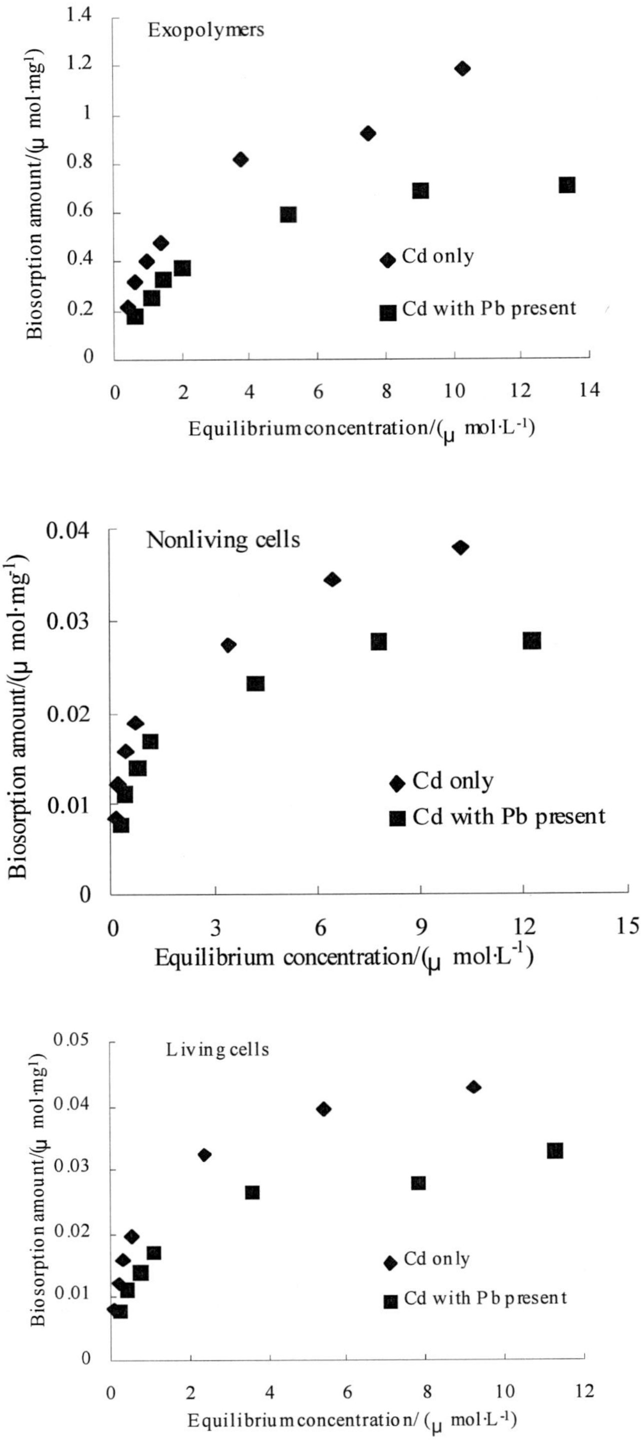

Figure 4.1. Adsorption isotherm curves of Cd^{2+} by the exopolymers, the living or the nonliving cells (25°C, pH=6).

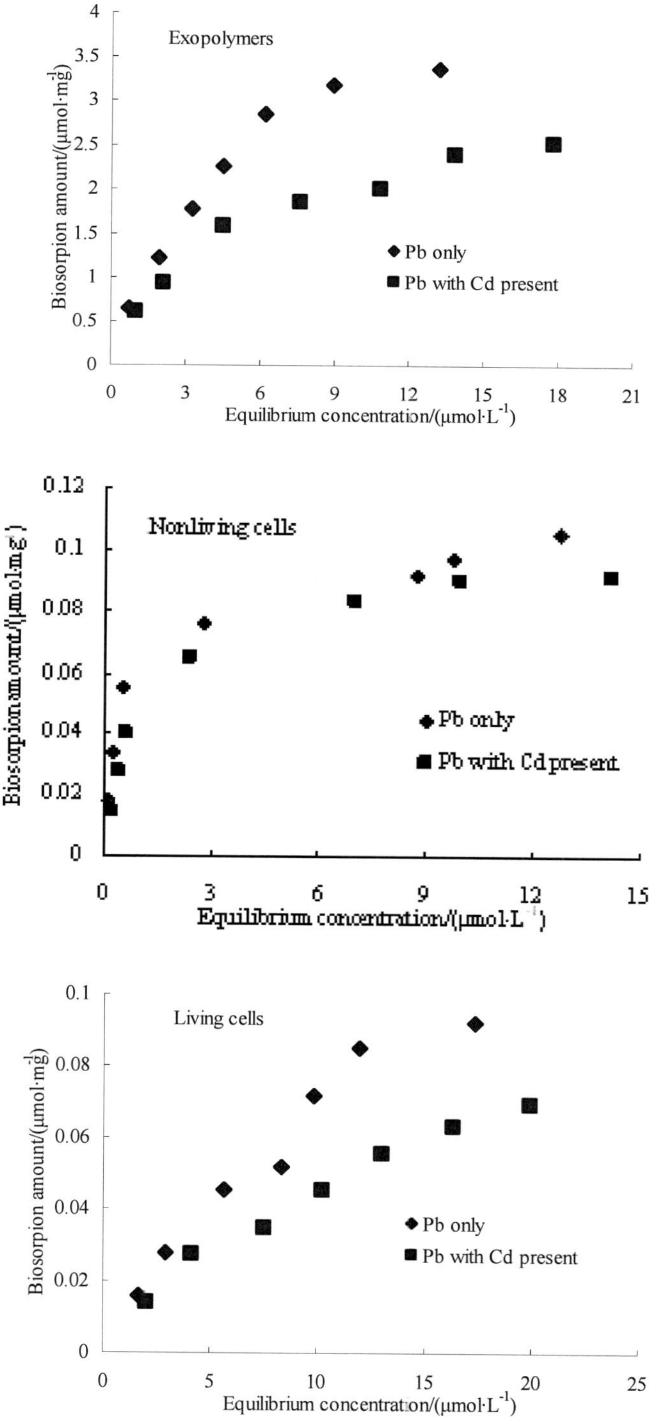

Figure 4.2. Adsorption isotherm curves of Pb^{2+} by the exopolymers, the living or the nonliving cells (25°C, pH=6).

Table 1.1. Freundlich and Langmuir adsorption parameters of Cd^{2+} adsorption by the exopolymers, the living or the nonliving cells (25°C, pH=6)

Organic component	Freundlich equation $\mathrm{Ln}\Gamma = \mathrm{Ln}\,K_d + (1/N)\cdot \mathrm{Ln}\,C$			Langmuir equation $1/\Gamma = 1/\Gamma_{max} + 1/K\Gamma_{max}\cdot 1/C$		
	K_d	N	r	Γ_{max}	K	r
Exopolymers	0.3824	2.0325	0.9890	1.2038	0.3824	0.9964
Nonliving cells	0.0500	2.8241	0.9453	0.0319	0.0028	0.9909
Living cells	0.0212	2.6875	0.9821	0.0386	0.0031	0.9934

Table 1.2. Freundlich and Langmuir adsorption parameters of Pb^{2+} adsorption by the exopolymers, living or nonliving cells (25°C, pH=6)

Organic component	Freundlich equation $\mathrm{Ln}\Gamma = \mathrm{Ln}\,K_d + (1/N)\cdot \mathrm{Ln}\,C$			Langmuir equation $1/\Gamma = 1/\Gamma_{max} + 1/K\Gamma_{max}\cdot 1/C$		
	K_d	N	r	Γ_{max}	K	r
Exopolymers	0.8385	2.0210	0.9866	4.3740	4.4160	0.9964
Nonliving cells	0.0500	2.8241	0.9654	0.1607	0.0182	0.9947
Living cells	0.0113	1.4667	0.9895	0.0105	0.0025	0.9961

Table 2.1. Freundlich and Langmuir adsorption parameters of Cd^{2+} adsorption with Pb^{2+} present by the exopolymers, living or nonliving cells (25°C, pH=6)

Organic component	Freundlich equation $\mathrm{Ln}\Gamma = \mathrm{Ln}\,K_d + (1/N)\cdot \mathrm{Ln}\,C$			Langmuir equation $1/\Gamma = 1/\Gamma_{max} + 1/K\Gamma_{max}\cdot 1/C$		
	K_d	N	r	Γ_{max}	K	r
Exopolymers	0.2523	2.2257	0.9797	0.8251	0.2970	0.9973
Nonliving cells	0.0139	3.0722	0.9716	0.0285	0.0011	0.9968
Living cells	0.0146	2.7917	0.9801	0.0304	0.0008	0.9957

Table 2.2. Freundlich and Langmuir adsorption parameters of Pb^{2+} adsorption with Cd^{2+}present by the exopolymer, living or nonliving cells (25°C, pH=6)

Organic component	Freundlich equation $\mathrm{Ln}\Gamma = \mathrm{Ln}\,K_d + (1/N)\cdot \mathrm{Ln}\,C$			Langmuir equation $1/\Gamma = 1/\Gamma_{max} + 1/K\Gamma_{max}\cdot 1/C$		
	K_d	N	r	Γ_{max}	K	r
Exopolymers	0.6566	1.657	0.9893	2.84	2.133	0.9957
Nonliving cells	0.0375	2.4925	0.9579	0.0847	0.0092	0.9901
Living cells	0.0093	1.2	0.9934	0.0081	0.0009	0.9956

It can be deduced from table 2-1and table 2-2 that Langmuir and Freundlich isotherms also be suitable to describe the adsorption of Cd^{2+} with Pb^{2+} present or Pb^{2+} with Cd^{2+} present。But both the adsorption amounts and the maximum adsorption amounts were decreased. This indicated that the adsorption mechanism of Cd^{2+} and Pb^{2+} was similar. In the adsorption process, the coexistent metal ions competed the surface binding sites and the definite quantities of binding sites resulted in the adsorption amounts decreased.

3.2. $Cd^{2+}$and Pb^{2+} Biosorption by the Exopolysaccharides and the Extracellular Proteins

3.2.1. Effect of Biosorption Time on the Adsorption by the Exopolysaccharides and the Extracellular Proteins

Pb^{2+} and Cd^{2+} biosorption dynamic processes by the exopolysaccharides and the extracellular proteins were showed in Fig5-1 and Fig.5-2.The concentrations of the metal ions in the solution decreased rapidly at first within 20 min and then slowly within 2h until reaching the adsorption equilibrium within 6h.

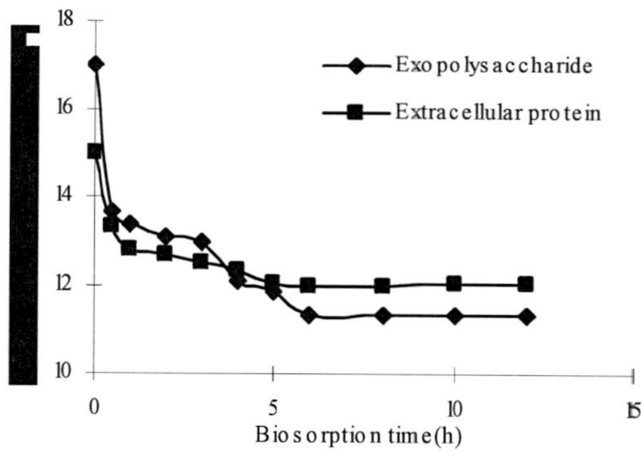

Figure 5.1. The process of Pb^{2+} adsorption by the exopolysaccharides and the extracellular proteins (20°C, pH=6).

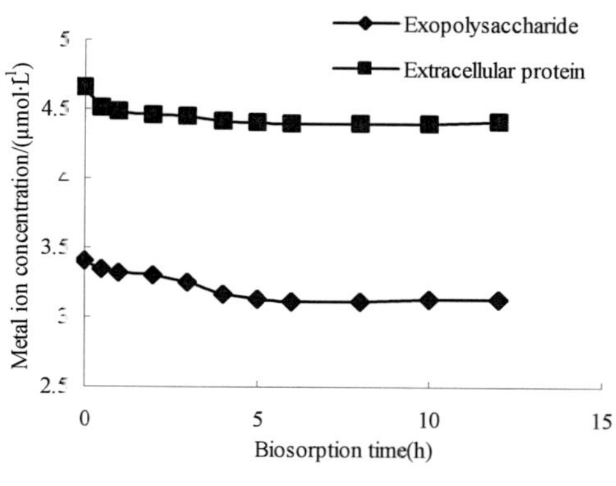

Figure 5.2. The process of Cd^{2+} adsorption by the exopolysaccharides and extracellular proteins (20°C, pH=6).

3.2.2. Effect of pH Value on the Adsorption by the Exopolysaccharides and the Extracellular Proteins

The maximum biosorption amounts by the exopolysaccharides and the extracellular proteins were also obtained at pH=6 (Fig.6-1and Fig.6-2). This result was the same with that of the exopolymers, living or nonliving cells and the reason also would be the same.

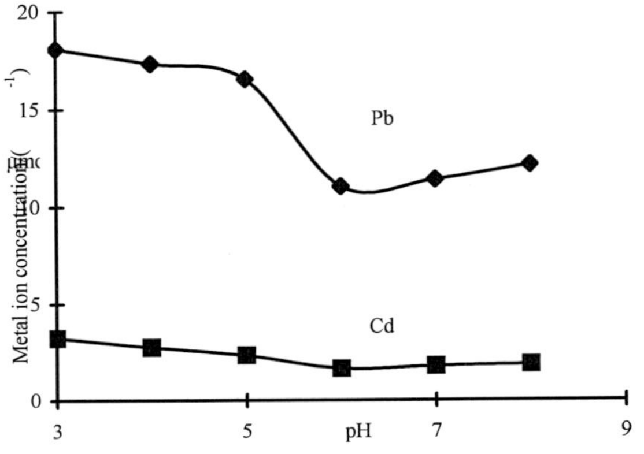

Figure 6.1. Effect of pH value on Pb^{2+} and Cd^{2+} adsorption by the exopolysaccharides (20°C).

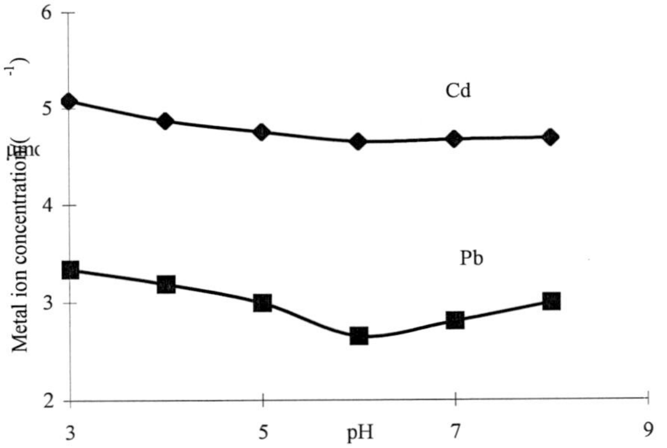

Figure 6.2. Effect of pH value on Pb^{2+} and Cd^{2+} adsorption by extracellular protein (20°C).

3.2.3. Effect of Temperature on the Adsorption by the Exopolysaccharides and the Extracellular Proteins

From Fig.7-1 to 7-2 it could be seen that in the range of 10-30 °C the biosorption amounts increased with the temperature raising and this was possibly caused by the higher adsorption reaction activity at higher temperature.

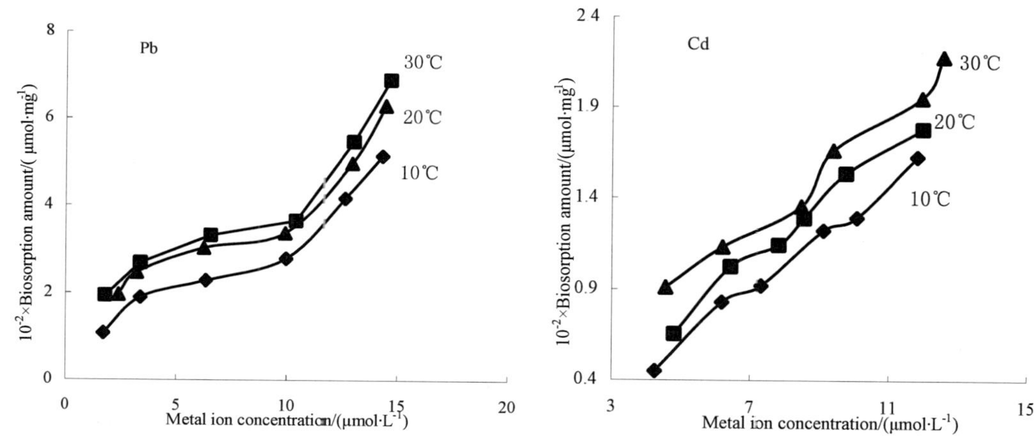

Figure 7.1. Effect of temperature on Cd^{2+} and Pb^{2+} adsorption by exopolysaccharides (pH=6).

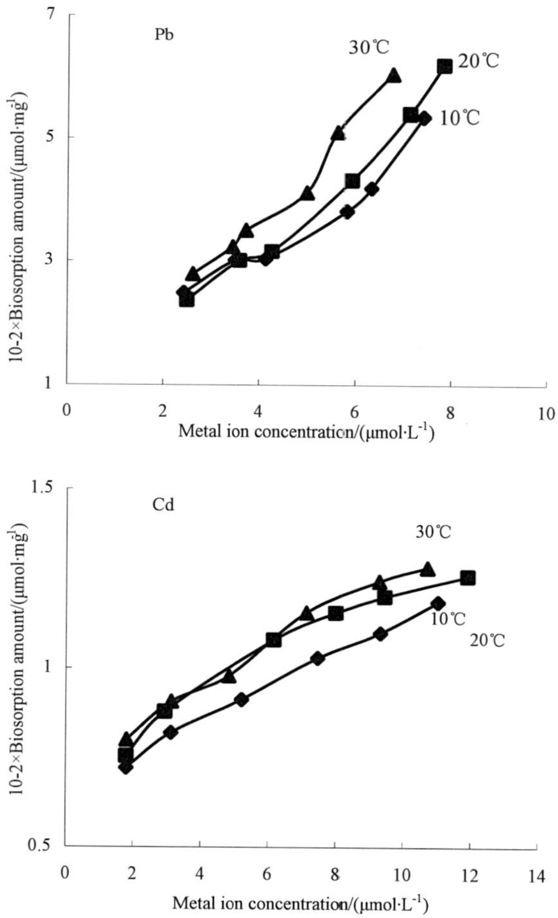

Figure 7.2. Effect of temperature on Cd^{2+} and Pb^{2+} adsorption by the extracellular proteins (pH=6).

3.2.4. The Cd^{2+} with Pb^{2+} Adsorption Isotherms by the Exopolysaccharides and the Extracellular Proteins

The adsorption isotherm curves of Cd^{2+}, Cd^{2+} with Pb^{2+} present, Pb^{2+} or Pb^{2+} with Cd^{2+} present were showed in Fig.8-1 to Fig.8-2 respectively. Both Langmuir and Freundlich isotherms were applied to estimate the adsorption parameters (Table 3) which indicated that both Langmuir and Freundlich isotherms were able to describe the adsorption and the stability of the Pb^{2+} after adsorbed is better than Cd^{2+}, the Pb^{2+} adsorption amounts were larger than Cd^{2+}. For both Cd^{2+} and Pb^{2+}, the adsorption amounts by the exopolysaccharides were larger than the extracellular proteins. Both Cd^{2+} and Pb^{2+} adsorption amounts were decreased with other metal ions present (Fig.8-1 to Fig.8-2) .This was also due to the definite quantities of the surface binding sites and other metal ions present together would possess some of them.

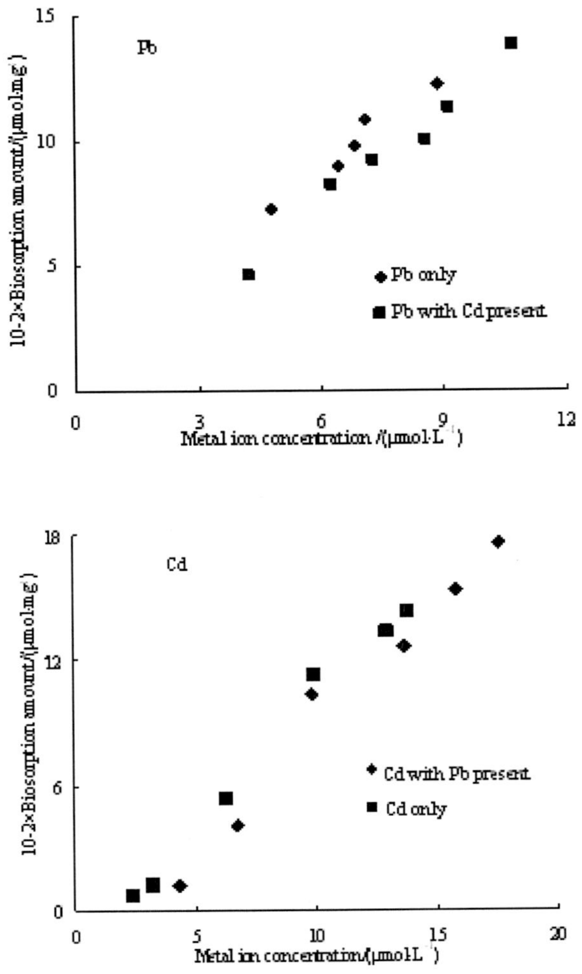

Figure 8.1. Adsorption isotherm curves of Cd^{2+} and Pb^{2+} by the exopolysaccharides (20°C, pH=6).

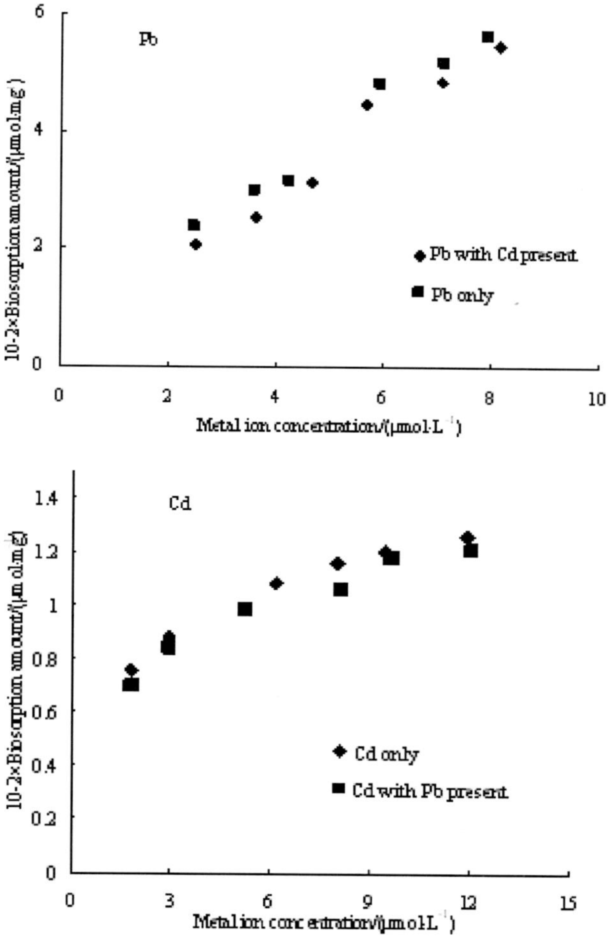

Figure 8.2. Adsorption isotherm curves of Cd^{2+} and Pb^{2+} by the extracellular proteins (20°C, pH=6).

Table 3. Freundlich and Langmuir adsorption parameters of Cd^{2+} and Pb^{2+} by the exopolysaccharides and the extracellular proteins (20°C, pH=6)

Organic component	Metal ion	Langmuir equation $\mathrm{Ln}\Gamma = \mathrm{Ln}\, K_d + (1/N) \cdot \mathrm{Ln}\, C$			Freundlich equation $1/\Gamma = 1/\Gamma_{max} + 1 / K\Gamma_{max} \cdot 1/C$		
		$1/(Q^0 b)$	$1/Q^0$	r	$1/n$	lg K	r
Exopolysaccharides	Pb^{2+}	0.3927	0.0633	0.9965	0.5025	0.4926	0.9936
	Cd^{2+}	0.1204	0.2775	0.9963	0.0699	0.4629	0.9758
Extracellular proteins	Pb^{2+}	0.1331	0.2869	0.9921	0.2417	0.0863	0.9785
	Cd^{2+}	1.1214	0.7276	0.9905	0.2445	0.0163	0.9860

4. Conclusion

$Cd^{2+}$and Pb^{2+} biosorption amounts by the five organic components all were increased with the adsorption time increasing and the adsorption equilibrium times respectively were 360min (the exopolymers), 100min (the living or the nonliving cells), 360 min (the exopolysaccharides and the extracellular proteins).Initial pH value took a great effect on the biosorption. The maximum biosorption amounts by the five organic components were all obtained at pH=6. The condition of pH<6 or pH > 6 generally prohibited the biosorption. For exopolymers, the biosorption amounts increased with the temperature increasing in the range of 15-30°C and then decreased. As for the nonliving cells and the living cells, the effect of the temperature on the adsorption was negligible in the range of 15~35 °C. While for the exopolysaccharide and the extracellular proteins, the biosorption amounts increased with the temperature raising in the range of 10-30°C.Both Langmuir and Freundlich isotherms were able to describe the $Cd^{2+}$and Pb^{2+} biosorption by the five organic components.Cd^{2+}or Pb^{2+} biosorption amounts were decreased with other metal ions present together. $Cd^{2+}$and Pb^{2+} biosorption ability followed the order: the exopolymers >> the nonliving cells > the living cells, the exopolysaccharides > the extracellular proteins.

Acknowledgments

This research was supported by the National Natural Science Foundation of China (NSFC, No. 20477014 and 20077011) and the Specialized Research Fund for the Doctoral Program of Higher Education (SRFDP, No. 20060183020) provided by the Ministry of Education, P. R. China.

References

Akiyoshi O, Hideki H.. A novel concept for evaluation of biofilm adhesion strength by applying tensile force and shear force. *Water Science Technology*. 1996, 34(5-6): 201-211

Andreas J, Per H N.. Cell biomass and exopolymer composition in sewer biofilms, *Water Science and Technology*. 1998, 37(1): 17-24

Dong De-ming, Zhang Bai-yu, Yue Bao-hua et al.. Separation of components of natural biofilms by chemical extraction, *Acta Scientiarum Naturalium Universitatis Jilinensis*. 2000, (4):71-74

Flemming H-C.. Sorption sites in biofilms, *Water Science and Technology*. 1995, 32(8): 27-33

Kang Chun-li, Guo Jing, Guo Ping et al.. Biosorption of Cd^{2+} by the organic components of surface coatings in natural waters, *Ecology and Environment*. 2005a, 14(5):636-639

Kang Chun-li, Guo Jing, Guo Ping et al.. Adsorption behavior of Pb^{2+} by organic components of natural biofilms in natural water and the influencing factors, *Chemical Journal of Chinese Universities*. 2005b, 26(11):2043-2045

Liehr S K, Effect of pH on metals precipitation in denitrifying biofilms, *Water Science and Technology*, 1995, 32 (8): 179-183

Lu Wen-xi, Zhu Yan-cheng, Tian Ye. Journal of Northeast Normal University, *Natural Science Edition*. 2002, 34 (3) : 103-107

Liu Yan-zhi, Tian SHheng-Yan, Shang Ping et al.. Adsorption of heavy metals Na-montmorillonite: effect of pH and organic acid, *Ecology and Environment*. 2005, 14(3): 353-356

Motoyuki S. Role of adsorption in water environment processes, *Water Science Technology*. 1997, 35(7): 1-11

Ridyan S, Adil D. Biosorption of cadmium (II), lead (II) and copper (II) with the filamentous fungus Phanerochaete chrysosporium, *Bioresearch Technology*. 1999, 70: 95—104

Volesky B, May H, Holan Z R.. Cadmium biosorption by saccharomyces, *Biotechnology Bioengineering*. 1993, 41: 826-829

Zhang Xiao-qi, Paul L Bishop, Margaret J Kupferle. Measurement of polysaccharides and proteins in biofilm extracellular polymers, *Water Science and Technology*. 1998, 37(4-5):345-348

In: Algae: Nutrition, Pollution Control and Energy Sources ISBN: 978-1-60692-008-4
Editor: Kristian N. Hagen, pp. 41-56

Chapter 4

DESICCATION TOLERANCE IN GREEN ALGAE: IMPLICATIONS OF PHYSIOLOGICAL ADAPTATION AND STRUCTURAL REQUIREMENTS

Andreas Holzinger

University of Innsbruck, Institute of Botany, Department of Physiology and Cell Physiology of Alpine Plants, Sternwartestr. 15, A-6020 Innsbruck

Abstract

Several green algal groups are known to tolerate transient desiccation in their vegetative state. This appears to be a substantial advantage when spreading to new habitats. Only recently the physiology of algae form arid habitats or man-made surfaces has been investigated extensively. The phylogenetic relationships of desiccation tolerant green algae have been determined at a molecular level, and substantial progress has been made in understanding avoidance or damage repair strategies. In addition to physiological adaptations, structural conditions like thickened cell walls covered with or containing substances that reduce evaporation and prevent desiccation are an important strategy in these organisms. Very limited information is available on the structure and ultrastructure of desiccated vegetative algal samples. This however appears to be important, when attempting to understand how a quick recovery from desiccation stress is achieved. In this chapter transmission electron microscopic data will demonstrate possible effects of desiccation in algal samples. Moreover it is demonstrated that in desiccation tolerant algae organelles like the chloroplast have the capacity to remain structurally integer during desiccation. These are new results that contribute to our understanding of desiccation tolerance in green algae.

1. Introduction

At a first view most dry habitats appear free of algae. As this chapter will show, this is definitely not the case. Desiccation tolerant green algae may not be very obvious, but it will be pointed out that the capacity of several groups of green algae to survive in a rather dry surrounding is extremely important since microalgal species account as basic food source for invertebrates (Lukesova and Frouz 2007). To some extent a better understanding of desiccation reactions in algae may help to control unwanted algal growth.

Limiting factors for terrestrial life of green algae are manifold (eg. Brown et al. 1964). Besides damage through high radiation (eg. Holzinger and Lütz 2006, Karsten et al. 2007a, Chockell et al. 2008) or insufficient supply with nutrients, the availability of liquid water appears to be one key restricting factor (eg. Oliver 2005a, Karsten et al. 2007b). Desiccation tolerance and functional adaptation to arid zones is well documented in the literature for cyanobacteria, lichens (eg. Lange 1989; 1997, Fletcher 2007) as well as green algae (eg. Gray et al. 2007, Lewis 2007). Most of these works are dealing with physiological aspects. Also in bryophytes and other "poikilohydric" organisms investigations on the behaviour and physiology upon drying have been performed (eg. Proctor and Smirnoff 2000, Heber et al. 2006, Oliver et al. 2005b). In contrast, only little information is available on the effects of desiccation by means of microscopic or transmission electron microscopic observations. For higher plants, where the phenomenon of desiccation tolerance is rarely realized for the vegetative state, only a small number of publications can be found. The situation is completely different eg. for seeds where a recent book summarizes the results on this issue (Jenks and Wood 2007). For the vegetative organs like leaves early studies focus on the maintenance of chloroplast structure, the ability to survive is explained with the ability to maintain certain structures eg. the thylakoids (Hallam and Gaff 1978). This is particularly interesting, as also in poikilohydric bryophytes the same "strategy" appears to be established (Proctor and Smirnoff 2000, Proctor et al. 2002; 2007), pointing out that at this level desiccation tolerance is linked with very early prokaryotic abilities to withstand drought. This appears to be conserved throughout the plant kingdom, regardless of the evolutionary level reached. An excellent overview on desiccation tolerance in a holistic way is given by Alpert (2006). There, the relations between organism size and desiccation tolerance is drawn and the generalized strategies also valid for the animal kingdom are pointed out. It is reported which genes could be involved in the desiccation relevant avoidance or protection strategies. Most of these genes are still present even in advanced eukaryotes, therefore it can only be speculated if these genes are either not needed in the vegetative state or recruited to other functions.

This chapter will be highlighting the situation in green algae. The moreover intention is to demonstrate strategies that allow and realize desiccation tolerance. Therefore, some information on prokaryotic desiccation tolerance is included as these strategies are regarded as "natures toolbox" (eg. Potts 1994, 2001; Fletcher 2007). Comparisons with lichens, bryohpyts and higher plants will be drawn, but the focus remains on desiccation tolerance in green algae. This topic will be discussed in a more general way – so the key question addressed will be: what is needed to keep green algae in their vegetative state alive in a "water-lacking" environment?

2. Desiccation Tolerant Green Algae

In this chapter selected examples of desiccation tolerant green algae will be presented. It should represent an overview on the existing literature in combination with own observations on desiccation tolerant green algae – but without claiming to be complete. For an elegant molecular based phylogenetic overview on desiccation tolerant green algae the reader is referred to Lewis (2007). These results are based mostly on desiccation tolerant algae collected in US deserts. A diversification of desert lineages is given, which is very useful in understanding their phylogeny. Some additional considerations on the systematics of this group is given by Lopez-Bautista et al. (2007). Systematic or taxonomic issues will not be further discussed here, but a profound knowledge of this background appears to be important for the understanding of desiccation tolerance. Morphologically difficult to distinguish unicellular or filamentous green algae might react extremely different in their physiology. This can be understood in relation to their different ancestors that may have conserved or evolved "desiccation strategies". So, when just sticking to a few established model organisms, we might very well miss important issues. As pointed out by Lewis (2007) more sophisticated strategies in molecular analysis like ITS rDNA sequence analysis will give important insights in strains of eg. *Klebsormidium*, *Cylindrocystis* or any other genera that can be isolated from distinct habitats. Especially these relationships promise insights in strategies of different algae.

Figure 1. Alpine soil alga Zygogonium ericetorum observed and prepared for TEM right after collection from it's natural habitat. bare surface of alpine a) Image of the bare soil with dried out algae at the sampling site near Obergurgl, Tyrol Austria at 2300 m a.s.l. b) light microscopic image of the coloration in rewetted state c) low magnification TEM image of an algal sample collected at the dry growing site d) TEM micrograph of desiccated sample with condensed cytoplasm and basically intact chloroplast, note the tremendous amount of plastoglobules (arrow) and the damaged cytoplasm. Bars b,c 10 μm; d 1 μm.

2.1. Green Algae in the Soil

Where do we expect to find desiccation tolerant green algae? The first "place to look" for green algae is on the bare soil (edaphic – eg. Trainor and Glydych 1995, Nienova 1996, Hoppert et al. 2004). These organisms are at least periodically exposed to desiccation stress and depending on the habitat this may last for prolonged time periods. Gray et al. (2007) give a nice comparison of the physiological effects when treating desiccation tolerant desert algae and closely related aquatic relatives. They report on the survival rates after a 4-week desiccation period in the chlorophycean genera *Bracteacoccus* sp., *Scenedesmus rotundus*, *Chlorosarcinopsis* sp., and the trebouxiophycean genera *Chlorella* sp. and *Myrmecia* sp. depicted from desert habitats. *Trentepohlia odorata* is another prominent candidate that tolerates desiccation (Ong et al. 1992).

Another example of pioneering desiccation tolerant organisms is taken from a sampling site in the Austrian Alps at 2300 m a.s.l. There soil algal communities are established (eg. Ettl and Gärtner 1995). These algal communities are dominated by a variety of different organisms including cyanobacteria and lichens, but also green algae like *Klebsormidium* sp., *Stichococcus* sp. (see also Büdel 2002). In a recent project these algae have been collected and prepared for TEM investigations directly after collection in the field. Here, the attention should be drawn to an organism which tolerates desiccation in the vegetative state: *Zygogonium ericetorum*. (Figure 1). The species is relatively easy to determine, even in the dried state – the filamentous algae form reddish-green/white sheets of dried cells directly on the soil (Figure 1a). These sheets have been collected and either rewetted for observations in the light microscope (Figure 1b) or directly subjected to TEM fixation in the dried state. At the TEM level these samples have a rather "destroyed" and dense appearance with high electron contrast in the cytoplasm (Figure 1c), but the chloroplasts and other large organelles like the nucleus remain structurally integer. Chloroplasts show the regular arrangement with thylakoid membranes, but are densely filled with plastoglobules (Figure 1d). This can be regarded as "stress-indicator", as the lipids of the thylakoid membranes can be "melted" into plastoglobules. *Vice versa*, plastoglobules are believed to be the source for new lipid like components of the thylakoid membranes and biosynthetic enzymes, when the chloroplasts recover (for further reading on plastoglobules see eg. Austin et al. 2006). The interpretation how organelles like chloroplasts can remain intact is not fully understood. Especially in *Zygogonium ericetorum* a reddish colored vacoular content (Figure 1b) may contribute to this phenomenon as well, or give protection against other stresses (UV irradiation) during the desiccation process. It has to be pointed out that the beneficial effect of intact chloroplasts is an extremely fast recovery of physiologic reactions like photosynthesis after rehydration (see above).

2.2. Algae on Anthropogenic "Open" Surfaces

Other "places to look" for green algae are the anthropogenic pendants to open surfaces, eg. the facades of buildings (Hofbauer et al. 2003, Häubner et al. 2006, Rindi 2007). These habitats may appear rather different to natural habitats from a human point of view, but from the algal demands there is not much of a difference (as long as no toxic components from paint etc. are involved). Indeed a high biodiversity of organisms, based on 18S rRNA sequence data, living on man-made surfaces has been found (eg. Karsten et al. 2007b). These authors also clearly state that the critical environmental factor ensuring ecological success of these organisms appears to be water availability (Häubner et al. 2006). But also nutrient availability might be an important factor, and organisms that tolerate limited water availability together with low nutrition are certainly the ecologically most fittest to allow a pioneering inhabitation of "new" open surfaces. The most extreme anthropogenic environments are worked stones or concrete surfaces. In the following two selected examples of own observation are described:

Green algae usually live directly on the open surface, but only when factors like periodic water availability and moderate light exposure are secured. This shall be illustrated at a prominent example: The *Basilique du Sacré-Cœur* (Paris, France) consists of lime-stone (Travertin). North-facing walls show a green coloration caused by green algal inhabitants (Figure 2a). The most abundant green algal species living on the surface are *Klebsormidium* sp. and *Apatococcus* cf. *lobatus*. (Figure 2b). These algae have been collected at ~ 2 m above the ground, so a local nutrition by eg. dog faeces is not expected there. However, a general nutrition level due to bird faeces washed from the top of the roof can not be excluded. It was easy to scratch off the green algae from the lime-stone, demonstrating that these algae live indeed on the surface. In contrast the west facing walls appear "black" due to endolithic cyanobacteria (Figure 2a). The cyanobacteria avoid "open" surfaces by becoming "endolithic" and thus creating a protected surrounding.

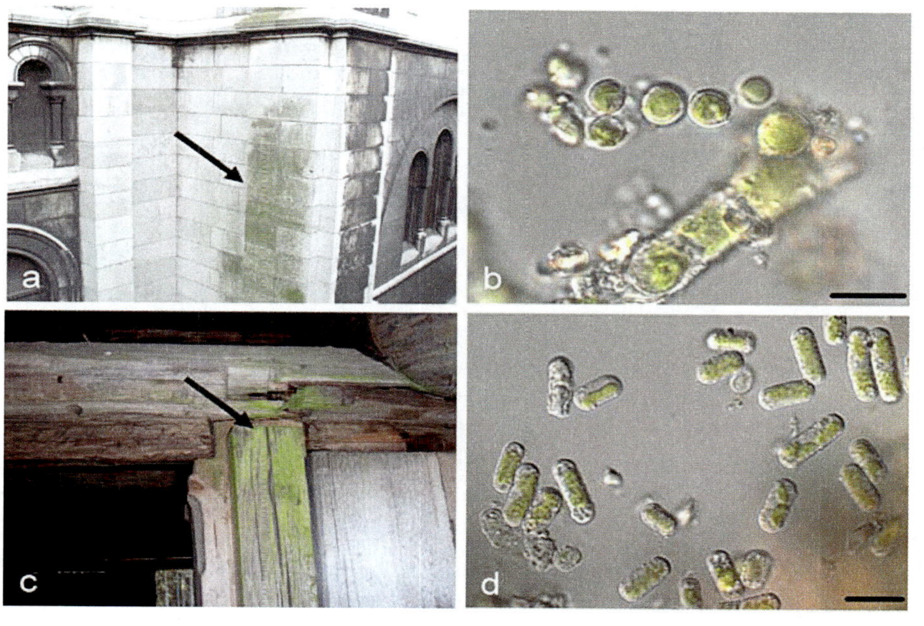

Figure 2. Examples for desiccation tolerant green algae in anthropogenic habitats. a) lime-stone (Travertin) surface of Basilique du Sacré-Cœur (Paris, France), west facing walls inhabited by endolithic cyanobacteria, smaller north facing walls are covered with green algae (arrow) b) DIC image of Klebsormidium sp. and Apatococcus cf. lobatus as the most abundant algae from the sampling site indicated in a, c) ~200-year-old Austrian wood construction building (near Salzburg, Austria) covered with a dense "powder" of aerophytic green algae d) the dominating species from this growing site is Stichococcus cf. bacillaris. All demonstrated examples are at least temporary desiccation tolerant. Bars 10 μm.

Elegant works of Häubner et al. (2006) and Karsten et al. (2007b) demonstrate the physiological effects of desiccation on these algae which have also become of economic relevance. The unfavourable coloration of facades, has to be removed periodically, which is a substantial cost factor in building maintenance.

Another, not so obvious example of green algal organisms growing on the surface of buildings are organisms that grow on wood eg. on old buildings like in the Austrian alpine region (Figure 2c). The demonstrated example was found on wood that is either directly under a roof extension, or in shaded rather moist parts of buildings. These organisms have the ability to utilize moisture for a positive net photosynthesis that remains either stored in the wood or is held by the special "micro climate" situation. Moreover they are virtually excluded from liquid water. The taxa found in these habitats are mostly "pure cultures" of *Stichococcus* cf. *bacillaris* (Figure 2d). In a dried out stage, the "green powder" formed by these algae can be removed easily from the wood surface. This powder remains green even during long dry periods, indicating that thylakoid architecture (and chlorophyll contents) remain intact.

2.3. Green Algae in Intertidal Zones

A third "place to look" for desiccation tolerant, or at least periodically drying out green algae are the intertidal zones of ocean shores. A vast variety of sea weed is living there and mostly brown algae have an astonishing ability to tolerate desiccation as summarized by Garbary (2007). From the green algae, one prominent example is *Ulva* sp. Despite living in these zones and dealing very well with UV irradiation (eg. Bischof et al. 2002) only limited desiccation tolerance capacity has been reported (Beer and Eshel 1983a). A critical factor especially for *Ulva* sp. might be, that different enzymes vary in their sensibility to water loss. For example carbonic anhydrase (the enzyme that converts the water dissolved HCO_3^- into photosynthetically useable CO_2), located at the cell's surface seems to be rather easily damaged when drying out (Beer and Eshel 1983b). As a result especially during spring time large colonies die. This phenomenon is also macroscopically visible by totally bleached thalli, which indicates total damage to the photosynthetic apparatus.

In contrast *Prasiola crispa* can be regarded as a very robust generalist, not only inhabitating intertidal zones, but all sorts of rocky environments. *Prasiola* has become a model organism for extensive studies concerning salt stress, photosynthetic performance and osmoregulation (for a summary see Holzinger and Lütz 2006). *Prasiola* is capable of producing MAAs (Karsten et al. 2005). It is well documented in the literature that *Prasiola* can withstand almost any unfavourable condition, as long as nitrogen availability is guaranteed (eg. Davey 1989, Rindi et al. 1999, Holzinger et al. 2006). This nitrogen dependence might be the critical factor for the synthesis of desiccation tolerance promoting substances. This is not only the case for *Prasiola*, also a nitrogen-dependency of photo acclimation in *Ulva rotundata* has been described (Henley et al. 1991).

3. Desiccation as a General Problem

There are some general problems arising when plants are drying out. Usually desiccation goes along with elevated temperatures, which cause a stress scenario in itself, which will not further be discussed here. One of the first reactions upon water loss is an osmotic change which can be tolerated to a certain extent by every "land plant". In order to avoid this stress plants have developed a variety of strategies to minimize water loss. Among these strategies are the development of polymeric compounds that prevent evaporation from the plant body (see Oliver 1996). In contrast, there is a large polyphyletic group of poikilohydric plants which do not employ such strategies, but instead developed mechanisms to cope with the desiccation event by withstanding it. This situation is given for example in the intertidal zones of oceans around the world, the stressful exposure of algae to this situation is summarized elegantly by Garbary (2007)

What is the "smaller scale" problem when a plant dries out? Indeed, proteins and membranes depend in their structure and function on an at least very thin water layer surrounding them. Poikilohydric plants obviously can survive drying out almost totally, but have the possibility to hold water in this smaller scale. Most likely this capacity allows to survive a lower water content than non-poikilohydric plants are able to tolerate. Alpert (2006) points out that the limit for metabolism is given when the water content is less than 10% of the dry mass. Below that threshold too little water is available to form this necessary monolayer around the proteins and membranes. The same author gives the rough estimate that a water content of 10% would be equivalent to the equilibration of a plant body with air of 50% relative humidity at 20°C, or expressed as a water potential of -100 MPa. The general trend can be drawn that plant species either die when dried to about 20% water content (non-desiccation tolerant plants) or are able to survive, further drying to about 10% water content with measurable metabolism (desiccation-tolerant plants). The cellular and physiological consequences of desiccation stress results in disturbance and/or inhibition of cytokinesis and a total loss of growth. As a prominent metabolic reaction, photosynthesis is maintained only to a certain threshold of water eg. in lichenized green algae (eg. Lange et al. 1997, Lange et al. 1989) or desert green algae (Gray et al. 2007). However, especially for small organisms it is difficult to determine how much water a cell can actually retain.

Water may also be unavailable when cells are subjected to temperatures that cause freezing eg. for alpine or tundra plants (eg. Benson et al. 2007). Extreme examples are snow and ice algae, which are continously exposed to low temperatures, but usually do not show freezing injuries which is likely due to "cryo-protection" (Remias et al. 2005). Strategies have been developed to "liquefy" the needed water in order to create a "waterfilm" around the organisms which is also in snow algae of substantial benefit for their survival. Additionally, algae in hyper variable environments like salt plains are exposed to desiccation phenomena as excellently review by Henley et al. (2007). In all cases the availability of at least a small proportion of water is essential for maintaining the physiological activity.

3.1. Strategies to Cope with Desiccation

It is generally recognized that a total avoidance of desiccation is not realized in the vegetative state of the algae subjected to evaporation. Instead, when exposed to air, algal cells or thalli fully dry out, but in desiccation tolerant groups this is achieved in a "non-destructive" way. In general, one of the key strategies to cope with drying out is the synthesis of several organic osmotically active compounds (eg. Davis 1972, Potts et al. 2005, Karsten et al. 2007b, Oren 2007). First to be mentioned are sugars, particularly trehalose is found to be involved.

Already in bacteria and cyanobacteria are able to osmoregulate via sugars (eg. Potts 1994, Wingler 2002, Fletcher 2007). Especially in green algae multi-functional sugar-alcohols are synthesized as summarized by Karsten et al. (2007b). Lichens have developed multiple biochemical traits for desiccation management (eg. Kranner et al. 2003). Sugars and sugar alcohols serve as organic osmolytes and compatible solutes and their function is basically to keep homeostasis as long as possible. When extended drought periods are expected, most green algae produce extremely persistent spores or "resting stages" which will not be discussed further in this chapter.

Once an algal cell has developed strategies to survive extremely dry conditions in the vegetative state it will be the first organism in a new habitat. Indeed pioneering organisms benefit from strategies that initially allow them to tolerate a scenario that is not life-supporting, but have the ability to recover quickly upon release from the repression. Most reports indicate that in the desiccated state these organisms are not able to perform positive net photosynthesis (Gray et al. 2007), however upon re-wetting a positive net photosynthesis is achieved within a few minutes. This chapter will summarize and consider possibilities why such a quick recovery is possible.

3.2. Function of Cell Walls

The function of several substances integrated or adhesive to the cell wall is to reduce water loss. However, most desiccation tolerant green algae follow this strategy only up to a certain threshold. However, when reaching a critical level the cells then allow drying out in an organized way and even their secondary cell walls remain dynamic and actively follow the shrinking process. This feature of the cell walls is mostly overseen, but realized eg. by containing expansins for wall extensibility that allows such shrinking processes without damage as reported from higher resurrection plant *Craterostigma* (eg. Franka et al. 2000, Jones and McQueen-Manson 2004). Aerophytic growing algae can be found totally "air-dry" in the vegetative state. To investigate the cell walls in this desiccated state microscopy has to be performed in a water-free medium (eg. directly in the immersion oil). The cells appear fully green, yet shrunk with bent in cell walls – the ability of these walls to follow the water loss is likely realized by the above mentioned mechanisms of "protecting proteins" within the walls. These cells are rather difficult to re-hydrate because applied liquid water forms spherical droplets on the dried "algal powder". In nature the uptake of liquid water may not even bee necessary, as these algae grow on places hardly if ever exposed to water droplets, but to air moisture. The physiological relevance of vapour eg. for *Apathococcus lobatus* is demonstrated below (Bertsch 1966).

3.3. Photosynthesis

Desiccation suppresses photosynthesis (eg. Bewley 1979, Häubner et al. 2006, Gray et al. 2007) and may result in inactivation of electron transport, cyclic photophosphorylation of isolated chloroplast membranes or generation of reactive oxygen species (ROS) as reported elsewhere (eg. Kranner and Birtic 2005). All of these reactions are very destructive to plants but strategies to deal with eg. ROS will not be discussed further here.

The CO_2 exchange and water relations have been investigated in the aerophytic green-alga *Apatococcus lobatus* by Bertsch (1966). The "unwettable" thalli achieve hydration equilibrium with the vapour pressure of the air only. The most favourable CO_2-balance has been measured at 97–98% r.h. (27-40 bar water potential as reported by the authors, which can be converted into -2.7 to -4 MPa – for a better comparison this conversion will be used in the following), 90% r.h. (-14.2 MPa) allows still fifty percent of the maximum CO_2-uptake capacity. The limit of measurable CO_2-uptake is at 68% r.h. (- 52 MPa) Despite moisture is necessary, liquid water has an unfavourable influence on the CO_2-balance, and the relation between apparent CO_2-uptake and dark-respiration of vapour-saturated thalli is almost optimal. The CO_2-exchange in these algae is extremely well adapted to their aerophilic way of life.

How is the protection of the photosynthetic apparatus achieved? It has been reported in rather early experiments that in the presence of sugars during water stress, the thylakoids may be partially or completely protected in higher plants (eg. Santarius 1973). This membrane stabilization has been found to depend on the concentration of sugars and their molecular mass. It has been described that the trisaccharide raffinose is more effective than the disaccharide sucrose and the latter more than the monosaccharide glucose (Santarius 1973). As green algae have the capacity to form a vast variety of sugars and sugar alcohols especially during desiccation (eg. Karsten et al. 2007b), this strategy can be regarded as extremely important for algal desiccation tolerance, not only in protecting photosynthetic membranes, but also relevant proteins and protein complexes.

4. Lessons from other Desiccation Tolerant Plants

Despite this chapter should emphasize on the situation in green algae, some other organisms, where research on desiccation tolerance has been conducted shall be discussed.

Most original work on desiccation tolerance in algae provides the reader with valuable information on physiological details, but omit cytological or ultrastructural information on the desiccation effect. This part will draw the attention to a more cell biological point of view in non-algal plants. There is obviously a demand for additional experimentation and visualization of desiccation effects.

4.1. Cyanobacterial Background

Cyanobacteria ("blue-green algae") had to face the stress scenario of drying out since their emergence in the early Precambrian. Desiccation tolerance of aerophytic cyanobacteria is often achieved through an "endolithic life-style" which frequently goes along with complete dryness of the rocks as their habitats. However, it is difficult to estimate how much water can be hold by endolithic cyanobacteria, and it is likely that the surface of the rock appears to be totally dry, whereas within the cells traces of water necessary to maintain living processes retmain. For sure desiccation was not the only limitation, especially high UV irradiation was an important factor as well in development of early oxygenic life (summarized eg. in Holzinger and Lütz 2006). As chloroplasts of green algae are regarded as primary endosymbiontic cyanobacteria (see Lewis and McCourt 2004, Keeling et al. 2005, Lewis

2007), it appears interesting to analyze which strategies cyanobacteria are following in terms of desiccation tolerance (Potts 1994, 2001). Moreover it has been know for some time that cyanobacteria have the ability to form several UV protecting substances like MAAs and carotenoides (eg. Karsten and Garci-Pichel 1996).

From the perspective of the invaded "chloroplasts", some strategies of holding water, especially through the cell walls or extracellular matrixes have either been given up or – what is more likely - are substituted by the cell walls of their permanent hosts, the green algal cells. In this chapter not a philosophical discourse of the benefits of endocytosis shall be launched, but for the understanding of desiccation tolerance some considerations are unambiguous. As *water* is "natures solvent" (Alpert 2006), it appears to be very critical especially for these ancient organisms to have the capacity to keep at least a minimum amount needed for physiological performance inside their cells. Viewed at a mechanistic way, it may simply be the beneficial effect of the "solvent" that allows the proteins maintaining their structures and functions.

What is the repertoire that cyanobacteria have brought into the eukaryotic world? Certainly the organization of a milieu consisting of polysaccharides that allow to osmotically "hold" water. This allows a life-supporting proportion of water inside the cells (Caiola et al. 1993, Caiola and Billi 2007). A detailed analysis of the membranes in *Synechocystits* (Liberton et al. 2006) gives insight into the structural basis. For *Nostoc commune* a "protein index" has been determined, and the changes induced by water stress are reported (Potts 1986). These early results have not been reinvestigated by means of modern molecular tools, which would be an important contribution to our understanding of the mechanisms involved in desiccation tolerance. There is, however, a report by Close and Lammers (1993) that an "osmotic stress protein" of cyanobacteria is immunologically related to higher plant dehydrins. The latter are well characterized proteins involved in desiccation tolerance of eg. seeds (eg. Rorat 2006) and have also been detected in fucoid algae (Li et al. 1998).

4.2. Bryophytes

In addition to physiological investigations of desiccation effects, ultrastructural data have been described for bryophytes. Attention shall be drawn to the studies by Proctor and coworkers (eg. Proctor and Smirnoff 2000, Proctor and Tuba 2002, Proctor et al. 2007). This is insofar important as theses studies are valuable contributions in a rather underrepresented field.

Net CO_2 uptake is reported to fall to zero at about 40% relative water content. It is interesting to see that even in samples that were desiccated for 9-18 d the initial negative net CO_2 uptake recovered within 10-30 min after rewetting. (Proctor et al. 2007). The full Fv/Fm values recovered to ~80% of the initial value within 10 min. This is tremendously fast and it suggests that the basic structures must have remained intact during the desiccation period. This time-frame would be way to short for a de-novo synthesis of membrane compounds or protein synthesis to recover destroyed elements. As expected, thylakoids, grana stacks and mitochondrial cristae remained intact during the drying-rewetting cycle giving the structural basis for this astonishing physiological effect. Some effort has been undertaken to generate defined drying scenarios in these experiments. Some of which would be worthwhile to be also employ on desiccation tolerant green algae, alongside with structural and ultrastructural evaluation.

4.3. Higher Plants

Desert plants, plants exposed to arctic and subarctic habitats and high alpine plants have developed effective strategies to cope with desiccation. The vegetative plant body, when exposed to arid or periodically freezing climates is adapted in the direction of maximal avoidance and efficient use of the smallest water resources (Vicre et al. 2004, Vander-Willigen 2004). Additionally these plants have put every effort in producing desiccation tolerant "permanent stages" in form of seeds, which are highly efficient and can be stored in "seedbanks" over years before finding appropriate conditions to germinate (eg. Jenks and Wood 2007).

There are not many desiccation tolerant higher plant species, that allow a complete drying of vegetative parts. This is likely due to the complexity of the organisms. Of the few "resurrection" plants *Craterostigma* is a reasonably well investigated example, especially concerning physiology (eg. Jones and McQueen-Manson 2004), but no structural or ultrastructural data do exist. In contrast TEM data have been reported for *Talbotia elegans*, a desiccation tolerant angiosperm (Velloziaceae) growing in extremely arid regions (Hallam and Gaff 1978, Hallam and Luff, 1980). This plant has the ability to retain chlorophyll during drying, and it is one of the few examples where the ultrastructure has been studied in the desiccated state (Hallam and Luff, 1980). It has been found that some thylakoid structures remain intact, even in a laboratory generated total desiccation (time period of 14 days for complete dry out). The authors report that when desiccation is complete, the only detectable remains of chloroplasts are intact thylakoid membranes organized in a loose array (Hallam and Gaff 1978). This is in contrast to desiccation-tolerant plants that lose chlorophyll during dehydration and reduce the chloroplast to pro-plastid like bodies containing vesicles. In *Talbotia*, additionally to these remaining thylakoid structures mitochondria with few cristae are observed and condensed nuclei with granular contents are viewed after the desiccation treatment (Hallam and Luff 1980). Very elegantly a solely aqueous fixation process is circumvented by adding dimethylsulphoxide in aqueous fixatives to match the water content of the dry plant tissue with that of the fixative. This shows better preservation of the ultrastructure, a smart idea that has not been followed by other investigators, but definitely could bring new insights.

5. Conclusion

In habitats with only periodical rain where organisms can not count on the frequent availability of water at certain times of the year, often the strategy of the formation of permanent stages is chosen. This is an extremely efficient strategy, as these permanent stages can withstand even long periods where water is unavailable. Moreover this strategy might be used for subjecting the permanent stages to a long distance transport, which might eventually result in finding new habitats, but on the short run may result in year long "dormancy", when such habitats are not reached. However, this process is possibly not fast enough for an establishment on new "open" habitats like freshly broken rocks or the man-made *pendent*, the facades of buildings. The faster strategy utilized is remaining in the vegetative state and having the ability to tolerate desiccation for a certain period of time. For these organisms often a higher air vapour is enough for their survival. There are some other abilities of these algae – they achieve a positive net photosynthesis even under very limited light conditions. Most likely these light conditions are physiologically necessary for desiccated algae, which would be damaged when exposed to higher light intensities.

Within this chapter the attention was drawn on strategies of various green algae to withstand or tolerate at least periodical desiccation stress. These organisms are characterized by an extremely quick recovery after desiccation, which is likely based upon their ability to retain the integrity of organelles like chloroplasts during desiccation for certain periods of time. When cells are able to resemble their "autotrophic production" after a desiccation period, everything else that might be damaged can obviously be repaired very quickly. Despite some protection strategies are known, this appears to be a wide field for more investigations. Further elucidating phylogenetic relationships, as well as physiological, structural and ultrastructural research will help to understand the cellular basis of desiccation tolerance.

Acknowledgement

I thank Univ. Prof. C. Lütz for critically reading the manuscript and for several helpful suggestions and discussions. Moreover I would like to thank Prof. G. Gärtner, University of Innsbruck for help in determination of some algal species and Prof. R. Türk, University of Salzburg for helpful discussions on desiccation tolerance. This work has been supported, in part, by a Research Grant from the "Universitätszentrum Obergurgl", University of Innsbruck, to the author.

References

Alpert, P. Constraints of tolerance: why are desiccation-tolerant organisms so small or rare? *J. Exp. Biol*, 2006, 209, 1575-1584.

Austin II JR; Frost, E; Vidi, P-A; Kessler, F; Staehelin LA. Plastoglobules are lipoprotein subcompartments of the chloroplast that are permanently coupled to thylakoid membranes and contain biosynthetic enzymes. *Plant Cell*, 2006, 18, 1693–1703.

Beer, S; Eshel, A. Photosynthesis of *Ulva* sp. I. Effects of desiccation when exposed to air. *J. Exp. Mar. Biol.* 1983a, 70, 91-97.

Beer, S; Eshel, A. Photosynthesis of *Ulva* sp. II. Utilization of CO_2 and HCO_3^- when submerged. *J. Exp. Mar. Biol.* 1983b, 70, 99-106.

Benson, EE; Harding, K; Day, JG. Algae at extreme low temperatures: the cryobank. In: J. Seckbach, editor. Algae and Cyanobacteria in Extreme Environments. Dordrecht, The Netherlands: Springer; 2007; 585-597.

Bertsch, A. CO_2-Gaswechsel und Wasserhaushalt der Aerophilen Grünalge *Apatococcus lobatus Planta* 1966, 70, 46-72.

Bewley, JD. Physiological aspects of desiccation tolerance. *Annu. Rev. Plant Physiol*, 1979, 30, 195-238.

Bischof, K; Kräbs, G; Wiencke, C; Hanelt, D. Solar ultraviolet radiation affects the activity of ribulose-1,5-bisphosphate carboxylase-oxygenase and the composition of photosynthetic and xanthophyll cycle pigments in the intertidal green alga *Ulva lactuca* L. *Planta* 2002, 215, 502-509.

Brown, RM; Larson, DA; Bold, HC. Airborne algae: Their abundance and heterogeneity. Science 1964, 143, 583-584.

Büdel, B. Diversity and ecology of biological crusts. *Prog. Bot*, 2002, 63, 386-404.

Caiola, MG; Billi, D. *Chroococcidiopsis* from desert to Mars. In: J. Seckbach, editor. Algae and Cyanobacteria in Extreme Environments. Dordrecht, The Netherlands: Springer; 2007; 555-568.

Caiola, MG; Ocampo-Friedmann, R; Friedmann, EI. Cytology of long-term desiccation in the desert cyanobacterium *Chroococcidiopsis* (Chroococcales). *Phycologia*, 1993, 32, 315-322.

Close, TJ; Lammers, PJ. An osmotic stress protein of cyanobacteria is immunologically related to plant dehydrins. *Plant Physiol*, 1993, 101, 773-779.

Cockell, CS; McKay, CP; Warren-Rhodes, K; Horneck, G. Ultraviolet radiation-induced limitation to epilithic microbial growth in arid deserts – Dorsimetric experiments in the hyperarid core of the Atacama Desert. *J. Photochem. Photobiol. B: Biol*, 2008, 90, 79-87.

Davey, MC. The effects of freezing and desiccation on photosynthesis and survival of terrestrial Antarctic algae and cyanobacteria. *Polar Biol.* 1989, 10, 29-36.

Davis, JS. Survival records in the algae, and the survival role of certain algal pigments, fat and mucilaginous substances. *Biologist*, 1972, 54, 52-93.

Ettl, H; Gärtner, G. Syllabus der Boden-, Luft- und Flechtenalgen. Stuttgart, Jena, and New York: G. Fischer Verlag; 1995.

Fletcher, VR. North American desert microbiotic soil crust communities: Diversity despite challenge. In: J. Seckbach, editor. Algae and Cyanobacteria in Extreme Environments. Dordrecht, The Netherlands: Springer; 2007; 539-551.

Franka, W; Munnikb, T; Kerkmannc, K; Salaminid, F; Bartelsd, D. Water deficit triggers phospholipase D activity in the resurrection plant *Craterostigma plantagineum. Plant Cell,* 2000, 12, 111-124.

Garbary, DJ. The margin of the sea: Survival at the top of the tides. In: J. Seckbach, editor. Algae and Cyanobacteria in Extreme Environments. Dordrecht, The Netherlands: Springer; 2007; 571-582.

Gray, DW; Lewis, LA; Cardon, ZG. Photosynthetic recovery following desiccation of desert green algae (Chlorophyta) and their aquatic relatives. *Plant Cell Environ,* 2007, 30, 1240-1255.

Hallam, ND; Gaff, DF. Regeneration of Chloroplast Structure in *Talbotia elegans*: A Desiccation- Tolerant Plant. *New Phytologist*, **81**, 1978, 657-662.

Hallam, ND; Luff, SE. Fine Structural Changes in the Leaves of the Desiccation-Tolerant Plant Talbotia elegans during Extreme Water Stress. Botanical Gazette, 1980, 141, 180-187.

Häubner, N; Schumann, R; Karsten, U. Aeroterrestrial algae growing on facades – response to temperature and water stress. *Microb. Ecol*, 2006, 51, 285-293.

Heber, U; Lange, OL; Shuvalov, VA. Conservation and dissipation of light energy as complementary processes: homoiohydric and poikilohydric autotrophs. *J. Exp. Bot*, 2006, 57, 1211-1223.

Henley, WJ; Levavasseur, G; Franklin, LA; Osmond, CB; Ramus, J. Photoacclimation and photoinhibition in *Ulva rotundata* as influenced by nitrogen availability. *Planta* 1991, 184, 235-243.

Hofbauer, KW; Breuer, K, Sedlbauer, K. Algen, Flechten, Moose und Farne auf Fassaden. *Bauphysik*, 2003, 25, 383-396.

Holzinger, A; Karsten, U; Lütz, C; Wiencke, C. Ultrastructure and photosynthesis in the supralittoral green macroalga *Prasiola crispa* (Lightfoot) Kützing from Spitsbergen (Norway) under UV exposure. *Phycologia*, 2006, 45, 168-177.

Holzinger, A; Lütz, C. Algae and UV irradiation: Effects on ultrastructure and related metabolic functions. *Micron*, 2006, 37, 190-207.

Hoppert, M; Beimer, R; Kemmling, A; Schroder, A; Gunzl, B; Heinken, T. Structure and reactivity of a biological soil crust from a xeric sandy soil in central Europe. *Geomicrobiol J*, 2004, 21, 183-191.

Jenks, M; Wood, A. Plant desiccation tolerance. Oxford: Blackwell Publishing; 2007.

Jones, L; McQueen-Mason, S. A role of expansins in dehydration and rehydration of the resurrection plant *Craterostigma plantagineum*. *FEBS Lett*, 2004, 559, 61-65.

Karsten, U; Friedl, T; Schumann, R; Hoyer, K; Lembcke, S. Mycosporine-like amino acids and phylogenies in green algae: *Prasiola* and its relatives from the Trebouxiophyceae (Chlorophyta). *J. Phycol*, 2005, 41, 557-566.

Karsten, U; Garcia-Pichel, F. Carotenoids and mycosporine-like amino acid compounds in members of the genus *Microcoleus* (Cyanobacteria): a chemosystematic study. *System. Appl. Microbiol*, 1996, 19, 285-294.

Karsten, U; Karsten, U; Lembcke, S; Schumann, R. The effects of ultraviolet radiation on photosynthetic performance, growth and sunscreen compounds in aeroterrestrial biofilm algae isolated from building facades. *Planta*, 2007a, 225, 991-1000.

Karsten, U; Schumann, R; Mostaert, AS. Aeroterrestrial algae growing on man-made surfaces: What are the secrets of their ecological success? In: J. Seckbach, editor. Algae and Cyanobacteria in Extreme Environments. Dordrecht, The Netherlands: Springer, 2007b; 585-597.

Keeling, PJ: Burger, G; Durnford, DG; Lang, BF; Lee, PW; Pearlman, RE; Roger, AJ; Gray, MW. The tree of eucaryotes. *Trends Ecol. Evol*, 2005, 20, 670-676.

Kranner, I; Birtic, S. A modulating role for antioxidants in desiccation tolerance. *Integr. Comp. Biol*, 2005, 45, 734-740.

Kranner, I; Zorn, M; Turk, B; Wornik, S; Beckett, RP; Batic, F. Biochemical traits of lichens differing in relative desiccation tolerance. *New Phytol*, 2003, 160, 167-176.

Lange, OL; Belnap, J; Reichenberger, H; Meyer, A. Photosynthesis of green algal soil crust lichens from arid lands in southern Utha, USA: role of water content on light and temperature responses of CO_2 exchange. *Flora,* 1997, 192, 1-15.

Lange, OL; Bilger, W; Schreiber, U. Chlorophyll fluorescence of lichens containing green and blue-green algae during hydration by water vapor uptake and by addition of liquid water. *Bot. Acta,* 1989, 102, 306-313.

Lewis, LA. Chlorophyta on land: Independent lineages of green eukaryotes from arid lands. In: J. Seckbach, editor. Algae and Cyanobacteria in Extreme Environments. Dordrecht, The Netherlands: Springer; 2007; 571-582.

Lewis, LA; McCourt, RM. Green algae and the origin of land plants. *Am. J. Bot*, 2004, 91, 1535-1556.

Li, R; Brawley, SH; Close, TJ. Proteins immunologically related to dehydrins in fucoid algae. *J. Phycol,* 1998, 34, 642-650.

Liberton, M; Berg, RH; Heuser, J; Roth, R; Pakrasi, HB. Ultrastructure of the membrane systems in the unicellular cyanobacterium *Synechocystis* sp. Strain PCC 6803. *Protoplasma* 2006, 227, 129-138.

Lopez-Bautista, JM; Rindi, F; Casamatta, D. The systematics of subaerial algae. In: J. Seckbach, editor. Algae and Cyanobacteria in Extreme Environments. Dordrecht, The Netherlands: Springer; 2007; 601-617.

Lukesova, A; Frouz, J. Soil and freshwater microalgae as a food source for invertebrates in extreme environments. In: J. Seckbach, editor. Algae and Cyanobacteria in Extreme Environments. Dordrecht, The Netherlands: Springer; 2007; 267-284.

Nienow, JA. Ecology of subaerial algae. Nova Hedwigia Beihefte, 1996, 112, 537-552.

Oliver, MJ. Desiccation tolerance in vegetative plant cells. *Physiol. Plant*, 1996, 97, 779-787.

Oliver, MJ; Tuba, Z; Mishler, BD. The evolution of vegetative desiccation tolerance in land plants. *Plant Ecol.* 2005a, 151, 85-100.

Oliver, MJ; Velten, J; Mishler, BD. Desiccation tolerance in byrophytes: A reflection of the primitive stratgy for plant survival in dehydrating habitats. *Integr. Compl. Biol*, 2005b, 45, 788-799.

Ong, B; Lim, M, Wee, Y. Effects of desiccation and illumination on photosynthesis and pigmentation of an edaphic population of *Trentepohlia odorata* (Chlorophyta). *J. Phycol*, 1992, 28, 768-772.

Oren, A. Diversity of organic osmotic compounds and osmotic adaptation in cyanobacteria and algae. In: J. Seckbach, editor. Algae and Cyanobacteria in Extreme Environments. Dordrecht, The Netherlands: Springer; 2007; 641-655.

Potts, M. Desiccation tolerance in prokaryotes. *Microbiol. Rev*, 1994, 58, 755-805.

Potts, M. Desiccation tolerance: A simple process? *Trends Microbiol,* 2001, 9, 553-559.

Potts, M. The protein index of *Nostoc commune* UTEX 584 (Cyanobacteria): Changes induced in immobilized cells by water stress. *Arch Microbiol*, 1986, 146, 87-95.

Potts, M; Slaughter, SM; Hunneke, F-U; Garst, JF; Helm, RF. Desiccation tolerance of prokaryotes: Application of principles to human cells. *Integr. Comp. Biol*, 2005, 45, 800-809.

Proctor, MCF; Ligrone, R; Duckett, JG. Desiccation tolerance in the moss *Polytrichum formosum*: Physiological and fine-structural changes during desiccation and recovery. *Annal. Bot,* 2007, 99, 75-93.

Proctor, MCF; Smirnoff, N. Rapid recovery of photosystems on rewetting desiccation tolerant mosses: chlorophyll fluorescence and inhibitor experiments. *J. Exp. Bot*, 2000, 51, 1695-1704.

Proctor, MCF; Tuba, Z. Poikilohydry and homoihydry: antithesis or spectrum of possibilities? *New Phytol.* 2002, 156, 327-349.

Reisser, W. The hidden life of algae underground. In: J. Seckbach, editor. Algae and Cyanobacteria in Extreme Environments. Dordrecht, The Netherlands: Springer; 2007; 79-58.

Remias, D; Lütz-Meindl, U; Lütz, C. Photosynthesis, pigments and ultrastructure of the alpine snow alga *Chlamydomonas nivalis*. *Eur. J. Phycol.* 2005, 40, 259–268.

Rindi, F. Diversity, distribution and ecology of green algae and cyanobacteria in urban habitats. In: J. Seckbach, editor. Algae and Cyanobacteria in Extreme Environments. Dordrecht, The Netherlands: Springer; 2007; 621-638.

Rindi, F; Guiry, MD; Barbiero, RP; Cinelli, F. The marine and terrestral Prasiolales (Chlorophyta) of Galway City Ireland: a morphological and ecological study. *J. Phycol.* 1999, 35, 469-482.

Rorat, T. Plant dehydrins – tissue location, structure and function. *Cell Mol. Biol. Lett,* 2006, 11, 536-556.

Santarius, KA. The protective effect of sugars on chloroplast membranes during temperature and water stress and its relationship to frost, desiccation and heat resistance. *Planta* 1973, 113, 105-114.

Trainor, FR; Glydych, R. Survival of algae in a desiccated soil: a 35-year study. *Phycologia* 1995, 34, 191-192.

Tschaikner, A; Ingolić, E; Holzinger, A; Gärtner, G. Phycobionts of some species of *Evernia* and *Ramalina*. Herzogia 2007, 20, 53-60.

Vander-Willigen, C; Pammenter, NW; Mundree, SG; Farrant, JM. Mechanical stabilization of desiccated vegetative tissue of the resurrection grass Eragrostis nindensis: does a TIP 3;1 and/or compartmentalization of subcellular components and metabolites play a role? *J. Exp. Bot,* 2004, 55, 651-661.

Vicre, M; Farrant, JM; Driouich, A. Insitghts into the cellular mechanisms of desiccation tolerance among angiosperm resurrection plant species. *Plant Cell Environ*, 2004, 27, 1329-1340.

Wingler, A. The function of trehalose biosynthesis in plants. *Phytochemistry* 2002, 60, 437-440.

In: Algae: Nutrition, Pollution Control and Energy Sources ISBN: 978-1-60692-008-4
Editor: Kristian N. Hagen, pp. 57-87

Chapter 5

UTILIZATION OF ALGAE FOR POLLUTION ELIMINATION

Jana Kadukova* and Miroslav Štofko
Technical University of Kosice, Faculty of Metallurgy, Department of Non-Ferrous Metals and Waste Treatment, Letna 9, 04200 Kosice, Slovakia

Abstract

Algae are present in most kinds of environments. They are adapted to saline and freshwater even to wetland. As they mostly live in an environment connected with water where concentration of essential nutrients is usually low, they had to develop mechanisms to concentrate nutrients in their body. For that reason they can accumulate several inorganic ions up to high levels. This capability can be used in the processes of pollutant removal. Metal ions represent an important group of hazardous contaminants in industrial wastewater, and algae can be effectively used to remove them. One of the possibilities represents bioaccumulation when metals are accumulated inside the cells as the result of metabolic activity. The other possibility is to use the ability of algae to adsorb metals onto their cell surface in the process called biosorption. In some situations the enzymatic apparatus of algae can cause the precipitation of metals such as gold or silver present in the environment. But the use of algae for the degradation of pollutants is much wider. They are able to decrease concentration of nutrients, especially nitrogen in wastewater. They assimilate NH_4^+ ions produced during the biodegradation of organic compounds and decrease their concentration more than twice compared with the biodegradation without algae. Their role in organic removal is also well known. They were applied to the dye concentration reduction in wastewater, olive oil mill wastewater treatment, water from paper industry treatment, etc. Because of their high sensitivity to high concentrations of pollutants, they are also used in toxicity tests or as sensitive bioindicators of ecological changes. It is obvious that algae have great potential in pollution control and in the processes applied to decrease the level of pollution in the environment.

* E-mail address: jana.kadukova@tuke.sk, kadukova@post.sk, tel. 00421-55-6022426, fax. 00421-55-6022428.

1. Introduction

Heavy metal contamination is one of the most serious environmental problems at the present time. They are major pollutants in marine, ground, industrial and even treated wastewater and soil. Relevant for algae are mostly the metals in soil and water. Rather little is taken up directly from the atmosphere. There are many sources, both natural and anthropogenic, from which a high concentration of metals can originate (Paulíková, 2002; Luptáková, 2004). Major sources of metal pollution are mining, different types of metal production and combustion of fossil fuels connected with the disposal of various wastes. Ever-intensifying human technological activities have resulted in the generation of large quantities of aqueous effluents that contain high levels of heavy metals and cause environmental disposal problems (Pichtel et al., 2000; Volesky, 2001).

Stringent regulations have increased the demand for new technologies for metal removal from wastewater to attain today's toxicity-driven limits (Nuhoglu et al., 2002). The use of biological processes for removing and recovering heavy metals from contaminated industrial effluents has emerged as a potential alternative method to conventional techniques, which may be expensive or ineffective (Kratochvil, Volesky, 1998). Algae are very often used in these biological processes. Their effectiveness is based on their ability to sequester and concentrate elements from the environment. This ability originates in the fact that concentration of essential elements that are necessary for normal development of organisms is often very low in an aquatic environment so organisms had to develop mechanisms able to sequester and concentrate these elements from the surrounding environment (Rubio et al., 1996; Butler, 1998).

Algae do not form a uniform group of plants but several parallel divisions and groups where it is sometimes difficult to determine phylogenetic relations among the individual divisions (Lane, Morel, 2000). Algae are traditionally characterized according to differences in pigmentation and cell complexity arising as a result of evolution (Hindák, 2001). They are included to the plant kingdom and are distinguished from other chlorophyllous plants on the basis of sexual reproduction (Davis et al., 2003).

Algae are photosynthetic, aquatic plants that utilize inorganic nutrients such as nitrogen and phosphorus. They belong among eukaryotic organisms in contrast to prokaryotic cell of cyanobacteria (cyanobacteria are, in English literature, often named blue-green algae). Many of the algae live in the sea, some in brackish water. Many species live in freshwater, some in wet soil, rocks or tree bark. Some species form with fungi lichens, some live in symbiosis with the cells of animals (Hindák et al., 1978; Urban, Kalina, 1980). The aquatic algae as the important elementary producers in marine and inland water plays a key role in the whole ecosystem (Zhou et al., 2008). The disturbance of aquatic ecosystems provoked by heavy metals pollution from industrial and domestic sources has as a consequence the loss of biological diversity, as well as increased bioaccumulation and magnification of toxicants in the food chain (Hindák, Makovinská, 1999). Algae are not only key-elements in the cycling of elements in aquatic ecosystems, often significantly enhancing bioavailability of toxic substances to higher consumers, but are in turn also adversely effected by such pollutants (Schlacher-Hoenlinger, Schlacher, 1998). Algae play an important role in metals biogeochemistry. On one hand, they can regulate the speciation and bioavailability of trace metals through the production and release of organic ligands. On the other hand they can mediate the aggregation of colloids, resulting in a faster removal of colloid-bound metals from surface waters to the sediments (Koukal et al., 2007).

Algae are ubiquitous in surface water but do not pose a problem to water treatment processes while their populations are relatively low. However, seasonal algal blooms can dramatically increase populations on relatively rapid timescales and as a result water treatment process efficiency can be impaired. It is required that algae be removed from drinking water, preferably during the initial stages to ensure minimal impact on subsequent processes (Henderson et al., 2008).

Metals represent a serious threat to the environment because of their toxicity and tendency to accumulate within particular trophic levels. Remediation of inorganic contaminants differs from the case with organic compounds. Because organic compounds can be mineralized, but the remediation of inorganic contamination must either physically remove the contaminant from the system or convert it into a biologically inert form. Removal of the inorganic pollutants can be accomplished by removing the biomass or, with certain inorganic contaminants, by contaminant volatilization (Cunningham, Ow, 1996). But organic pollution or an increasing amount of nutrients in natural waters causing eutrophication are also serious problems. Algae because of their metabolic activity and abilities to transform organic compounds are often used in their treatment.

2. Behaviour of Algae in the Presence of Inorganic Stressors

Stress caused by the presence of inorganic chemicals, including heavy metals, affects algae at biochemical, cellular, population and community levels of biological organization. Algae differ in tolerance to stress owing to the activity of inorganic chemicals on intracellular and cell-surface binding sites. Algae concentrate inorganic ions to amounts several thousand fold greater than in external dilute solutions (Genter, 1996; Muse et al., 2006). The whole process involves a variety of biological, chemical and physical mechanisms which are described in the chapters of this work.

While living in a polluted environment, algae adopted several mechanisms on how to cope with pollution. Even at the beginning, several physiological changes appear but after a while the population recovers without visible changes. For example alga *Phymatolithon calcareum* was initially negatively affected by an exposition to heavy metals at all concentrations, but after one week they appeared to have assimilated all the contamination and recovered normal photosynthetic values (Wilson et al., 2004). Algal populations can exist in a polluted environment due to:

- *acclimation* – physiological changes which occur during exposure to particular chemicals resulting in tolerance increasing,
- *adaptation* – in which natural selection acts on genetically based individual variation so that populations evolve increased resistance,
- a change in specific composition to a less sensitive species resulting in decrease of biodiversity in the environment (Genter, 1996).

Metals and their compounds represent the biggest group of inorganic chemicals so the next section will be mostly dedicated to the influence of metals and their compounds on the algal populations. Some metals influence living organisms positively. They are present in the organisms in low concentrations. These metals belong among essential or trace metals (e.g. Zn). Other are not known to be important for organisms. They are called nonessential such as Pb, Cd. The effects of nonessential and essential metals are shown on the Figure 1.

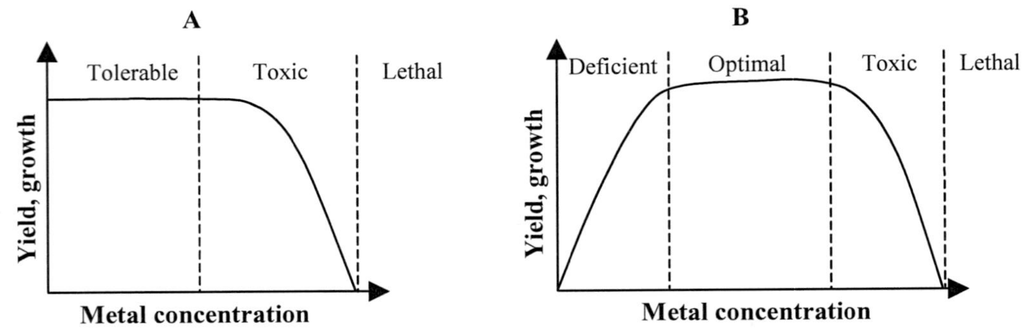

Figure 1. The effect of nonessential (A) and essential (B) metals on the population growth (Förstner, 1983; Wood, Wang, 1983).

In the case of nonessential element (Fig. 1a) low concentration already has negative effects on the organism but organisms can tolerate them and cope with them within its detoxification mechanisms. Higher concentration causes severe damages, develop acute toxicity and organisms are dying at high concentrations. In the case of lack of the essential metal (Fig. 1b) organisms does not develop normally, its growth is retarded or stopped, pathological changes are present. For normal function of organism is necessary particular "optimal" metal concentration but in its excess again normal physiological activities are decreased, different pathological changes are present and develop acute toxicity symptoms and organism is dying (Förstner, 1983, Clemens, 2006). Limit concentrations vary for different organisms and different metals.

In general metals represent serious polluting compounds at very low concentration and they can evoke damages and illnesses of organisms (Tölgyessy, 1984). Heavy metal exposure can cause the disturbance of normal metabolism and biological function, inhibition of photosynthesis, reduction of cytochrome, cellular mutation, putrescence, even death in algae. As an example, a purified strain of algae (*Chlorella ellipsoidea*) in a pond near the plating factory was reported to exhibit growth inhibition due to Cu, Zn, Ni and Cd exposure. Negative correlation existed between the contents of chlorophyll *a* and heavy metal levels (Zhou et al, 2008, Rai, Chandra, 1989). Partial inhibition of the epoxidation step of the xanthophyll cycle in the marine diatom *Phaeodactylum tricornutum* by Cd was recorded (Bertrand et al, 2001). Extremely dangerous are metals which block activity of enzymes containg SH groups (Hg, Pb, Cd, As, Se, Cu, V etc.) (Tölgyessy, 1984).

Not only high concentrations of metal causing acute toxicity represent a problem for algal population. Chronic pollution may affect algae even more than a single event (Wilson et al, 2004). Except of heavy metals other inorganic chemicals can adversely influence the algal population. Fluoride was found to inhibit or enhance the population growth of algae and affect nucleotide and nucleic acid metabolism (Camargo, 2003).

Presence of the inorganic chemicals alone does not inform about their toxicity for cells. Toxicity can be influenced by other chemical, physical or biological factor or season. These factors are shortly described in the Figure 2. Except of it, biological effect of inorganic chemical is influenced by its concentration, way, frequency and duration of exposition (Tölgyessy, Fargašová, 1991, Ostapczuk et al, 1997).

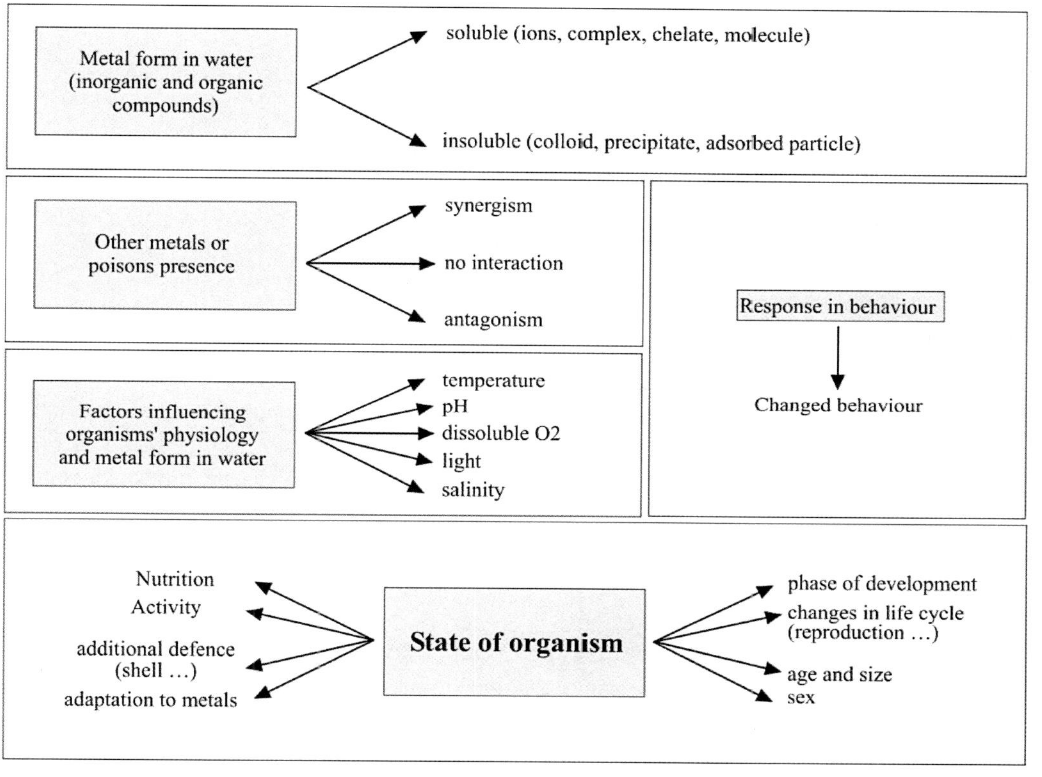

Figure 2. Factors influencing the metal toxicity in aquatic environment (Förstner, 1983).

Chemical Factors

An insufficient supply of CO_2 leads to high pH levels and can be a growth limiting factor. On the other hand, excess of CO_2 can inhibit metal-dependent enzyme systems. Some metals are for algae more toxic in specific pH range, e.g. at low pH values are Cd, Cu and Zn ions less toxic in contrast with Pb ions which are more toxic at these conditions. Excess of hydrogen ions may decrease the toxicity of inorganic ions (and induce acid stress) by completively excluding them from binding to cell-surface ligands. Content of Ca, Mg and Mn ions can significantly influence the toxicity. At their high concentration the toxicity of many metal ions decreases. Metals may be more toxic outside the normal salinity ranges for freshwater and marine algae. Other significant factor influencing metal toxicity is the content of basic nutrients. For example the concentration of PO_4^{3-} ions directly influences metal toxicity to algae. Metal ions can be precipitated together with phosphate ions in external solution or bound on intracellular polyphosphate. Organic compounds such as amino acids, humic acids or fulvic acid work as chelating agents and bind inorganic ions what decreases their toxicity (Genter, 1996).

Presence of other metals can also change the behaviour of algal population. It was found that algae previously tolerant to Cd were also able to live in the environment with elevated concentration of Zn. Co-tolerance or multitolerance to metals is a common phenomenon for algae (Paulsson et al, 2000).

Physical Factors

It seems that temperature increasing causes decreasing of the stress activated by inorganic ions what can be the results of metabolism inhibition. It is not known if light has any direct effect on toxicity. Probably its effect is mediated by influencing of algal growth and metabolisms (Genter, 1996).

Biological Factors

With increasing the amount of biomass decrease the toxicity because larger algal populations lead to more membrane binding sites being available, which leads to less metal being bound per alga and, hence, less toxicity. Except of it toxicity is influenced by biotransformation (Genter, 1996). Biotransformation is the enzyme-catalyzes conversion of one chemical into another that may be more or less toxic (good example is methyl-mercury or organotin as more toxic compounds and methylarsenic as less toxic compound (Cullen, Reiner, 1989, Gadd, 2000).

Communities formed by many species are usually more tolerant to high concentration of metals or organic pollution. For example, high input of organic waste may stimulate heterotrophs to secrete protective mucus or detoxify metals to such a degree that benthic organisms surviving in this habitat are protected by their dense packing. Toxic effects of zinc (and cadmium) are prevented by metal binding in the biofilm, stimulated by the simultaneous pollution with organic waste water and precipitation of iron in dense layers of cells, mucus and detritus (Admiraal et al, 1999).

Season

The influence of the season to toxicity is given probably by its influence to particular above mentioned factors. In summer is higher temperature, more nutrients what is the opposite in winter. For example, dimethylarsenic acid concentration grows with increasing of water temperature and its maximum copies temperature maximum. Concentration maxima of this acid differ in dependence on depth. Similar is the behaviour of monomethylarsenic acid (Hasegawa, 1997). Concentration of trivalent compounds of arsenic is probably independent of temperature changes (Hasegawa, 1996).

3. Uptake of Metal Ions from the Environment

Inorganic ion present in the environment has several ways how to influence the organisms. It depends on its speciation in the environment and its biological availability which is connected with many factors. Any chemical present in the environment and toxic for aquatic organisms must first dissolve in water, it must be transported across the external membranes and then internally dissolve in a body fluid (Escuder-Gilabert et al, 2001). Uptake of inorganic ions by algal cell can be basically divided into two stages (Cho et al, 1994, Genter, 1996):

- the first is fast and metabolism independent (the basis of the biosorption),
- the second is slow and dependent on metabolism (the basis of the bioaccumulation).

The first includes inorganic ion accumulation in cell wall components and takes approximately 5 – 10 minutes. The second is connected with the ion transport to cell. It is slower and it is inhibited by metabolic inhibitors such as low temperature, lack of energetic sources (light) etc. Metabolism can also affect the rate at which substance arrives at the receptor. Electronic and steric factors rule interactions with the corresponding metabolizing enzymes (Escuder-Gilabert et al, 2001). Schematic model of pathways by which metals enter the cell and influence its particular functions is on the Figure 3.

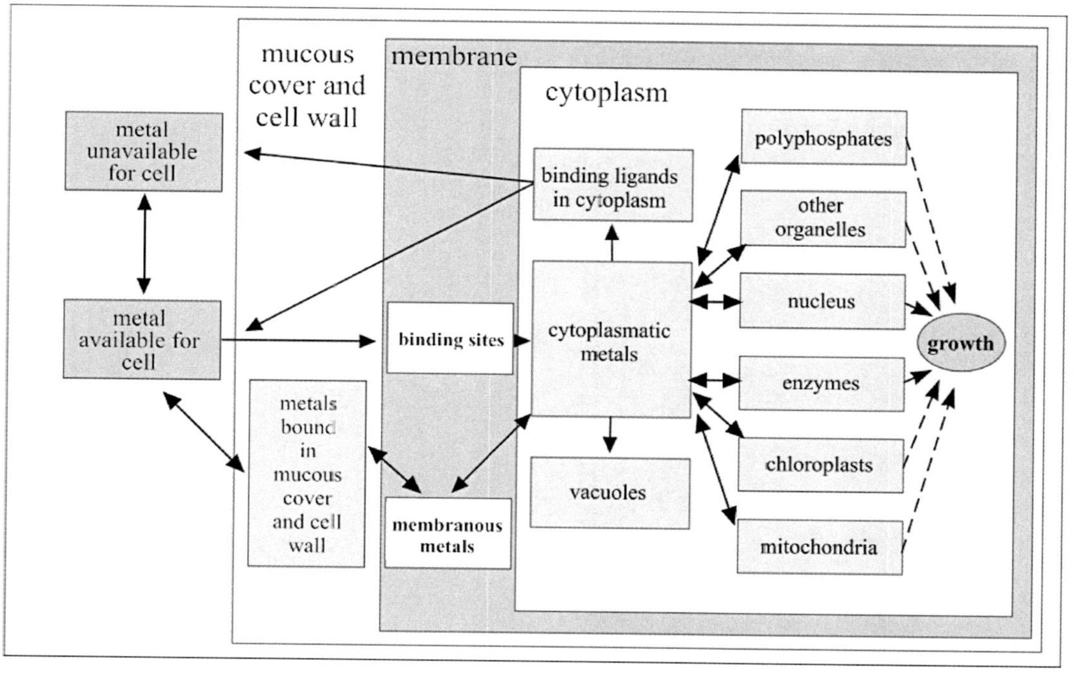

Figure 3. Schematic model of pathways by which metals enter the cell and influence its particular functions (Genter, 1996).

The ability of organisms to survive higher concentration of toxic chemicals depends (except of other factors) on their ability to cope with the characteristics of pollutants present in the environment. Organisms during their evolution developed various mechanisms which enable them to maintain low concentration of toxic compounds inside the cells. This adaptation to resist toxic substances became the component part of their genome, coded by nuclear chromosomes. Other microorganisms have acquired a transferred resistance to the polluted environment relatively recently (after industrial revolution). Extra chromosomal DNA molecules (plasmids) often determine these resistances (Irgolic, Martell, 1985, Wood, Wang, 1985). According to the resistance mechanism location, it is possible to divide them into three groups:

1. Mechanisms localized on the cell wall
   - cell surface sorption,
   - extra-cellular precipitation.

2. Mechanisms localized on the cell wall and cytoplasmic membrane
   - ion efflux mechanisms.

3. Intracellular localized mechanisms
   - enzymatic oxidation and reduction,
   - intracellular polymers (intracellular accumulation),
   - methylation.

Most of these mechanisms are involved in transport of metal into the cell and inside the cell as well as their uptake and storage (Wood, Wang, 1985, Kaduková, Štofko, 2006). In the case of living algal cells usually more mechanisms are responsible for the accumulation. Inorganic ion can be in first step taken up by only physico-chemical interactions but then it can be transported into the cell by several transport systems.

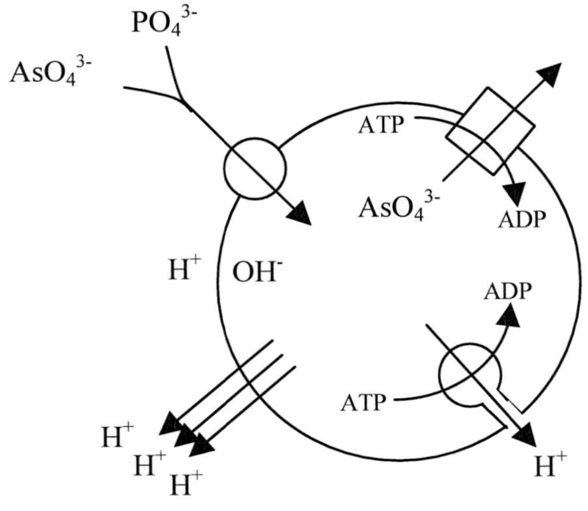

Figure 4. System of the phosphate and arsenate ions transport into the cell and system of the arsenate exclusion from the cell (Silver, 1985).

Essential elements are accumulated because cells need them, e.g. phosphorus in the form of phosphates or nitrogen in the form of nitrates. But the same systems are able to take up arsenate or arsenite on the basis of their similar physico-chemical properties with phosphate ions (Beceiro-González et al, 2000). Due to this reason cell sequesters arsenic ions as well as cadmium and antimony via normal transport systems by chance. Accumulation of these elements is non selective (Wood, Wang, 1985, Elbaz-Poulichet et al, 2000). But the ability of cells to exclude these elements out from their organism is highly selective and it is connected with the presence and functioning of ion efflux mechanisms. As an example sequestration and excluding of arsenic ions is shown on the Figure 4.

Presence of the specific type of ion efflux mechanisms is linked with the presence of specific ions. For example organisms living in the environment with arsenic owe arsenic efflux mechanism (Cervantes, 1995). Similar mechanisms based on the existence of ion efflux mechanisms were found for heavy metal ions as well.

Biosorption

Adsorption of the inorganic ions on the algal cell wall is the first step in their uptake. This process is often called biosorption or bioadsorption. It is the passive immobilization of metal or other inorganic ions. Biosorption mechanism by cell surface does not depend on cell metabolisms. It is based on physical – chemical interactions among ions and functional groups of cell wall (Kaduková, Virčíková, 2003).

Algae biosorption is mainly attributed to their cell wall properties where both electrostatic attraction and complexation can play a role (Davis et al, 2003). The most common types of algae (*Chlorophyta, Phaeophyta* and *Rhodophyta*) showed different sorption behaviour, due to the different structures of the cell wall polysaccharides. The association of some metals with polyphenolic fraction of algal cells was found to be responsible for sequestering of high concentration of heavy metals by some brown algae. It was in some part found for *Padina gymnospora* accumulating high Zn and Cd concentrations (Amado Filho et al, 1999).

The role that any given group of the biomass plays depends on factors such as the number of sites in the biosorbent material, its accessibility, its chemical state (i.e. availability), and the affinity between site and metal (i.e. binding strength) (Herrero et al, 2008). Knowledge of the cell wall composition is very important for the study of binding possibilities and mechanisms responsible for uptake of inorganic ions. Algal cell wall is formed by multilayer microfibrillar matrix which contains celluloses with dispersed amorphous particles. Sometimes it is incrusted SiO_2 or carbonates. Cell wall of some algae consists of ten layers. In dependence on species cell wall represents from 1 to 90 % of total mass, amorphous material is formed by different glycoproteins. Basic building material of cell wall of some algal species can be mannan or xylan not celluloses. Very often mucilaginous cover is an external surface which consists of uronic acids important in metal ions sorption (Volesky, 1990, Schiewer, Volesky, 2003).

Metal ions bind to specific functional groups. On the algal surface are mainly present following groups: hydroxyl (-OH), phosphoryl ($-PO_3H_2$), amino ($-NH_2$), carboxyl (-COOH) and thiol (-SH) groups (Genter, 1996). Presence of particular groups and their number is species specific for that different algae have the ability to bind different ions depending on the environmental conditions. For example alga *Chlorella vulgaris* is better biosorbent for zinc ions than cadmium ions (Cho et al., 1994). Rijstenbil et al. (1994) reported that stress caused by metal presence can increase surface and roughness of cell wall and so increase adsorption capacity of algae. Amado Filho et al. (1999) after studying Zn and Cd accumulation by *Padina gymnospora* confirmed that cell walls play a main role in metal accumulation.

Since biosorption is independent of the metabolism, binding of metals is very quick, up to 1 min (Delgado et al, 1998, Bustard et al, 1997) and mostly reversible (Veglio, Beolchini, 1997). The biosorption mechanisms can be different but as major mechanisms are considered:

1. physical adsorption
2. chemisorption
3. ion exchange.

Minor mechanisms which can also participate in biosorption are microprecipitation or complexation (Davis et al, 2003, Volesky, 2003).

The main mechanism responsible for the metal biosorption is probably ion exchange (Kratochvil, Volesky, 1998). It was found (Volesky, 1990) that bivalent ions are exchanged with ions with the same charge from polysaccharides. The process can be described by the equation:

$$2\ NaAlg + Me^{2+} \rightarrow Me(Alg)_2 + 2\ Na^{2+},$$

where Alg^- represents rest of the alginic acid and Me^{2+} bivalent ion.

Alginate in marine algae is present in the form of K^+, Na^+, Ca^{2+} and Mg^{2+} salts. These ions can exchange with the metal ions such as Co^{2+}, Cu^{2+}, Cd^{2+} and Zn^{2+} resulting in their removal from solution (Veglio, Beolchini, 1997).

In the case of biosorption of heavy metals by algal biomass mechanism can be described as being extracellular or occurring discretely at the cell wall. Intracellular sorption would normally imply bioaccumulation by living organisms (Davis et al, 2003).

Bioaccumulation

Bioaccumulation is a process of accumulation of chemicals from the outside environment inside the living organisms. Due to the bioaccumulation chemicals enter food chain, circulate within the individual trophic levels and directly threat the human beings. Mechanisms of this process do not distinguish between essential and toxic chemicals (Volesky, 2003).

Bioaccumulation is the second part of the metal sequestration process by living cell. Sometimes it is also called active biosorption. It is the opposite of the passive physico-chemical sorption called biosorption. It depends on many factors such as the individual species, its state, inorganic ion characteristics. Absorption, transport, and distribution of the substance in a biological system are governed largely by its partitioning behaviour between lipid and aqueous phase, which depends on molecular properties such as hydrophobicity, polarity, degree of ionization, molecular shape, and size (Escuder-Gilabert et al, 2001).

One of the basic differences between the two processes is their kinetics and the values of activating energy. The activation energy for biosorption is about 21 kJ/mol what is in agreement with physical nature of the process (Raraz, 1995). The activation energy for bioaccumulation is about 63 kJ/mol corresponding to biochemical processes. Biosorption is a fast process independent of the presence of specific nutrients while bioaccumulation is slow and nutrient dependent. The extent of bioaccumulation is lower than that of the biosorption. Bioaccumulation represents only 0.5 – 2% dry cell weight (Wood and Wang, 1985, Raraz, 1995, Kaduková, Virčíková, 2005). Intracellular accumulation is very complicated processes with many mechanisms responsible for it but many of them are not known till now. The most common mechanisms of intracellular accumulation can be divided into four groups:

1. the bind of metal ions with intracellular compounds
2. precipitation of metals inside the cell
3. methylation
4. volatilization

Biosynthesis of intracellular compounds represents temporal action of cell which helps it to prevent increasing of the toxic ion concentration over tolerable limits. It is often followed by the exclusion of metal from the cell. The example is the synthesis of metalothioneins – low-molecular proteins or polypeptides which form clusters with SH-groups (Winge et al, 1998, Malmström, Leckner, 1998).

In some cases metals can be stored in the cell in the form of free precipitates in cytoplasm, e.g. forming of intracellular granules of Cu and Ni (Wood, Wang, 1985) or CdS precipitates (Raskin et al, 1994).

Biomethylation belongs among basic mechanisms responsible for the detoxification. It is the ability of organisms to produce alkyl and aryl-alkyl compounds. Sometimes volatile compounds are formed and sometimes compounds which are stored in the organisms. In dependence on the type of organism and type of metal biomethylation can decrease or increase the toxicity of the metal (Brickman, 1985, Cullen, Reiner, 1989). In the case of mercury methylated compounds are more toxic, because HgS as the most common Hg compound in nature is very little soluble but methylated mercury is soluble in the lipids, thus, available for organisms (Šírl, 1995). In contrast, methylation of arsenic by higher organisms, including algae, produces different compounds which are less toxic as inorganic arsenic. The final product of methylation in marine and freshwater algae is arsenic built in arsenosugars (Cullen, Reiner, 1989). As marine algae are often part of the human diet it was necessary to find if arsenosugars are toxic for people. After experiments with volunteers slight cytotoxicy, 10000-times lower in comparison with inorganic arsenic was confirmed (Le et al, 1994, Sakurai et al 1997).

Biovolatilization is the forming of volatile compounds by cells. The basic mechanisms are methylation and reduction (Michalke et al, 2000). Most of volatile metal compounds is more toxic than their inorganic forms except of As and Se (Thayer, 1995).

Bioprecipitation

Another very important mechanism of metal detoxification is bioprecipitation. Some organisms, with the aim to avoid metal enter their cell, excrete different molecules into the environment which form insoluble compounds such as organometallic complexes, sulphides or oxides (Raraz, 1995, Luptakova et al, 2002). This process is known as extracellular precipitation which can be dependent but also independent of metabolism (Veglio, Beolchini, 1997) usually is slow, irreversible and dependent on temperature (Kotrba et al, 1994). Holan et al. (1993) suggested that precipitation is additional mechanisms in Cd removal by marine algae. Similarly gold is reduced to the metallic form on the surface of the cells *Chlorella vulgaris* (Borovec, 1989; Yazawa, Kuwabara, 1996). Extracellular organic exudates excreted by *Pseudokirchneriella subcapitata* interact with metals present in water forming complexes with Cd, Cu and Zn (Koukal et al, 2007).

4. Removal of Inorganic Pollution

Pollution by metal ions in industrial waste waters is one of the most important causes of contamination to humans, as they can accumulate throughout the food chain, and environment, due to its great persistence and high toxicity (Herrero et al, 2008, Fečko et al, 2004). Above described mechanisms which algae use to cope with high toxicity can be used also for decreasing of the amount of pollutants in the ecosystems and serve as the basis for new technologies.

4.1. Bioaccumulation as the Pollutant Removal Technology

Algae offer great potential for the cleaning up aquatic environment within phytoremediation techniques. They grow in a wide spectrum of water qualities, sometimes in really extreme environment. Their ability to accumulate high quantities of metals represents possibility to take up pollutants from natural ecosystems, ranging from heavy metals (Rai, Chandra, 1989, Peña Castro et al, 2004), metalloids to radionuclides (Zhou et al, 2008).

Most of the elements biding in brown algae may be attributed to ion exchange on cell walls which contain a large number of compounds with anionic groups including proteins, polysaccharides and polyphenols. Once elements are bound to the surface they are transported, actively or passively, through the cell membrane into the intracellular liquid (Ostapczuk et al, 1997). Aquatic organisms do not accumulate all elements in the same way. Algae accumulate some elements such as As, Mn, Co, Ni and Ba in higher extent than for example mussels, Hg and Se in lower and now differences were observed for P, S, Na, K, Mg, Cu, Cd, Pb and Zn accumulation (Ostapczuk et al, 1997).

Except of direct removal of metals by bioaccumulation algae can be the part of ecosystems used for wastewater cleaning. Scholz and Xu (2002) used passive vertical flow filter containing different macrophytes such as *Phragmites* and *Typha* and granular media with different adsorption capacities where also algae were present for Cu and Pb removal from wastewater. The role of algae was not to accumulate metals but to produce O_2 and increase the pH values.

Freshwater alga *Chlorella vulgaris* accumulated arsenate ions and transformed them into dimethylarsenate compounds (Kaise et al, 1997). Alga *Chara canescens* is able to accumulate Se and methylate it, however, it is not known how big part of methylated compounds is stored inside the algal body and how big part is released into the atmosphere in the form of volatile compounds. Concentration of Se in the environment by means of *Ch. canescens* was decreased by 72% (Lin et al., 2002). Microalga *Scenedesmus incrassatulus* removed copper, cadmium and chromium in continuous system with the efficiency from 25-78% (higher efficiency was for chromium) (Peña Castro et al, 2004).

Travieso et al. (2002) constructed a reactor – BIOALGA – where immobilized microalga *Scenedesmus obliquus* was used. Alga was immobilized in polyurethane foam. About 82% of Co from model solution was removed within first three days; equilibrium was achieved after 10 days with the efficiency of Co removal 94.5%.

Alga *Scenedesmus quadricauda* decreased Ni concentration from 30 mg/l to 0.9 mg/l within 5 minutes and to 0.4 mg/l after 90 minutes after inoculation. It represented 97% of Ni removal. When Ni and Zn mixture was used alga accumulated 98% of both metals within 5 minutes (Chong et al, 2000).

Technology which uses bioaccumulation for arsenic removal was described by Borovec (1989).They used algae *Ceratophyllum demursum* and *Largosiphon major* to treat 24 700 litres of water per day. Biomass was after bioaccumulation filtered, dried and burned. As_2O_3 was taken up together with hot gas and arsenic after cooling condensated. Concentration of As in water after treatment was less than 0.04 mg/l.

Bioprecipitation caused by algae is rarely considered as the technology but it was confirmed that growing of algae in shallow lagoons can contribute to the treatment of wastewater by transforming solved nutrients into particle aggregates (Steinmann et al, 2003).

Similarly algae which grew on the surface of the bricks were able to oxidize Mn^{2+} ions to Mn^{4+} and so formed micronodules of MnO_2 (López-Arce et al, 2003) or in the case of Zn removal by *Padina gymnospora* Zn granules were found in cell wall (Amado Filho et al, 1999).

4.2. Biosorption as the Technology

The aim of the biosorption is to remove dissolved ions from solutions using the abilities of organisms or their parts to take up metals on their surface. It is a passive immobilization of metals (Kaduková, Štofko, 2007). In biotechnology is used to call biosorption the process when metal ions are bound on the dead biomass (Volesky, 1990) but in wider sense biosorption takes place also in living cells as the first step of the bioaccumulation. Difference with the technological point of view is that using dead biomass means that metal is bound only at the surface and does not enter the cell so it can be desorbed. New (biological) waste contaminated with metals is not created then. Biosorption has several advantages over the conventional methods: the process does not produce chemical sludge (i.e. non-polluting), it could be highly selective, more efficient, easy to operate and hence cost effective for the treatment of large volumes of wastewaters containing low pollutant concentrations (Dönmez, Aksu, 2002).

Table 1. Biosorption of metal by algal biomass. N-undetermined

Metal	Alga	Algal treatment	Initial metal concentration (mg/l)	Highest observed q (mg/g)	Ref.
Cu	*Caulerpa lentillifera*	drying at 80°C	10	5.6	Pavasant et al, 2006
	Ulothrix zonata	drying at 100°C	N	176.2	Nuhoglu et al, 2002
	Padina sp.	drying at 60°C and pre-treating with $CaCl_2$	127	50.8	Kaewsarn, 2002
	Sargassum vulgare	drying at 45°C	10 - 250	59	Volesky et al, 1999
	Padina sp.	drying at 60°C	N	72.4	Sheng et al, 2004
	Chlorella kessleri	drying at 55°C	0 - 5000	1260	Kaduková Virčíková, 2005
	Cymodocea nodosa	drying at 60°C	25 - 300	52.7	Sánchez et al, 1999
	Scenedesmus obliquus	drying	N	10	Mattuschka et al, 1993
Cr	*Chlorella vulgaris*	drying	25 - 400	3.5	Nourbakhsh et al, 1994
	Chlorella vulgaris	protonating in H_2SO_4 and drying at 60°C	100	50	Chovancová, 2001
	Cladophora crispata	drying	25 - 400	3	Nourbakhsh et al, 1994
	Laminaria japonica	pre-treating with $CaCl_2$ and drying at 60°C	N	1.81	Lee et al, 2004
	Spirogyra sp.	sun drying	5	14.7	Gupta et al, 2001
Cd	*Padina sp.*	sun drying and pre-treating with $CaCl_2$	225	59.6	Kaewsarn, Yu, 2001
	Sargassum baccularia	drying in owen	50 - 400	83.2	Haskim, Chu, 2004
	Sargassum vulgare	drying at 45°C	10 - 450	87	Volesky et al, 1999
	Sargassum polycystum	drying at 80°C and 100°C	10 - 1200	103.4	Srikrajib et al, 1999
	Sargassum sp.	drying at 70°C	20	5.04	Cruz et al, 2004
	Chaetomorpha linum	drying in owen	50 - 400	53.9	Haskim, Chu, 2004
	Chlorella kessleri	drying at 55°C	26.5	8.7	Kaduková, Virčíková, 2002
	Ecklonia maxima	pre-treating with $CaCl_2$ and drying at 100°C	30 - 300	88.5	Feng, Aldrich, 2004

Table 1. Continued

Metal	Alga	Algal treatment	Initial metal concentration (mg/l)	Highest observed q (mg/g)	Ref.
Pb	*Ecklonia maxima*	pre-treating with $CaCl_2$ and drying at 100°C	30 - 300	232.6	Feng, Aldrich, 2004
	Padina sp.	drying at 60°C	N	133	Sheng et al, 2004
	Sargassum hystrix	drying at 50°C	500	280	Jalali et a., 2002
Zn	*Cymodocea nodosa*	drying	25 - 300	46.6	Sánchez et a., 1999
Ni	*Chlorella sorokiniana*	immobilization in loofa sponge	2.5 - 200	60.4	Akhtar et al, 2004
	Chlorella vulgaris	drying at 60°C	250	60.2	Aksu, 2002
	Sargassum sp.	soaking with acetone and drying at 40°C	250	70	Kalyani et al, 2004
Al	*Laminaria japonica*	pre-treating with $CaCl_2$ and drying at 60°C	N	2.79	Lee et al, 2004
Hg	*Chlorella vulgaris*	immobilization in silica gel	0.02 - 1	$0.4.10^{-3}$ – $6.6.10^{-3}$	Tajes-Martínez et al, 2006
	Ulva lactuca	protonating in H_2SO_4 and drying at 50°C	N	84.74	Zeroual et al, 2003
Cs	*Hypnea valentiae*	sun drying	20 - 500	71.9	Jalali-Rad et al, 2004
La	*Sargassum fluitans*	sun drying	N	33.3	Palmieri et al, 2002

As in biosorption dead biomass is used no nutrients are necessary and biosorbents can be used to treat industrial wastewater with high metal concentrations. Advantage of biosorption by algal biomass is that dead immobilized algal biomass is good biosorbent for heavy metals at Ca and Mg concentration higher than 10 000 mg.kg^{-1}. Higher concentration of Ca^{2+}, Mg^{2+}, Na^{+} and K^{+} did not interfere with the biosorption of heavy metals (Kaewsarn, 2002); also organic materials do not interfere with bound ions. These advantages are very useful for ground water cleaning where Ca and Mg concentration are often very high (Darnall, Hyde, 1989). Other advantages are high rate of the process, normal pressure and temperature required for performance, metal concentrate as the final product of biosorption can be used for metal recovery by conventional hydrometallurgical methods and biomass can be used repeatedly (Volesky, 2003). When microscopic algae are used for biosorbents they are often immobilized in solid carriers. This way beads with right size, mechanical resistance, rigidity and porosity are formed that are more suitable for industrial practice (Stams, Oude Elferink, 1997, Shah et al, 1999).

The most effective biosorption set-up in practice is in packed-bed columns (Volesky, 2001). Biosorption column works in cycles consisting of filling, regeneration and washing. Whole process starts with filling of column by biosorbent particles then metal contaminated water flow through the column. Biosorbent particles take up metal ions and after whole column is saturated it is taken out of the process, metal is eluted by the acid or base solutions and finally the column is washed by water to removed rest of regenerating agent or suspended solids. Metal concentrate obtained after regeneration can be treated by conventional hydrometallurgical processes (Kratochvil, Volesky, 1998).

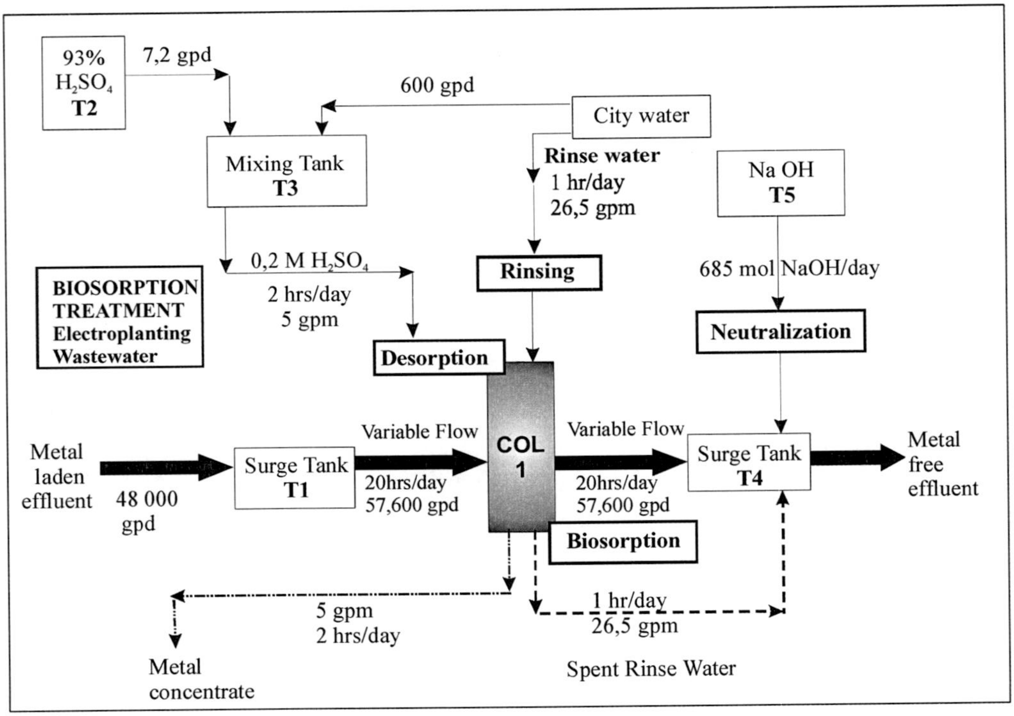

Figure 5. The design of the wastewater treatment plant using biosorption (Volesky, 2003).

Algae with high biosorption abilities are often used for biosorbents preparation. Some of algae studied for metal biosorption are listed in the Table 1.

Some of the processes using biosorption have already been commercialized. Biosorbent AlgaSORB prepared from alga *Chlorella* immobilized in silica gel or polyacryl gel was developed in USA in Bio-Recovery Systems, New Mexico (Volesky, 1990, Smith et al, 1994). Biosorbent BIO-FIX formed by algal, moss or other organisms was developed in U.S. Bureau of Mines and Licencees, USA (Kotrba et al, 1994). Other commercialized biosorbents are prepared from bacterial or other biomasses. Company BV Sorbex, Inc. Canada also successfully works in the field of biosorption application. They applied biosorption for wastewater originating from various metal processing plants. Their design for biosorption plant is in Figure 5 (Volesky, 2003).

5. Removal of Organic Pollution

The bioaccumulation of persistent organic compounds in aquatic food chains is undesirable because of their adverse effect on aquatic organisms and top predators as well as the humans (Binelli, Provini, 2003). The use of naturally occurring microorganisms in the soil and water to actively degrade toxic or unwanted organic compounds and transferring them into harmless substances is very effective way to clean up the environment (Swaminathan, 2003). Application of algae for the removal of nutrients and organics from wastewater is called phycoremediation (Olguín, 2003). Very important feature of algae is their ability to change from autotrophic to mixotrophic growth which enables them to metabolize organic chemicals (Semple, 1998, Tarlan et al, 2002).

Interesting and promising method offers using of macroalgae for seawater cleaning from oil. Usual problem connected with the open sea bioremediation is that any seeded microorganisms would be washed out and diluted. Radwan et al. (2002) found that macroalgae are often associated with several species of hydrocarbon utilizing bacteria. Each gram fresh alga from the Arabian Gulf, which they have studied, was associated with about two to about 30 million cells of bacteria predominantly belonging to the nocardioforms and the genus *Acinetobacter*. These bacteria consumed big part of hydrocarbons from the media (ranging from about 64 to 98% of n-octadecane and about 38– 56% phenanthrene) after 2 weeks. Bacteria adhering to macroalgae were firmly immobilized on their surface and very rarely were released into the sea water.

The role of algal communities formed mostly by *Volvox aureus* in phenol bioremediation in small forest pond water in Siberia (Russia) was also reported (Gladyshew et al, 1998). Algae did not degrade phenol directly but they provided phenol-degrading bacteria coexisting in the same ecosystem by inorganic nutrients (mostly PO_4^{3-} and NO_3^-) so bacteria were not limited by these nutrients. In comparison with the pond where not algae but cyanobacteria were present phenol degradation was significantly lower during the period of their "bloom" because of massive nutrients decrease.

Also green alga *Ochromonas danica* was found to remove within 3 days phenol totally from the cultivation media when it was the sole carbon source. This alga was able to remove phenols preferentially from media with other organic compounds only the rate was slower than in the first case (Semple, 1998).

Chlorella fusca var. *vacuolata* was used in the removal of phenols. It removed more than 90% of o- and m-nitrophenol, 85% of bisphenol A and 90% of dinitrophenol. No addition of organic carbon sources was necessary during treatment (Hirooka et al, 2003).

Potential of algae to metabolize DDT and transform it to relatively biodegradable DDD was reported suggesting the role of algae in its bioremediation (Megharaj et al, 2000).

Metabolic degradation of organics by green algae (*Chlorella sp.*) and diatom species was the main mechanism of organics removal in wood-based pulp and paper industry wastewater. Within 42 days algae removed up to 58% of COD, 84% of colour and 80% of absorbable organic xenobiotics. The main mechanism of colour and organics removal from pulping effluents was partly metabolism and partly metabolic conversion of coloured and chlorinated molecules to non-coloured and non-chlorinated molecules. Adsorption onto algal biomass was not so effective (Tarlan et al, 2002).

Biosorption can be used to remove metal ions as well as some organic compounds. Recently it is often studied to treat waters containg synthetic dyes. Khalaf (2008) has successfully applied freshwater green alga *Spirogyra sp.* to remove biological reactive dye Synazol. Efficiency of the process was 85% after 18 hours. Alga *Caulerpa lentillifera* proved high biosorption capacity for three basic dyes Astrazon Blue FGRL, Astrazon Red GTLN and methylene blue. Biosorption capacities were even higher that capacities observed for activated carbon. Maximum biosorption capacities for Astrazon Blue FGRL, Astrazon Red GTLN and methylene blue were 38.9, 47.6 and 417 mg/g, respectively (Marungrueng, Pavasant, 2007).

Very big area of pollution treatment is a treatment of hydrocarbons from oils. In general, bacteria are widely used in this process. But an example that algae can be used to degrade oils was provided by Suzuki and Yamaya (2005). They applied achlorophyllous microalga *Prototheca zopfii* to remove n-alkanes - C_{14}, C_{15} and C_{16}. After 2 days of operation about 95% of hydrocarbons were removed proving that microalgae can be used also in petroleum hydrocarbons biodegradation.

6. Treatment of Domestic Wastewater and Nutrient Removal

The increase in nitrogen and phosphorus from various sources of wastewater has resulted in serious eutrophication of natural bodies of water, and pollution of the water environment has become a serious global problem (Martínez et al, 2000, Feng et al, 2003).

Algae are often part of ecosystems constructed to treat domestic wastewater. Microalgae play an important role during tertiary treatment of domestic wastewater in maturation ponds, treatment of small-middle-scale municipal wastewater in facultative aerobic ponds or waste stabilization ponds (Van Luijn et al, 1999, Arauzo et al, 2000, Zimmo et al, 2004, Beran, Kargi, 2005). They enhance the removal of nutrients, heavy metals and pathogens and furnish O_2 to heterotrophic aerobic bacteria to mineralize organic pollutants using in turn the CO_2 released from bacterial respiration (Mashauri, Kayombo, 2002, Kayombo et al, 2003, Muñoz, Guieysse, 2006).

Use of algae in nutrient removal is comparable to other conventional technologies (Olguín, 2003, Jiménez-Pérez et al, 2004). Algae are usually found to be very effective in treatment of domestic wastewater in stabilization ponds but probably their combination with higher plants such as *Lemna sp. Spirodela sp. Wolffia sp.* gives better results in the treatment of domestic wastewater according to study of Dalu and Ndamba (2003).

Biological treatment is the most effective means of N removal, converting biologically available N compounds into N_2 gas. The most important mechanism for N removal by algae was found nitrification/denitrification (Hamersley, Howes, 2002). Zimmo et al. (2003) confirmed that ammonia volatilisation rates in algae-based ponds were higher than in duckweed-based ponds, but ammonia volatilization in both kinds of treatment did not exceed 1.5% of total ammonium nitrogen removal.

Algal species have been successfully used in the treatment of municipal sewage and effluents from livestock operations. They are able to metabolize dissolved nitrogen and phosphorus. Lincoln et al. (1996) found that in laboratory condition algal uptake of ammonia nitrogen was rapid and complete, lowering concentrations by more than 24 mg/l/day with minimum density of algal population about 250 mg/l. The advantage of this process was that at least half of the dry weight of algal biomass is edible protein.

In the case of the degradation of N-containing contaminants such as acetonitrile application of algae in biodegradation batch process decreased the concentration of produced NH_4^+ twice in comparison with bacterial biodegradation without algae (Muñoz, Guieysse, 2006).

Alga *Chlorella vulgaris* applied to a nutrients reduction in municipal sewage removed more than 90% of NH_4^+ and 80% of PO_4^{3-} within 10 days treatment. The higher was algal concentration in inoculum the higher was the treatment efficiency. Decrease of COD was more than 50% and the decrease of total organic nitrogen was more than 60% but it was mostly caused by indigenous bacteria not by algae (according to comparison with controls). Interaction between algal and bacterial cells seems to be important in simultaneous removal of N, P and organic matter from primary settled sewage (Lau et al, 1995, Lau et al, 1998). Alga *Phormidium laminosum* was found to remove NO_3^-, NO_2^- and PO_4^{3-} within 14 hours (Garbisu et al, 1994, Sawayama et al, 1998). Mixture of alga *Chlorella vulgaris* and bacterium *Azospirillum brasilense* immobilized in alginate beads removed NH_4^+ and P with high efficiency (De-Bashan et al, 2002). 99% ammonium removal by alga *Chlorella vulgaris* was observed when L-glutamic acid was added into media (Khan, Yoshida, 2008).

To remove nutrients from stillage via the growth on the treated effluents several algae were successfully applied such as *Spirulina platensis*, *Chlorella vulgaris* and *Chlamydomonas reinhardii* (Wilkie et al, 2000).

Algae play an important role in the nutrient removal from animal wastewater when growing in the integrated recycling system (IRS). Within such systems, animal waste is the first input and several by-products and high added value products such as algae are the overall output. Nutrients originally contained in the waste are recycled and used to produce new products. An assessment of an IRS for pig waste with recuperation of *Spirulina* has been carried out recently (Olguín, 2003).

Treatment of piggery slurry, effluents from fisheries and high BOD-containing wastewaters was successfully carried out in high rate algal ponds – HRAP. HRAP are shallow (0.3-0.6 m) in order to allow maximum light penetration and promote algal growth. They can operate at short hydraulic retention time in the range of 4 to 10 days depending on climatic conditions reducing the required surface area (Olguín, 2003).

In sub-tropical conditions a pilot-scale engineered ecosystem where algae were integral part was built and tested in Australia. Although algae and plants were shown to remove 5% of the influent nitrogen and 6% of the influent phosphorus whole ecosystem removed 75% of ammonium and 53% of total inorganic nitrogen. Effluent produced by ecosystem did not exceed limits required by the government (Kavanagh, Keller, 2007).

7. Biomonitoring Using Algae

Pollution has the potential to affect the biological integrity of aquatic systems, decreases the quality of waters and may directly affect human health (Graça et al, 2002). Traditionally the evaluation of toxicants in water relies on the extraction of chemicals from the water, followed by chemical analysis (Schreiber et al, 2007). Conventional analyses provide information about the amount of pollutants but they do not provide information about their influence on ecosystems (Zhou et al, 2008). The effects of anthropogenic disturbances in the receiving systems can be assessed by measuring changes in community structural parameters including biodiversity indicators (diversity, species and taxonomic richness), biotic indices, community structure or changes in functional parameters such as energy allocated to growth (Graça et al, 2002). Using of organisms sensitive to the presence of particular pollutant in the environment to quantitative and qualitative evaluation of its negative effect on biological system is called biomonitoring. It is important tool of environmental monitoring and inseparable part of chemical assays (Kadukova, 2006). The algae species and amounts can directly reflect the water quality and because of their sensitivity to environmental pollution they are especially suitable bioindicators for bio-tests (Millán de Kuhn et al, 2006, Swaminathan, 2003, Nassar et al, 2003). Algae are very often used in standard test to evaluate water quality and toxic affect of inorganic and organic pollutants (Staples, Davis, 2002, Tare et al, 2003, Tišler et al, 2004).

The advantages of using algae are numerous: access to large populations, ease of culture, use of simple inorganic culture media and rapid growth rate. The progress has been made in diversifying the algal toxicity tests to include different taxonomic groups; e.g. brown, green and red algae, and uni- and multi-cellular species. Another major advance has been the miniaturization of the tests (Koukal et al, 2007). Although significant results have been obtained based on algae biomonitoring some disadvantages has been also found. For example the size of microalgae makes it difficult to isolate enough amount of purebred strain. The complexity of phytoplankton communities makes the monitoring data sinuous for the actual evaluation (Zhou et al, 2008). Except of it, heavy metal levels in algae species do not depend only on metal concentration in water but also on environmental parameters (salinity, temperature, pH, light, oxygen, nutrient concentrations, complexing agents) as well as on the structural differences among the algae species (Topcuoğlu et al, 2003). But respecting these limitations algal biomonitoring is very important and useful for the evaluation of metal pollution in aquatic ecosystem including harbours, continental waters, heavy metal mining areas etc. It may offer the effective precaution system based on these biomonitoring data. The performance of the wastewater treatment can be evaluated as well (Zhou et al, 2008). To evaluate the evolution of pollutants over time or to compare pollution level among different sites the concentration factor CF is often used. CF represents the ratio between the content of given metal in an organism and its concentration in the surrounding environment (Muse et al, 2006).

Very important tool is the use of living organisms for monitoring the concentration of metals occurring in water in very low amount undetectable by chemical test but in spite of it causing toxic effects in organisms. Gosavi et al. (2004) has studied several algal species growing in contaminated water. They found that essential metals were accumulated to a concentration greater than all non-essential metals, except for Al. But what is more interesting, they found that trace metals As, Cd and Pb, though undetectable in water samples, accumulated to significant levels in all macroalgae studied.

Except for direct monitoring of the environmental pollution, algae can be used as the monitors of the other organisms' damage. For example, symbiotic algae can give an early indication of copper-induced damage in coral *Plesiastrea versipora* before the damage becomes visible or lethal (Grant et al., 2003).

Well chosen algal species and methods of biomonitoring can not only give information about the state of the environment, the range of the damage and effect on ecosystems but also can be used to distinguish among the sources of pollution. Regel et al. (2002) studied the esterase activity of *Microcystis aeruginosa* and *Selenastrum capricornutum* as a rapid measure of the biological effects of acid mine drainage. Their study demonstrated the successful application of algal esterase activity bioassays, in combination with flow cytometry, to rapidly assess the toxicity of AMD-affected waters and to differentiate this response from the effects of other pollutants. Also a source of the pollution can be identified. For example studying of metal content in algae growing in the lake Nainital (India) indicated that pollution input in the lake was mainly through domestic waste and municipal sewage (Ali et al., 1999).

Alga *Selenastrum capricornutum* was used also for the monitoring of bioavailable dissolved organic nitrogen (Pehlivanoglu, Sedlak, 2004).

Measuring of chlorophyll fluorescence using algae is a rapid and sensitive method to determine the toxic effect of pollutants on plants. Alga *Closterium ehrenbergii* was used in this measurement to detect toxicity of sewage treatment plant effluent (Juneau et al., 2003).

8. Conclusion

Algae are part of most ecosystems on the Earth. They are forced to gain nutrients and essential elements for their growth, to adapt to various kinds of environments and conditions and finally, with the intensifying human industrial activities, to cope with the excess of undesirable or even toxic chemicals. They developed different mechanisms which help them to survive. Many of these mechanisms can be effectively used for the benefit of the whole ecosystem and people when they are reasonably intensified and applied to pollution removal. They are very efficient for metal removal from the environment.

Biosorption as the method of metal ions uptake from solutions on the surface of dead biomass is in the present highly studied. Biosorbents prepared from algae compete with classic sorptive materials such as active carbon or ion-exchangers. Their capacity is comparable with commercial materials. The advantage is that biosorbents prepared from algae do not change their sorption capacity for metals in the presence of Ca and Mg ions and they can sequester metals selectively. But the greatest advantage is their low price as they are mostly prepared from waste or naturally abundant biomass. According to published studies, bioaccumulation and bioprecipitation can also be successfully used for metal removal.

Algae are also often used in biological systems to decrease nutrient concentration or to degrade organic compounds. In these systems their ability to change from autotrophy to mixotrophy and metabolically degrade organic chemicals is used. It seems that in systems which contain more species of different organisms including bacteria, plants or even invertebrates, algae are more efficient in comparison with monocultures. These systems are exclusively biological, and thus, do not disturb the environment. No new waste is formed during their performance and they do not require addition of new chemicals. Their big advantage is low operational cost.

Using algae to monitor environmental pollution is a common practice. But there are still new possibilities of how to improve methods of biomonitoring such as using algae not only to monitor the level of pollution but also the type of pollution and its source.

It is obvious that algae have a great potential in pollution control as well as in processes applied to decrease the pollution level in the environment.

Acknowledgement

This work was supported by the Slovak grant agency – project VEGA1/3220/06.

References

Admiraal, W; Blanck, H; Buckert-De Jong, M; Guasch, H; Ivorra, N; Lehmann, V; Nyström, BAH; Paulsson, M-M; Sabater, R. Short-Term Toxicity of Zinc to Microbenthic Algae and Bacteria in a Metal Polluted Stream, *Water Research*, 1999, 33, 9, 1989-1996.

Akhtar, N; Iqbal, J; Iqbal, M. Removal and recovery of nickel(II) from aqueous solution by loofa sponge-immobilized biomass of *Chlorella sorokiniana*: characterization studies, *Journal of Hazardous Materials*, 2004, B108, 85-94.

Aksu, Z. Determination of the equilibrium, kinetic and thermodynamic parameters of the batch biosorption of nickel(II) ions onto *Chlorella vulgaris*, *Process Biochemistry*, 2002, 38, 89-99.

Ali, MB; Tripathi, RD; Rai, UN; Pal, A; Singh, SP. Physico - chemical characteristics and pollution level of lake Nainital (U.P., India): role of macrophytes and phytoplankton in biomonitoring and phytoremediation of toxic metal ions, *Chemosphere*, 1999, 39, 12, 2171-2182.

Amado Filho, GM; Andrade, LR; Karez, CS; Farina, M; Pfeiffer, WC. Brown algae species as biomonitors of Zn and Cd at Sepetiba Bay, Rio de Janeiro, Brazil, *Marine Environmental Research*, 1999, 48, 213-224.

Arauzo, M; Colmenarejo, MF; Martínez, E; García, MG. The Role of Algae in a Deep Wastewater Self-Regeneration Pond, *Water Research*, 2000, 34, 14, 3666-3674.

Beceiro-González, E; Taboada-de la Calzada, A; Alonso-Rodrígez, E; López-Mahía, P; Muniategui-Lorenzo, S; Prada-Rodrígez, D. Interaction between metallic species and biological substrates: approximation to possible intraction mechanisms between the alga *Chlorella vulgaris* and arsenic(III), *Trends in Analytical Chemistry*, 2000, 19, 8, 475-480.

Beran, B; Kargi, F. A dynamic mathematical model for wastewater stabilization ponds, *Ecological Modelling*, 2005, 181, 39–57.

Bertrand, M; Schoefs, B; Siffel, P; Rohacek, K; Molnar, I. Cadmium inhibits epoxidation of diatoxanthin to diadinoxanthin in the xanthophyll cycle of the marine diatom *Phaeodactylum tricornutum*, *FEBS Letters*, 2001, 508, 153-156.

Binelli, A; Provini, A. The PCB pollution of Lake Iseo (N. Italy) and the role of biomagnification in the pelagic food web, *Chemosphere*, 2003, 53, 143–151.

Borovec, Z. Bioakumulace kovů a dočisťování vod baktériemi, houbami a řasami, *Rudy*, 1989, 12, 37, 375-382.

Brinckman, FE. Environmental and Inorganic Chemistry of Main Groups Elements with Special Emphasis on Their Occurence as Methyl Derivatives, In: Environmental Inorganic Chemistry (proceedings), VCH Publishers, Inc. Deerfield Beach, Florida, 1985.

Bustard, M; Donnellan, N; Rollan, A; McHale, A.P. Studies on the biosorption of uranium by a thermotolerant, ethanol-producing strain of *Kluyveromyces marxianus, Bioprocess Engineering*, 1997, 17, 45-50.

Butler, A. Acquisition and utilization of transition metal ions by marine organisms, *Science*, 1998, 281, 207-210.

Camargo, JA. Fluoride toxicity to aquatic organisms: A review, *Chemosphere*, 2003, 50, 251–264.

Cervantes, C. Bacterial resistance to arsenic compounds (Resistencia bacteriana a compuestos de arsenico), *Revue of Latinoamerican Microbiology*, 1995, 34, 4, 387-395.

Cho, DY; Lee, ST; Park, S. W., Chung, A. S.: Studies on the biosorption of heavy metals onto *Chlorella vulgaris*, *Jouranl of Environmental Science and Health*, 1994, A29, 389-409.

Chong, AMY; Wong, YS; Tam, NFY. Performance of different microalgal species in removing nickel and zinc from industrial wastewater, *Chemosphere*, 2000, 41, 251-257.

Chovancová, I. Uptake of trivalent chromium by biosorbent of algae of *Chlorella vulgaris,* New Trends in Mineral Processing, *Proccedings from international conference,* Ostrava, 2001, 493-499.

Clemens, S. Toxic metal accumulaiton, responses to exposure and mechanisms of tolerance in plants, *Biochimie*, 2006, 88, 1707-1719.

Cruz, CCV; da Costa, AC; Henriques, CA; Luna, AS. Kinetic modeling and equilibrium studies during cadmium biosorption by dead *Sargassum sp.* biomass, *Bioresource Technology*, 2004, 91, 249–257.

Cullen, RW; Reiner, KJ Arsenic Speciation in the Environment, *Chemical Review*, 1989, 89, 713-764.

Cunningham, SD; Ow, DW. Promises and Prospects of Phytoremediation, *Plant Physiology*, 1996, 110, 715-719.

Dalu, JM; Ndamba, J. Duckweed based wastewater stabilization ponds for wastewater treatment (a low cost technology for small urban areas in Zimbabwe), *Physics and Chemistry of the Earth,* 2003, 28, 1147–1160.

Darnall, DW; Hyde, LD. Removal of heavy metal ions from groundwaters using an algal biomass, In: Proceedings of the 2nd National Conference, November 27-29, Washington, USA, 1989, 41-45.

Davis, TA; Volesky, B; Mucci, A. A review of the biochemistry of heavy metal biosorption by brown algae, *Water Research*, 2003, 37, 4311-4330.

De-Bashan, LE; Moreno, M; Hernández, JP; Bashan, Y. Removal of ammonium and phosphorus ions from synthetic wastewater by the microalgae *Chlorella vulgaris* coimmobilized in alginate beads with the microalgae growth-promoting bacterium *Azospirillum brasilense*, *Water Research*, 2002, 36, 2941–2948.

Delgado, A; Anselmo, AM; Novais, JM. Heavy metal biosorption by dried powdered mycelium of Fusarium flocciferum, *Water Environment Research*, 1998, 70, 3, 370-376.

Dönmez, G; Aksu, Z. Removal of chromium(VI) from saline wastewaters by *Dunaliella* species, *Process Biochemistry*, 2002, 38, 751-762.

Elbaz-Poulichet, F; Dupuy, C; Cruzado, A; Velasquez, Z; Achterberg, EP; Braungardt, CB. Influence Of Sorption Processes By Iron Oxides And Algae Fixation On Arsenic And Phosphate Cycle In An Acidic Estuary (Tinto River, Spain), *Water Research,* 2000, 34, 12, 3222-3230.

Escuder-Gilabert, L; Martín-Biosca, Y; Sagrado, S; Villanueva-Camañas, RM; Medina-Hernández, MJ. Biopartitioning micellar chromatography to predict ecotoxicity, *Analytica Chimica Acta*, 2001, 448, 173–185.

Fečko, P; Kušnierová, M; Čablík, V; Pečtová, I. *Environmentální biotechnologie,* VŠB-TU, Ostrava, 2004.

Feng, Ch; Sugiura, N; Shimada, S; Maekawa, T. Development of a high performance electrochemical wastewater treatment system, *Journal of Hazardous Materials*, 2003, B103, 65–78.

Feng, D; Aldrich, C. Adsorption of heavy metals by biomaterials derived from the marine alga *Ecklonia maxima*, *Hydrometallurgy*, 2004, 73, 1-10.

Förstner, U; Wittmann, GTW. *Metal Pollution in the Aquatic Environment*, Springer-Verlag, Berlin, 1983.

Gadd, GM. Microbial interactions with tributyltin compounds: detoxification, accumulation, and environmental fate, *The Science of the Total Environment*, 2000, 258, 119-127.

Garbisu, C; Hall, DO; Llama, MJ; Serra, JL. Inorganic nitrogen and phosphate removal from water by free-living and polyvinyl-immobilized *Phormidium laminosum* in batch and continuous-flow bioreactors, *Enzyme Microbiology and Technology*, 1994, 16, 395–401.

Genter, RB. Ecotoxicology of Inorganic Chemical Stress to Algae, *Algal Ecology*, Academic Press, 1996, 403-468.

Gladyshew, MI; Sushchik, NN; Kalachova, GS; Shchur, LA. The Effect of Algal Blooms on the Disappearence of Phenol in a Small Forest Pond, *Water Research*, 1998, 32, 9, 2769-2775.

Gosavi, K; Sammut, J; Gifford, S; Jankowski, J. Macroalgal biomonitors of trace metal contamination in acid sulfate soil aquaculture ponds, *Science of the Total Environment*, 2004, 324, 25–39.

Graça, MAS; Rodrígez-Capítulo, A; Ocón, C; Gómez, N. In situ tests for water quality assessment: a case study in Pampean rivers, *Water Research*, 2002, 36, 4033–4040.

Grant, AJ; Graham, K; Frankland, S; Hinde, R. Effect of copper on algal-host interactions in the symbiotic coral *Plesiastrea versipora, Plant Physiology and Biochemistry,* 2003, 41, 383–390

Gupta, VK; Shrivastava, AK; Jain, N. Biosorption of Chromium(VI) from Aqueous Solutions by Green Algae *Spirogyra* Species, *Water Research*, 2001, 35, 17, 4079-4085.

Hamersley, MR; Howes, BL. Control of denitrification in a septage-treating artificial wetland: the dual role of particulate organic carbon, *Water Research*, 2002, 36, 4415–4427.

Hasegawa, H. Seasonal changes in methylarsenic distribution in Tosa Bay and Uranouchi Inlet, *Applied Organometallic Chemistry*, 1996, 10, 9, 733-740.

Hasegawa, H. The behaviour of trivalent and pentavalent metylarsenicals in Lake Biwa, *Applied Organometallic Chemistry*, 1997, 11, 4, 305-311.

Haskim, MA; Chu, KH. Biosorption of cadmium by brown, green, and red seaweeds, *Chemical Engineering Journal*, 2004, 97, 249–255.

Henderson, R; Parsons, SA; Jefferson, B. The impact of algal properties and pre-oxidation on solid–liquid separation of algae, Water Research, 2008 (article in press).

Herrero, R; Lodeiro, P; Rojo, R; Ciorba, A; Rodríguez, P; Sastre de Vicente, ME. The efficiency of the red alga *Mastocarpus stellatus* for remediation of cadmium pollution, *Bioresource Technology*, 2008, 99, 10, 4138-4146.

Hindák F; Makovinská J. Use of algae for monitoring rivers in Slovakia: recentdevelopments. In: Prygiel J. et al.: *Use of algae for monitoring rivers III,* Agence de l′ Eau Artois-Picardie, 1999, 35-38.

Hindák, F. et al.: *Sladkovodné riasy*, SPN Bratislava, 1978, 530-541.

Hindák, F. *Fotografický atlas mikroskopických siníc*, Veda, vydavateľstvo SAV, Bratislava, 2001.

Hirooka, T; Akiyama, Y; Tsuji, N; Nakamura, T; Nagase, H; Hirata, K; Miyamoto, K. Removal of Hazardous Phenols by Microalgae under Photoautotrophic Conditions, *Journal of Bioscience and Bioengineering*, 2003, 95, 2, 200-203.

Holan, ZR; Volesky, B; Prasetyo, I. Biosorption of cadmium by biomass of marine algae, Biotechnology and Bioengineering, 1993, 41, 819-825.

Irgolic, KJ; Martell AE.Environmental Inorganic Chemistry, Workshop summary, In: *Environmental Inorganic Chemistry (proceedings),* VCH Publishers, Inc. Deerfield Beach, Florida, 1985, 23-39.

Jalali, R; Ghafourian, H; Asef, Y; Davarpanah, SJ; Sepehr, S. Removal and recovery of lead using nonliving biomass of marine algae, *Journal of Hazardous Materials*, 2002, B92, 253–262.

Jalali-Rad, R; Ghafourian, H; Asef, Y; Dalir, ST; Sahafipour, MH; Gharanjik, BM. Biosorption of cesium by native and chemically modified biomass of marine algae: introduce the new biosorbents for biotechnology applications, *Journal of Hazardous Materials*, 2004, B116, 125–134.

Jiménez-Pérez, MV; Sánchez-Castillo, P; Romera, O; Fernández-Moreno, D; Pérez-Martínez, C. Growth and nutrient removal in free and immobilized planktonic green algae isolated from pig manure, *Enzyme and Microbial Technology*, 2004, 34, 392–398.

Juneau, P; Sumitomo, H; Matsui, S; Itoh, S; Kim, S-G; Popovic, R. Use of chlorophyll fluorescence of *Closterium ehrenbergii* and *Lemna gibba* for toxic effect evaluation of sewage treatment plant effluent and its hydrophobic components, *Ecotoxicology and Environmental Safety,* 2003, 55, 1–8.

Kaduková, J. Aplikácia biotechnológií na riešenie environmentálnych problémov v huntíctve (Application of biotechnologies to environmental problems in metallurgy solving), Habilitation Thesis, HF TU, Košice, 2006.

Kadukova, J; Štofko, M. Biosorption of Heavy Metals Ions from Aqueous Solutions, Chapter 4, *Environmental Research Trends* (Ed. Marvin A. Cato), Nova Publishers, 2007, 133-152.

Kaduková, J; Štofko, M. *Environmentálne biotechnológie pre hutníkov*, TU Košice, 2006.

Kaduková, J; Virčíková, E Cadmium and Silver Removal by Biosorption onto Alga *Chlorella kessleri* and Moss *Sphagnum sp.*, *Proceedings from 6th International*

Conference on Environment and Mineral Processing, Part I, Ostrava, Czech Republic, June 2002, 299-306.

Kaduková, J; Virčíková, E. Comparison of differences between copper bioaccumulation and biosorption, *Environment International*, 2005, 31, 2, 227-232.

Kaduková, J; Virčíková, E. Minerálne biotechnológie III, Biosorpcia kovov z roztokov, VŠB-TU Ostrava, 2003.

Kaewsarn, P. Biosorption of copper(II) from aqueous solutions by pre-treated biomass of marine algae Padina sp., *Chemosphere*, 2002, 47, 1081–1085.

Kaewsarn, P; Yu, Q. Cadmium(II) removal from aqueous solutions by pre-treated biomass of marine alga *Padina sp.*, *Environmental Pollution*, 2001, 112, 209-213.

Kaise, T; Ogura, M; Nozaki,T; Saitoh, K; Sakurai, T; Matsubara, Ch; Watanabe; Ch; Hanaoka, K. Biomethylation of arsenic in an arsenic - rich freshwater environment, *Applied Organometallic Chemistry*, 1997, 11, 4, 297-304.

Kalyani, S; Srinivasa Rao, P; Krishnaiah, A. Removal of nickel (II) from aqueous solutions using marine macroalgae as the sorbing biomass, *Chemosphere*, 2004, 57, 1225–1229.

Kavanagh, LJ; Keller, J. Engineered ecosystem for sustainable on-site wastewater treatment, *Water Research*, 2007, 41, 1823-1831.

Kayombo, S; Mbwette, TSA; Katima, JHY; Jorgensen, SE. Effects of substrate concentrations on the growth of heterotrophic bacteria and algae in secondary facultative ponds, *Water Research*, 2003, 37, 2937–2943.

Khalaf, MA. Biosorption of reactive dye from textile wastewater by non-viable biomass of *Aspergillus niger* and *Spirogyra sp.*, *Bioresource Technology*, 2008, article in press.

Khan, M; Yoshida, N. Effect of L-glutamic acid on the growth and ammonium removal from ammonium solution and natural wastewater by *Chlorella vulgaris* NTM06, *Bioresource Technology*, 2008, 99, 575-582.

Kotrba, P; Ruml, T; Macek, T. Vazba kadmia buňkami mikroorganismů a rostlin, *Chemické listy*, 1994, 88, 642-649.

Koukal, B; Rossé, P; Reinhardt, A; Ferrari, B; Wilkinson, KJ; Loizeau, JL; Dominik, J. Effect of *Pseudokirchneriella subcapitata* (Chlorophyceae) exudates on metal toxicity and colloid aggregation, *Water Research*, 2007, 41, 63–70.

Kratochvil, D; Volesky, B. Advances in the biosorption of heavy metals, *Trends in Biotechnology*, 1998, 16, 291-300.

Lane, T; Morel, A. A biological function for cadmium in marine diatoms, *Proceedings of the National Academy of Science in the USA,* 2000, 97, 4627-4631.

Lau, PS; Tam, NFY; Wong, YS. Effect of Algal Density on Nutrient Removal from Primary Settled Wastewater, *Environmental Pollution*, 1995, 89, 59-66.

Lau, PS; Tam, NFY; Wong, YS. Operational optimization of batchwise nutrient removal from wastewater by carrageenan immobilized *Chlorella vulgaris*, *Water Science and Technology*, 1998, 38, 1, 185–192.

Le, XC; Cullen, WR; Reimer, KJ. Human urinary arsenic excretion after one-time ingestion of seaweed, crab and shrimp, *Clinical Chemistry*, 1994, 40, 4, 617-624.

Lee, HS; Suh, JH; Kim, IB; Yoon, T. Effect of aluminum in two-metal biosorption by an algal biosorbent, *Minerals Engineering*, 2004, 17, 487-493.

Lin, Z-Q; de Souza, M; Pickering, IJ; Terry, N. Evaluation of the Macroalga, Muskgrass, for the Phytoremediation of Selenium-Contaminated Agricultural Drainage Water by Microcosms, *Journal of Environmental Quality* 2002, 31, 2104-2110.

Lincoln, EP; Wilkie, AC; French, BT. Cyanobacterial Process for Renovating Dairy Wastewater, *Biomass and Bioenergy*, 1996, 10, 1, 63-68.

López-Arce, P; García-Guinea, J; Fierro, JLG. Manganese micro-nodules on ancient brick walls, *The Science of the Total Environment*, 2003, 302, 267–274.

Luptáková, A; Kušnierová, M. Bioremediation of heavy metals contaminated industrial waste waters, *Acta Metallurgica Slovaca*, **10**, 2, 2004, 183-188.

Luptáková, A; Kušnierová, M; Fečko, P. Minerálne biotechnlógie II., *Sulfuretum v prírode a v priemysle*, VŠB-TU Ostrava, 2002.

Malmström, BG; Leckner, J. The chemical biology of copper, *Current Opinion in Chemical Biology*, 1998, 2, 286-292.

Martínez, ME; Sánchez, S; Jiménez, JM; El Yousfi, F; Muñoz, L. Nitrogen and phosphorus removal from urban wastewater by the microalga *Scenedesmus obliquus*, *Bioresource Technology*, 2000, 73, 263-272.

Marungrueng, K; Pavasant, P. High performance biosorbent (*Caulerpa lentillifera*) for basic dye removal, *Bioresource Technology*, 2007, 98, 1567–1572.

Mashauri, DA; Kayombo, S. Application of the two coupled models for water quality management: facultative pond cum constructed wetland models, *Physics and Chemistry of the Earth*, 2002, 27, 773–781.

Mattuschka, B; Straube, G. Biosorption of metals by waste biomass, *Journal of Chemical Technology and Biotechnology*, 1993, 58, 57-63.

Megharaj, M; Kantachote, D; Singleton, I; Naidu, R. Effects of long-term contamination of DDT on soil microflora with special reference to soil algae and algal transformation of DDT, *Environmental Pollution*, 2000, 109, 35-42.

Michalke, K; Wickenheiser, EB; Mehring, M; Hirner, AV; Hensel, R. Production of Volatile Derivatives of Metal(loid)s by Microflora Involved in Anaerobic Digestion of Sewage Sludge, *Applied and Environmental Microbiology*, 2000, 66, 7, 2791-2796.

Millán de Kuhn, R; Streb, Ch; Breiter, R; Richter, P; Neeße, T; Häder, DP. Screening for unicellular algae as possible bioassay organisms for monitoring marine water samples, *Water Research*, 2006, 40, 2695 – 2703.

Muñoz, R; Guieysse, B. Algal-bacterial processes for the treatment of hazardous contaminants: A review. *Water Research*, 2006, 40, 2799-2815.

Muse, JO; Carducci, CN; Stripeikis, JD; Tudino, MB; Fernández, FM. A link between lead and cadmium kinetic speciation in seawater and accumulation by the green alga *Ulva lactuca*, *Environmental Pollution*, 2006, 141, 126-130.

Nassar, CAG; Salgado, LT; Yoneshigue-Valentin, Y; Amado Filho, GM. The effect of iron-ore particles on the metal content of the brown alga *Padina gymnospora* (Espírito Santo Bay, Brazil), *Environmental Pollution*, 2003, 123, 301–305.

Nourbakhsh, M; Sağ, Y; Özer, D; Aksu, Z; Kutsal, T; Çağlar, A. A comparative study of various biosorbents for removal of chromium (VI) ions from industrial waste waters, *Process Biochemistry*, 1994, 29, 1-5.

Nuhoglu, Y; Malkoc, E; Gürses, A; Canpolat, N. The removal of Cu(II) from aqueous solutions by *Ulothrix zonata*, *Bioresource Technology*, 2002, 85, 331–333.

Olguín, EJ. Phycoremediation: key issues for cost-effective nutrient removal processes, *Biotechnology Advances*, 2003, 22, 81–91.

Ostapczuk, P; Burow, M; May, K; Mohl, C; Froning, M; Süßenbach, B; Waidmann, E; Emons, H. Mussels and Algae as Bioindicators for Long-Term Tendencies of Element Pollution in Marine Ecosystems, *Chemosphere*, 1997, 34, 9110, 2049-2058.

Palmieri, MC; Volesky, B; Garcia Jr., O. Biosorption of lanthanum using *Sargassum fluitans* in batch system, *Hydrometallurgy*, 2002, 67, 31–36.

Paulíková, A. Zariadenie na ekologické čistenie odpadových vôd. Proceedings of the 4th DAAAM Workshop, „Human Factor and Environmentalist" 5th December 2002, Košice, 85-86.

Paulsson, M; Nyström, B; Blanck, H. Long-term toxicity of zinc to bacteria and algae in periphyton communities from the river Göta Älv, based on a microcosm study, *Aquatic Toxicology*, 2000, 47, 243–257.

Pavasant, P; Apiratikul, R; Sungkhum, V; Suthiparinyanont, P; Wattanachira, S; Marhaba, TF. Biosorption of Cu^{2+}, Cd^{2+}, Pb^{2+}, and Zn^{2+} using dried marine green macroalga *Caulerpa lentillifera*, *Bioresource Technology*, 2006, 97, 2321–2329.

Pehlivanoglu, E; Sedlak, DL. Bioavailability of wastewater-derived organic nitrogen to the alga *Selenastrum Capricornutum*, *Water Research*, 2004, 38, 3189–3196.

Peña Castro, JM; Martínez-Jerónimo, F; Esparza-García, F; Cañizares-Villanueva, RO. Heavy metals removal by the microalga Scenedesmus incrassatulus in continuous cultures, *Bioresource Technology*, 2004, 94, 219–222.

Pichtel, J; Kuroiwa, K; Sawyerr, HT. Distribution of Pb, Cd and Ba in soils and plants of two contaminated sites, *Environmental Pollution*, 2000, 110, 171-178.

Radwan, SS; Al-Hasan, RH; Salamah, S; Al-Dabbous, S. Bioremediation of oily sea water by bacteria immobilized in biofilms coating macroalgae, *International Biodeterioration & Biodegradation*, 2002, 50, 55–59.

Rai, UN; Chandra, P. Removal of Heavy Metals from Polluted Waters by *Hydrodictyon Reticulatum* (Linn.) Lagerheim, *The Science of the Total Environment*, 87-88, 1989, 509-515.

Raraz, AG. Biological and Biotechnological Waste Managment in Materials Processing, *JOM*, 1995, 56-62.

Raskin, I; Kumar, PBAN; Dushenkov, S; Salt, DE. Bioconcentration of heavy metals by plants (Review article), *Current Opinion in Biotechnology*, 1994, 5, 285-290.

Regel, RH; Ferris, JM; Ganf, GG; Brookes, JD. Algal esterase activity as a biomeasure of environmental degradation in a freshwater creek, *Aquatic Toxicology*, 2002, 59, 209–223.

Rijstenbil, JW; Derksen, JWM; Gerringa, LJA; Poortvliet, TCW; Sandee, A; van den Berg, M; van Drie, J; Wijnholds, JA. Oxidative stress induced by copper: Defense and damage in the marine planktonic diatom *Ditylum drightwellii* grown in continuous cultures with high and low zinc levels, *Marine Biology*, Berlin, 1994, 119, 583-590.

Rubio, J; Schneider, IAH; Aliaga, W. New processes for heavy metals separation from waste water streams in the mining industry, Clean Technology for the Mining Industry, *Proceedings of the III International Conference on Clean Technology for the Mining Industry*, Santiago, Chile, May 15-17, Ed.: M.A. Sánchez, F. Vergara, S.H. Castro, University of Concepcion, 1996, 85-97.

Sakurai, T; Kaise, T; Ochi, T; Saitoh, T; Matsubara, Ch. Study of in vitro cytotoxicity of a water soluble organic arsenic compound, arsenosugar, in seaweed, *Toxicology*, 122, 3, 1997, 205-212.

Sanchez, A; Ballester, A; Blázques, ML; Muñoz, GJ; Hammaini, A. Biosorption of copper and zinc by *Cymodocea nodosa, FEMS Microbiology Reviews*, 1999, 23, 527-536.

Sawayama, S; Rao, KK; Hall, DO Nitrate and phosphate ion removal from water by *Phormidium laminosum* immobilized on hollow fibres in a photobioreactor, *Applied Microbiology and Biotechnology*, 1998, 49, 463– 468.

Schiewer, B; Volesky, B. Biosorption by marine algae, Remediation (proceedings), Valdes, J.J. (ed.), Dordrecht, The Netherlands, Kluwer Academic Publishers, 2000, 139-169.

Schlacher-Hoenlinger, MA; Schlacher, T. Differential Accumulation Patterns of Heavy Metals among the Dominant Macrophytes of a Mediterranean *Seagrass Meadow, Chemosphere*, 1998, 37, 8, 1511-1519.

Scholz, M; Xu, J. Performance comparison of experimental constructed wetlands with different filter media and macrophytes treating industrial wastewater contaminated with lead and copper, *Bioresource Technology*, 2002, 83, 71–79.

Schreiber, U; Quayle, P; Schmidt, S; Escher, BI; Mueller, JF. Methodology and evaluation of a highly sensitive algae toxicity test based on multiwell chlorophyll fluorescence imaging, *Biosensors and Bioelectronics*, 2007, 22, 2554–2563.

Semple, KT. Heterotrophic growth on phenolic mixtures by *Ochromonas danica, Research in Microbiology*, 1998. 149, 65-72.

Shah, MP; Vora, SB; Dave, SR. Evaluation of potential use of immobilzed *Penicillium griseofulvum* in bioremoval of copper, Biohydrometallurgy and the Environment Toward the Mining of the 21st Century, *Proceedings of the International Biohydrometallurgy Symposium IBS'99,* Ed.: R.Amils, A.Balester, Elsevier, 1999.

Sheng, PX; Ting, Y-P; Chen, J.P; Hong, L. Sorption of lead, copper, cadmium, zinc, and nickel by marine algal biomass: characterization of biosorptive capacity and investigation of mechanisms, *Journal of Colloid and Interface Science*, 2004, 275, 131–141.

Silver, S. Bacterial transformations of and resistances to heavy metals, In: *Environmental Inorganic Chemistry (proceedings),* VCH Publishers, Inc. Deerfield Beach, Florida, 1985, 513-537.

Šírl, M. Jedy a některé jiné látky zdraví škodlivé, (vyd. Ing. J. Loudín, RoVS), 1995, 25-32.

Smith, LA; Alleman, BC; Copley-Graves, L. Biological Treatment Options, In: *Emerging Technology for Bioremediation of Metals*, Ed: Means, JL and Hinchee, RE, Lewis Publishers, Ohio, 1994, 1-12.

Srikrabij, S; Tongta, A; Thiravetyan, P; Sivabovorn, K. Cadmium removal by the dry biomass of *Sargassum polycystum,* Biohydrometallurgy and the Environment Toward the Mining of the 21st Century, *Proceedings of the International Biohydrometallurgy Symposium IBS'99,* Ed.: R.Amils, A.Balester, Elsevier, 1999, 419-427.

Stams, AJM; Oude Elferink, SJWH. Understanding and advancing wastewater *treatment, Current Opinion in Biotechnology*, 1997, 8, 328-334.

Staples, ChA; Davis, JW. An examination of the physical properties, fate, ecotoxicity and potential environmental risks for a series of propylene glycol ethers, *Chemosphere*, 2002, 49, 61–73.

Steinmann, ChR; Weinhart, S; Melzer, A. A combined system of lagoon and constructed wetland for an effective wastewater treatment, *Water Research*, 2003, 37, 2035–2042.

Suzuki, T; Yamaya, S. Removal of hydrocarbons in a rotating biological contactor with biodrum, *Process Biochemistry*, 2005, 40, 3429–3433.

Swaminathan, MS. Bio-diversity: an effective safety net against environmental pollution *Environmental Pollution*, 2003, 126, 287–291.

Tajes-Martínez, P; Beceiro-González, E; Muniategui-Lorenzo, S; Prada-Rodríguez, D. Micro-columns packed with *Chlorella vulgaris* immobilized on silica gel for mercury speciation, *Talanta*, 2006, 68, 1489–1496.

Tare, V; Yadav, AVS; Bose, P. Analysis of photosynthetic activity in the most polluted stretch of river Ganga, *Water Research*, 2003, 37, 67–77.

Tarlan, E; Dilek, FB; Yetis, U. Effectiveness of algae in the treatment of a wood-based pulp and paper industry wastewater, *Bioresource Technology*, 2002, 84, 1–5.

Thayer, JS. *Environmental chemistry of the heavy elements: hydrido and organo compounds,* VCH Publishers, Inc., Weinheim, Germany, 1995.

Tišler, T; Zagorc-Končan, J; Cotman, M; Drolc, A. Toxicity potential of disinfection agent in tannery wastewater, *Water Research*, 2004, 38, 3503–3510.

Tölgyessy, J. et al. *Chémia, biológia a toxikológia vody a ovzdušia*, Veda, Bratislava, 1984.

Tölgyessy, J; Fargašová, A. *Základy ekológie a toxikológie*, Edičné stredisko SVŠT, 1991.

Topcuoğlu, S; Güven, KC; Balkıs, N; Kırbaşoğlu, Ç. Heavy metal monitoring of marine algae from the Turkish Coast of the Black Sea, 1998–2000, *Chemosphere*, 2003, 52, 1683–1688.

Travieso, L; Pellón, A; Benítez, F; Sánchez, E; Borja, R; O'Farrill, N; Weiland, P. BIOALGA reactor: preliminary studies for heavy metals removal, *Biochemical Engineering Journal*, 2002, 12, 87-91.

Urban, Z; Kalina, T. *Systém a evoluce nižších rostlin*, SPN Praha, 1980, 151-153.

Van Luijn, F; Boers, PCM; Lijklema, L; Sweerts, J-P. Nitrogen Fluxes and Processes in Sandy and Muddy Sediments from a Shallow Eutrophic Lake, *Water Research*, 1999, 33, 1, 33-42.

Veglio, F; Beolchini, F. Removal of metals by biosorption: a review, *Hydrometalurgy*, 1997, 44, 301-316.

Vieira, MGA; Oisiovici, RM; Gimenes, ML; Silva, MGC. Biosorption of chromium(VI) using a *Sargassum sp*. packed-bed column, *Bioresource Technology*, **99**, 2008, 3094–3099.

Volesky, B. *Biosorption of Heavy Metals*, CRC Press, 1990.

Volesky, B. Detoxification of metal-bearing effluents: biosorption for the next century, *Hydrometallurgy*, 2001, 59, 203-216.

Volesky, B. *Sorption and Biosorption*, BV Sorbex, Inc. Montreal, Canada, 2003.

Volesky, B; Weber, J; Vieira, R. Biosorption of Cd and Cu by different types of Saragassum biomass, *Process Metallurgy*, 1999, 9, 2, 473-482.

Wilkie, AC; Riedesel, KJ; Owens, JM. Stillage characterization and anaerobic treatment of ethanol stillage from conventional and cellulosic feedstocks, *Biomass and Bioenergy*, 2000, 19, 63-102.

Wilson, S; Blake, Ch; Berges, JA; Maggs, ChA. Environmental tolerances of free-living coralline algae (maerl): implications for European marine conservation, *Biological Conservation*, 2004, 120, 279-289.

Winge, DR; Jensen, LT; Srinivassan, Ch. Metal-ion regulation of gene expression in yeast, *Current Opinion in Chemical Biology*, 1998, 2, 216-221.

Wood, JM; Wang, H-K. Microbial Resistance to Heavy Metals. *In Environmental Inorganic Chemistry (Proceedings)*, VCH Publishers, Inc. Deerfield Beach, Florida, 1985, 487–512.

Yazawa, A; Kuwabara, T. Bio-recycle metallurgy by utilization of algae, *Mineralia Slovaca*, 1996, 5, 28, 313-317.

Zeroual, Y; Moutaouakkil, A; Dzairi, FZ; Talbi, M; Chung, PU; Lee, K; Blaghen, M. Biosorption of mercury from aqueous solution by *Ulva lactuca* biomass, *Bioresource Technology*, 2003, 90, 349–351.

Zhou, Q; Zhang, J; Fu, J; Shi, J; Jiang, G. Biomonitoring: An appealing tool for assessment of metal pollution in the aquatic ecosystem, *Analytica Chimica Acta*, 606, 2008, 135–150.

Zimmo, OR; van der Steen, NP; Gijzen, HJ. Comparison of ammonia volatilisation rates in algae and duckweed-based waste stabilisation ponds treating domestic wastewater, *Water Research*, 2003, 37, 4587–4594.

Zimmo, OR; van der Steen, NP; Gijzen, HJ. Nitrogen mass balance across pilot-scale algae and duckweed-based wastewater stabilisation ponds, *Water Research*, 2004, 38, 913–920.

In: Algae: Nutrition, Pollution Control and Energy Sources
Editor: Kristian N. Hagen, pp. 89-125
ISBN: 978-1-60692-008-4

Chapter 6

ULTRASONIC CONTROL AND REMOVAL OF CYANOBACTERIA

Guangming Zhang[,1], Panyue Zhang[2] and Hongwei Hao[3]***
[1]Harbin Institute of Technology, China
[2]Beijing Forestry University, China
[3]Tsinghua University, China

Abstract

Overgrowth of cyanobacteria has caused great serious water environment problems and threatens the drinking water supply worldwide. Many methods have been researched and practiced for its control. This paper investigates thoroughly the ultrasonic control of cyanobacteria growth in water, the ultrasonic enhancement of algal cell removal via coagulation, and the degradation of algal toxins by sonication. The paper also examines the important operational parameters and aqueous matrix factors.

The results showed that proper ultrasonic irradiation could effectively control the growth of some cyanobacteria species. Gas vacuole in the algal cells can be easily destroyed during ultrasonic irradiation by acting as the 'nuclei' for acoustic cavitation and collapse during the "bubble crush" period, which results in the cyanobacteria settlement and growth inhibition. Sonication can also instantly decrease the antenna complexes like chlorophyll *a* and phycocyanins. Direct damage on cell surface or even cell fracturing was also observed. But the dominant mechanism is the acoustic cavitation of gas vacuole. The same mechanism also explains the enhancement of coagulation by sonication. As a result, cyanobacteria with gas vacuoles can be easily controlled using ultrasonic waves, but algae without gas vacuoles are virtually immune to sonication.

On the other hand, the reason for ultrasonic degradation of algal toxins is because of the chemical changes caused by sonication, especially the formation of hydroxyl free radicals in water. Ultrasonic frequency, intensity, and duration play important roles in the algal growth control and ultrasonic-coagulation removal, and ultrasonic toxin degradation.

Keywords: Ultrasonic cavitation; Cyanobacteria; Kinetics; Weak spot; Microcystins

1. Introduction

Surface cyanobacterial blooms in eutrophic lakes and rivers have been recognized as a serious environmental problem globally in recent years (Smith et al., 1999). Autotrophic algae, especially cyanobacteria, become preponderant in eutrophic waters because of abundant nutriment like N and P, and even converge to algal slurry about several centimeters, forming algal blooms (Carr and Whitton, 1982; Chorus and Bartram, 1999). Cyanobacterial bloom increases water turbidity and is a source of serious taste and odor concerns. During catabolism, algae stink, consume a great deal of dissolved oxygen, which can cause the death of fish, and release metabolize products like biological toxins harming the safety of human and other organisms (Carmicheal, 2001; Dawson, 1998; Lian et al., 2000; Peng, 2001; Repavich et al., 1990). Reports showed that typical algal toxins such as microcystins have caused many instances of fish kill and death of domestic animals and patients. Field investigations in China have reported that microcystins in drinking water were one major factor that resulted in locally high incidence of liver cancer (Lian et al., 2000). To guarantee the security of drinking water, the newly emended national standards of drinking water regulated the concentration of microcystin-LR (MCLR), the commonest algae toxin, as 1.0 μg/L.

When the eutrophic water is used to produce drinking water, algae cells consume high amount of coagulants and disinfectant, jam filters, generate disinfection byproducts and oxidation byproducts, form tasty molecules, and deteriorate the water quality. In November 11, 2004, algae bloom increased the source water algae concentration to $\sim 8\times10^{7}L^{-1}$, which increased the coagulant (polychlorinated aluminum) consumption from 2 mg/L to 10 mg/L and jammed the filter within 3.5 h in a water plant in Shenzhen, China. Cyanobacteria overgrowth in Tai Lake, Cao Lake, and Dianchi Lake of China has long become a threat to local water plants. In the summer of 2007, a few cities around Lake Tai were forced to close their drinking water supply plants for several days due to the outbreak of algal bloom in Lake, which threatened the daily life of more than 3 million citizens. Similar accidents have been reported in other cities as well.

Many countries have strict regulation on the maximum level of algal toxins in the final water and some water plants set 90% or higher removal ratio for algae cells at all times as a goal for water treatment though there is no law regarding the number of algae cells in the drinking water. However, algae removal from water treatment process is hard because of their small size and their low specific gravity. The present control strategies in water-bodies as well as in water-plants include the use of microorganisms (Rao et al., 1999) and chemicals such as chlorine, ClO_2, $KMnO_4$, ozone, copper, and coagulants (Chen et al., 1998; Donati et al., 1994; Jiao, 2004; Ma and Liu, 2002; Nichlolson et al., 1994; Peng, 2001; Ryding and Rast, 1989; Shi et al., 2003; Simmons, 1997). These methods are costly, complex and may cause secondary pollution. Despite all efforts, algal bloom, especially cyanobacterial overgrowth, remains a major problem in many lakes and reservoirs, and keeps threatening drinking water supply. New engineering methods are needed to counteract excessive cyanobacteria growth in eutrophic water-bodies and excessive algal concentration in water plants.

Ultrasound, or sound of frequency >20 kHz, is inaudible to the human ear. Irradiation with ultrasound is widely used in medical imaging, sonochemical processing, ultrasonic cleaning of surfaces, and as the basis for underwater sonar ranging. Ultrasonic irradiation has been known to provide various chemical, physical and biological effects due to cavitation effect (Wu, 1992; Suslick, 1988; Rott, 1998; Price, 1992; Phillips et al., 1998; Mason and Lorimer, 1990). When applied to water, power ultrasound causes acoustic cavitation, in which millions of microbubbles form at various nucleation sites in the medium and grow during the rarefaction phase of the sound wave. Then, in the compression phase, the bubbles implode within μs. The collapsing bubbles release a violent shock wave that propagates through the medium. Cavitation causes intense local heating to reach temperatures as high as 5000 K and pressures as high as 100 MPa, so called 'hot spots', generates highly active hydroxyl free radical, and causes high-speed micro-jets (Suslick, 1988). Local temperature in the vicinity of a forming or collapsing bubble can change extremely rapidly at > 100 °C s^{-1}. Such extreme conditions induce chemical reactions and enforce mechanic damages on substances in water, which has been widely employed to accelerate chemical reactions. Since 1990, there has been an increase of interest in the use of ultrasound in environmental pollution control, and the main application was focused on destroying organic contaminants present in water, including halogenated hydrocarbon, phenol, aromatic compounds, pesticide and dyestuff (Cum et al, 1992; Entezari, 1997; Gondrexon et al., 1999; Huang et al., 1995; Kang and Hoffmann, 1998; Petrier and David, 1996; Petrier et al., 1998; Zhang and Hua, 2000).

Power ultrasound is known to have harmful effects on the structure and functional state of organisms (Rott, 1998; Phull, 1997). At sufficiently high acoustic power inputs, ultrasound is known to rupture cells and sonication is a well-established laboratory technique of cell disruption. A cell can be inactivated by ultrasound at intensities less than those needed to cause disruption. Intense ultrasound is known to damage macromolecules such as enzymes, probably from unfolding and scrambling the native protein and breaking the chain into radicals or small peptides (Chisti and Moo, 1986). Therefore, high doses of ultrasonic exposure provide a suitable method for reducing cyanobacteria in waterworks, with environmentally significant advantages, such as safety, cleanness and energy conservation. A few studies have been developed on cyanobacterial bloom control by ultrasound in lakes. Since it was reported that ultrasonic irradiation was used in cyanobacterial bloom control firstly in 2001, more and more researches have been carried out due to its environmentally significant advantages such as safety, cleanness and energy conservation (Ahn et al., 2003; Hao et al., 2004a, 2004b; Lee et al., 2000, 2001, 2002; Nakano et al., 2001; Wang, 2003). It was reported that the extent of algal growth reduction was impacted by ultrasonic parameters like frequency, intensity and time (Petrier and David, 1996; Zhang and Hua, 2000; Cum et al., 1992). Low frequency ultrasound effectively settled a naturally growing cyanobacterial suspension or reduced the growth rate of cyanobacteria by collapsing the gas vesicles and immediately damaged the photosynthetic activity (Lee et al., 2000, 2001, 2002; Nakano et al., 2001; Hao et al., 2004a). Although exciting inhibition results have been acquired, but the mechanism, kinetics and influencing factors have not been investigated in detail. Besides, ultrasound is also known to be able to lyse cells and release the intracellular materials. When applied to algae solution, ultrasound waves may release algae toxins into water, which is highly undesirable and has not been investigated.

In this paper, we studied the feasibility and kinetics of ultrasonic irradiation for inhibiting cyanobacterial bloom, for removing the algal cells directly and in combination with coagulation, for degradation of algal toxins, and tried to determine correct conditions to minimize the release of toxins while still rupturing gas vesicles and promoting the settling of cells. We investigated the impacts of some important sonochemical parameters such as sound frequency, power, and time, and studied the potential changes of algal cells and photo-activities during sonication. We also compared the effects of sonication with those of the more conventional ultraviolet light (UV) irradiation. The purpose of this investigation was to develop a safe and economical method for inhibiting the overgrowth of algal cells in surface water and for eliminating cyanobacteria from the source water in drinking water treatment plants with ultrasound and to understand the underlying mechanisms for cyanobacterial control by ultrasonic irradiation.

2. Materials and Methods

2.1. Algae Culture

Three prevailing cyanobacteria species were cultured and used in the experiment, namely the filamentous gas-vacuolate *Spirulina* (*Arthrospira*) *platensis*, the unicellular gas-vacuolate *Microcystis aeruginosa,* and the unicellular gas-vacuole negative *Synechococcus.* All three are bloom-formers and enjoy fast growth. Pure culture of the three species were purchased from the Institute of Hydrobiology, Chinese Academy of Sciences and then cultured in the lab before usage.

Spirulina platensis samples were cultured in Erlenmeyer flasks with Zarrouk's medium (Rao et al., 1999, Table 1), and were continuously agitated on a shaker at 150 rpm in a growth chamber (28°C), with the incandescent light at 32 μmol of photons m^{-2} s^{-1}.

Table 1. Contents of Zarrouk's culturing solution (based on 1 L solution)

Culturing medium					
$NaHCO_3$	$NaNO_3$	K_2SO_4	NaCl	$MgSO_4·7H_2O$	
16.8 g	2.5 g	1.0 g	1.0 g	0.2 g	
$CaCl_2·2H_2O$	$K_2HPO_4·3H_2O$	$FeSO_4·7H_2O$	Na-EDTA	A5 solution	B6 solution
40 mg	0.5 g	10 mg	80 mg	1.0 ml	1.0 ml
A5					
H_3BO_3	$MnCl_2·4H_2O$	$NaMoO_4·2H_2O$	$CuSO_4·5H_2O$	$ZnSO_4·7H_2O$	
2.86 g	1.81 g	39 mg	74 mg	0.222 g	
B6					
NH_4VO_3	$K_2Cr_2(SO_4)_4·2H_2O$	$NiSO_4·7H_2O$	$Na_2WO_4·2H_2O$	$Ti_2(SO_4)_3$	$Co(NO_3)_2·6H_2O$
0.23 g	0.96 g	0.48 g	0.18 g	0.40 g	0.44 g

Both *Microcystis aeruginosa* and *Synechococcus* PCC 7942 were cultivated at 25°C in axenic BG-11 medium (Table 2), under continuous illumination from general electric cool white fluorescent lamps at an average intensity of 15 μmol of photons m^{-2} s^{-1}. Aeration was provided by bubbling air at regular pressure.

Table 2. *Microcystis aeruginosa* and *Synechococcus* PCC 7942 culturing solution

Culturing medium (Based on 1L)					
BG-11	Glucose	K_2HPO_4	Na_2CO_3	Ammonia citrate iron	
10 ml	1.5 g	30.5 mg	20 mg	6.0 mg	
BG-11 (Based on 1L)					
$MgSO4{\cdot}7H_2O$	A1	$CaCl_2{\cdot}2H_2O$	Citric acid	Na-EDTA (pH 8.0, 0.25 M)	
7.49 g	100 ml	3.60 g	0.60 g	1.12 ml	
A1 (Based on 1L)					
H_3BO_3	$MnCl_2{\cdot}4H_2O$	$NaMoO_4.2H_2O$	$CuSO_4{\cdot}5H_2O$	$ZnSO_4{\cdot}7H_2O$	$Co(NO_3)_2{\cdot}6H_2O$
2.86 g	1.81 g	0.39 g	79 mg	0.222 g	49 mg

Spirulina platensis, Microcystis aeruginosa, and the unicellular gas-vacuole negative *Synechococcus* all followed the S shape growth. It reached the exponential growth stage after 3-4 days, which lasted around 10 days. All experiments were done with cells in the exponential growth stage unless stated otherwise. After grown for 14 days, algal concentration in the culturing solution reached almost $10^9\ L^{-1}$. The solution pH was around 7.3-7.9.

Surface water samples were collected from local reservoirs and used without further algal culturing.

Figure 1 shows the shapes and configurations of the three algal species used.

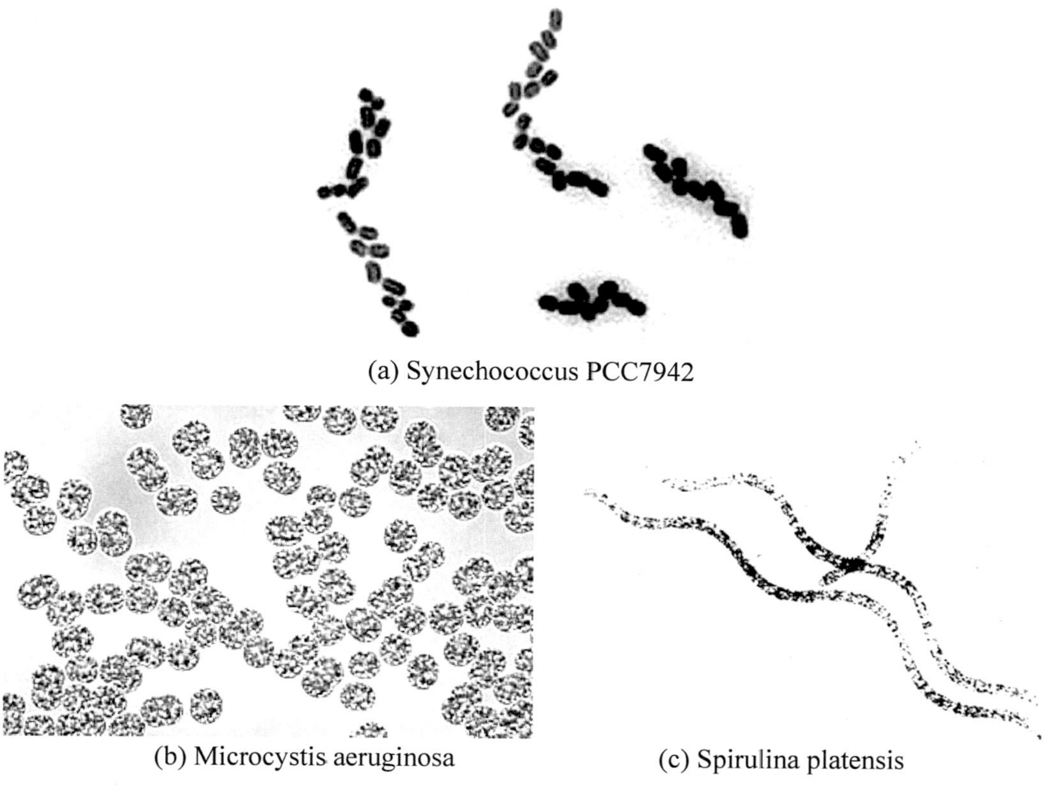

(a) Synechococcus PCC7942

(b) Microcystis aeruginosa

(c) Spirulina platensis

Figure 1. Three major algal species used in this study.

2.2. Chemicals Used

LIVE/DEAD® BacLightTM molecular probe and dyes were purchased from Molecular Probes Inc., USA. Microcystins standards were purchased from Sigma Aldrich, USA. All other chemicals were obtained from Ameta Chemical, China. All chemicals were used as received without further purification. Home made de-ioned water (SYNS50001 generator, Millipore Inc.) was used for all experimental trials.

2.3. Ultrasound Devices

Various ultrasonic devices were employed in this study. The 25 kHz horn system was provided by Xinzhi Inc. (JY90-II) that had a tip surface area of 2.12 cm^2. A home made 20 kHz horn system had a tip surface area of 7.5 cm^2. Ultrasound at higher frequencies (150kHz, 410 kHz, and 1.7 MHz) was emitted from the piezo-electric discs of lead zirconate titanite fixed on the underside of the beaker reactors with epoxy. The ultrasound generator designed in our laboratory consists of a voltage controlled oscillator (VCO), power amplifier and matched impedance. The emitting surface area was 4.5 cm^2. The volume of the ultrasonic reactor was 500 ml. Each frequency required a specific emitter.

The measurement of the ultrasonic field inside the beaker was carried out by dipping into the solution a standard calibration ultrasound-needle-hydro-phone (CS-3, Acoustic Academy of China) connected with a TDS 3000 oscillograph (Tektronix Inc.). The hydrophone had a diameter of 1.5 mm, and a receiving sensitivity of 31.3 nV/Pa for the frequencies involved. The output signal from the hydrophone was transmitted to the oscilloscope, digitized, and stored in a computer. Cavitation spectrum was analyzed using the Origin software (Version 6.1, Originlab Corp.), in which the reference acoustic pressure was10^{-6} Pa, to find out the working ultrasonic frequency and power. Figure 2 shows a typical cavitation pressure spectrum captured by the hydrophone.

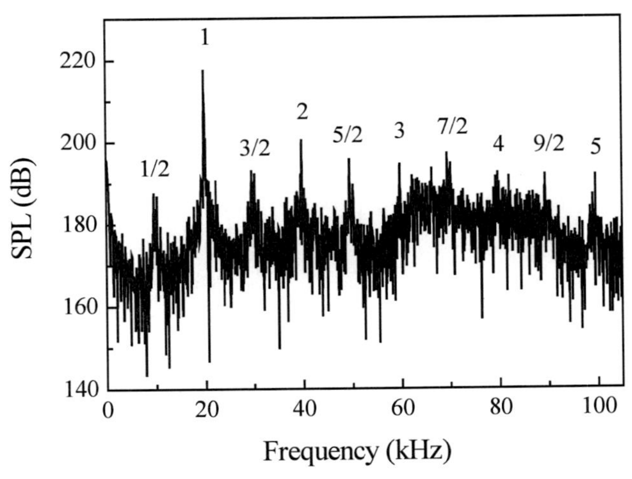

Figure 2. Typical cavitation spectrum, SPL = sound pressure level.

The standard method of KI dosimetry was used to calibrate the ·OH free radicals produced in cavitation, in which the liberated iodine was measured by the absorbance at 355 nm using an UV/Visible spectrophotometer (Ultra-Spec 2000, Amersham Biosciences AB, Uppsala, Sweden).

2.4. Ultrasonic and Ultraviolet Light (UV) Irradiation

The cultured algae solutions were diluted to a concentration of around $1.5\times10^8 L^{-1}$ for sonication and UV irradiation. The algal cell concentration in the surface water samples was not adjusted.

In the ultrasonic algal growth control and cell removal experiments, 400 ml cyanobacterial suspension was filled in the sono-reactor for each sonication trial. When horn system was used, the horn was dipped 1-3 cm into the suspension. The sono-reactor was water-jacketed and the temperature of solution was maintained during sonication ~23 °C using water bath (THZ95, SBL Ltd., China). After 10 min of sonication, the solution pH dropped from ~7.8 to ~7.2 and then stabilized.

In the UV irradiation experiments, 400 ml inoculated sample was filled in reactor and irradiated by an UV lamp at 254 nm (Ultra-Lum Corp., USA) each time. After that, samples were filled back to Erlenmeyer flasks and cultured in illumination incubator. Everyday samples were taken for analyses, and the intensity of UV light was measured by ultraviolet irradiation meter. The pH dropped a little in the UV irradiation trials as well.

For sonication-coagulation trials, 1 L algal suspension was used each time. Larger amount of sample, 10 L each time, was used for the pilot study. Standard jar tests were conducted for the coagulation experiments in a mixer equipped with six-paddle jar test apparatus. Samples were stirred at 150 rpm for 2 min and then at 50 rpm for 10 min.

For the algal toxin degradation experimental trials, 400 ml *Microcystis aeruginosa* suspension was used each time.

2.5. Sampling

For the algal cell removal experimental trials, samples were allowed to be settled quiescently for 30 min after the ultrasonic/UV irradiation stopped. Too long settling might allow gas vesicles get repaired and the cells could again become buoyant. 30 min was sufficient for settling and no refloatation of algae cell was observed. Thereafter, the upper 100 ml of the water sample was siphoned 1 cm below the water surface and taken for analysis.

For the algal growth control experimental trials, partial solution siphoned 1 cm below the water surface and taken for analyses of the immediate influences of sonication. The remaining solution was then fed back to Erlenmeyer flasks and cultured in the illumination incubator. Blank samples without irradiation were cultured under exactly the same conditions.

For the algal toxin degradation experimental trials, sampling and analysis were performed immediately after the irradiation stopped.

2.6. Analysis

The number of algae cells was counted with the aid of an Olympus 18-ZSX microscope, which was very time consuming. The alternative way was to monitor the optical transparency of the water sample using an Ultra-Spec 2000UV/Visible spectrophotometer (Pharmacia Inc.) since there is good linearity between dry weight of cyanobacterial cells and absorbency of suspended liquid. The absorbency value (OD) was linear with the counted cell number with R^2 of 0.99 within the tested cyanobacteria concentration range. Therefore, absorbency at 560 nm (OD560) is expressed for live cell biomass of *Spirulina platensi,* and absorbency at 684 (OD684) for live cell biomass of *Microcystis aeruginosa,* and absorbency at 730 nm (OD730) for live cell biomass of *Synechococcus* PCC 7942 (MacKinney, 1941; Hao et al., 2004a; Zhang et al., 2006a, 2006b). Both methods were utilized simultaneously and frequently cross-checked. Reported data were the average of three or more repeated experiments.

Note that when the algae cell concentration was below 5×10^5 L^{-1}, the cell numbers could not be counted accurately. Therefore, all experiments were designed to guarantee algae cell concentrations higher than 5×10^5 L^{-1}.

The alive condition of algae was monitored using fluorescence reflection. Algae cells were dyed with LIVE/DEAD® BacLightTM molecular probe and dyes purchased from Molecular Probes Inc., USA and then observed with a IX71 microscope (Olympus Inc., Japan). Alive cells show green and dead cells show red under certain wavelength.

The surface of cyanobacterial cells was observed with a differential interference microscope (Leica DM RE, Leica Microsystems AG, Wetzlar, Germany).

Chlorophyll *a* was determined following methods described by MacKinney (MacKinney, 1941) and refined by Hao (Hao et al., 2004a). 3 ml cell suspension was pipetted and centrifuged at 8000×g for 10 min (Sigma Laboratory Centrifuges, 3K30). 100% methyl alcohol (3 ml) was added to the pellet in a darkened room. After being kept for 3 min, the samples were centrifuged at 7000×g for 10 min. The absorbance of the supernatant was measured at 660 nm and the chlorophyll *a* concentration was calculated by the following formula: chlorophyll *a* concentration ($mg\cdot g^{-1}$) = $(13.9\times OD_{660}\times10^{-3})\times(0.5036\times OD_{560})$ (MacKinney, 1941).

Oxygen evolution rate was measured following the method of Hao (Hao et al., 2004a).

Microcystins was measured using competitive enzyme-linked immunosorbent assay (ELISA) method since it is quick, sensitive and can measure the total microcystins concentration. An ELISA plate reader (DG5031, Huadong, China) was used to measure the microcystins concentrations colorimetrically at 450 nm. The method has a low detection limit of 0.16 μg L^{-1}. To measure the concentration of extracellular microcystins, water sample was filtered through a 0.45 μm fiber glass filter to remove cyanobacteria cells, the filtrate was then extracted using a Waters C18 solid extracting column according to Zhang (Zhang et al., 2003) and then measured following the methods of Chorus (Chorus and Bartram, 1999).

To extract PC, the cyanobacteria sample was centrifuged at 12,000×g for 10 min (3K30 Sigma Laboratory Centrifuge, Sigma Inc.), the supernatant was discarded, and then 10 ml KH_2PO_4–K_2HPO_4 cushion solution (final concentration of 0.001 $molL^{-1}$, pH of 7.0, with 0.001 $molL^{-1}$ BME, 0.001 $molL^{-1}$, NaN_3 and 0.2 $molL^{-1}$ NaCl) was added to the remaining solids. The so-prepared solution was then wrapped with black paper, frozen at -4□ and then de-frozen at 4□, which was repeated three times; the solution was then centrifuged at 15,000 g^{-1} for 45 min at 4□, the supernatant was then used for the measurement of pycocyanins.

Pycocyanins absorbed strongly at 620 nm, the absorbance of the supernatant at 620 nm was thus used to quantify the PC concentration.

The pH and turbidity of samples were measured with a Hach pH meter and a Hach turbidity meter respectively.

3. Ultrasonic Algae Control

3.1. Ultrasonic Cyanobacterial Growth Inhibition

In eutrophic water-bodies, algae, especially cyanobacteria will over grow. The priority choice is to control the overgrowth of algal cells to such a level that it does not harm the utilization of water-bodies. Ultrasound irradiation was explored for this purpose and a more conventional method, ultraviolet light (UV) irradiation, was tested for comparison. *Spirulina platensis*, *Microcystis aeruginosa,* and *Synechococcus* solutions, as well as water samples collected from two water reservoirs were used for this study.

3.1.1. Feasibility of Ultrasonic Cyanobacterial Growth Inhibition

One *Spirulina platensis* suspension was irradiated with high frequency ultrasound waves (1.7 MHz) for 5 min. The ultrasonic power was 40 W. The color of the algal suspensions changed from blue-green to yellow-green during sonication and gas bubbles were observed during sonication. After 30 min settling, fractions of the algal cells obviously settled to the bottom of the sono-vessel. The treated suspension and a blank suspension without ultrasonic irradiation were then cultured under the same culturing conditions for 7 days and the color of the treated sample was much lighter than that of the control sample. Figure 3 reports the change of algal cells immediately after sonication and 4 days after sonication. The data showed that ultrasonic irradiation effectively inhibited the growth of the *Spirulina platensis* cells. The initial cell concentration decreased by 63%, then grew very slowly and had not shown significant growth for 3 days. After 3 days' culturing, the algal cell concentration of the treated sample was only 29% of that of the untreated sample.

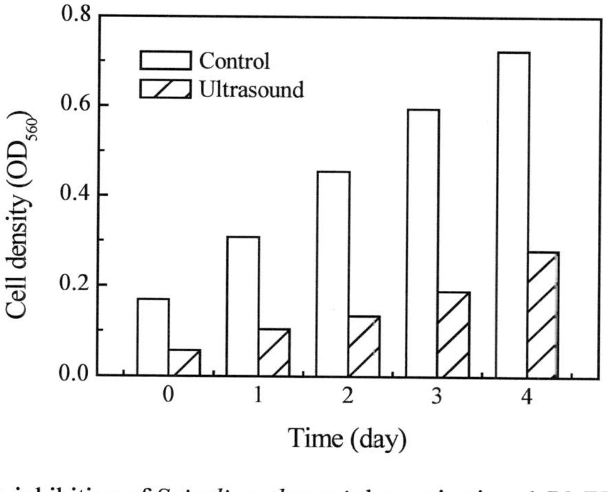

Figure 3. Growth inhibition of *Spirulina platensis* by sonication, 1.7 MHz, 40 W, 5 min.

3.1.2. Impact of Algal Species

Suspensions of *Spirulina platensis*, *Microcystis aeruginosa,* and *Synechococcus* were treated with 1.7 MHz ultrasound waves for 5 min respectively. The ultrasonic power was 40 W. The treated samples and control samples without sonication were then cultured together for 3 days and the suspension OD was monitored. Figure 4 showed clearly that algal species had great impact on the ultrasonic inhibition effects. The two species with gas-vacuole (*Spirulina platensis* and *Microcystis aeruginosa*) are both prone to ultrasonic irradiation and 5 min sonication inhibited the cell growth (after 3 days) by 64% and 58% respectively, but the one specie without gas-vacuole (*Synechococcus*) was much less sensitive to sonication and 5 min sonication only reduced the cell growth (after 3 days) by 11.5%. The significant difference showed clearly that the existence of gas-vacuoles was crucial for the ultrasonic irradiation to control the algal growth, which agreed with the hypothesis that gas-vacuoles inside the algal cells collapse during sonication and thus damages the cell growth (Lee et al., 2000; 2001; 2002). The difference between *Spirulina platensis* and *Microcystis aeruginosa* showed that the shape of algal species also impacts the sonication results. The filaceous *Spirulina platensis* is more prone to sonication than the spherical *Microcystis aeruginosa*, which may be explained by the mechanical effects of sonication, i.e., acoustic cavitation tends to damage long-chain chemicals. Further experiments were performed using algal samples collected from Xili Reservior in Southern China and from Daqing Reservior in Northern China. Analysis showed that more than twenty different algal species existed in the samples and the dominant species were different for the two samples. The ultrasonic irradiation tests showed that the algal growth of both samples could be inhibited by sonication. The final algal concentration, measured by chlorophyll *a* concentration, was 52% and 46% of that of control sample, showing the effectiveness of this method for control of surface water algal blooming.

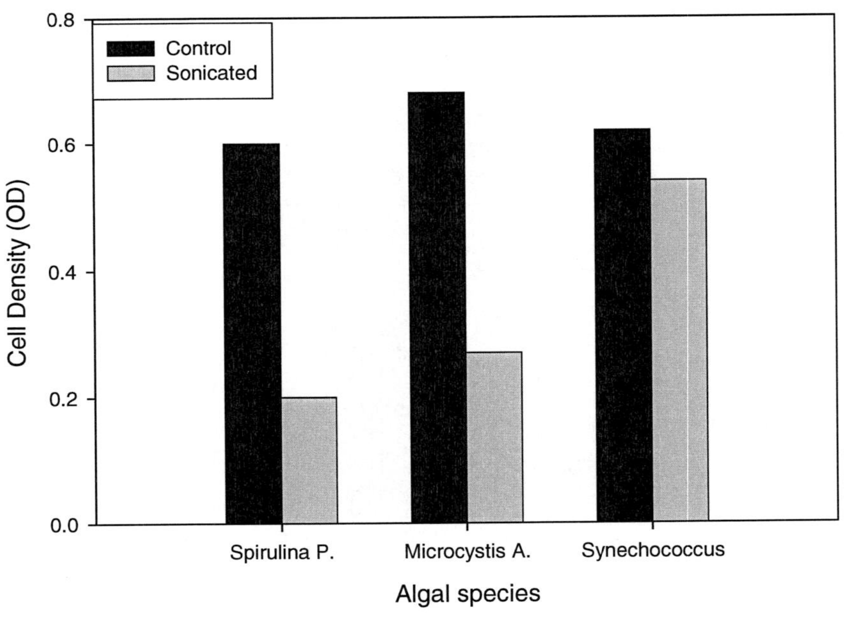

Figure 4. Impact of algal species on ultrasonic growth inhibition, 1.7 MHz, 40 W, 5 min.

3.1.3. Effluence of Algal Growth Stage

Two suspensions of *Spirulina platensis*, one in the initial arrearage growth stage and the other in the exponential growth stage, were treated for 5 min using ultrasound at 1.7 MHz. The ultrasonic power was 20 W. The two suspensions had different initial cell concentrations, the one in the arrearage stage had an initial OD of 0.085 and thus marked as 'Low' and the one in the exponential stage had an initial OD of 0.306 and thus marked 'High'. Figure 5 reported the growth of the two sonicated suspensions under the same culturing conditions. Obviously, the 'High' suspension grew much faster than the 'Low' suspension, and achieved a much higher final cell concentration than the 'Low' suspension after 3 days culturing. The reason was explained by that the 'High' suspension was in the exponential growth stage and was most health, could recover from the ultrasonic damage more easily. Previous research also supports the hypothesis that cells in the arrearage stage are 'weaker' to damages (Ahn et al., 2003).

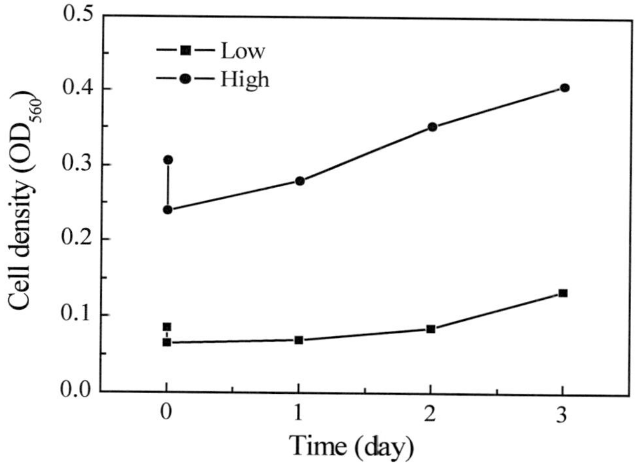

Figure 5. Influence of algal growth stage on ultrasonic algal growth inhibition, 1.7 MHz, 20 W, 5 min.

3.1.4. Influence of Sonication Duration

Five suspensions of *Spirulina platensis* were treated by 1.7 MHz, 20 W ultrasound waves for 1 min, 3 min, 5 min, 7 min, and 9 min respectively, and then cultured with the control sample without sonication for 5 days under the same culturing conditions. Figure 6 recorded the algal growth after sonication. Clearly, the sonication duration had great impact on the effect of algal growth inhibition. When 1 min of sonication was applied, the algal cell concentration decreased by 29% immediately and then grew rapidly; the slope of the growth curve was almost the same as that of the control sample, showing little growth inhibition. When 3 min of sonication was applied, the algal cell concentration decreased by 56% immediately; the algal growth was retarded by 2 days and then recovered slowly after 3 days. When 5 min and longer sonication durations were applied, the algal cell concentration decreased by 63%-67% immediately; the algal growth was markedly slower than that of the control sample. In summary, when the sonication duration increased from 1 min to 5 min, the ultrasound inhibition effect of algal growth built up; but further time increase brought in little benefit and thus was undesirable.

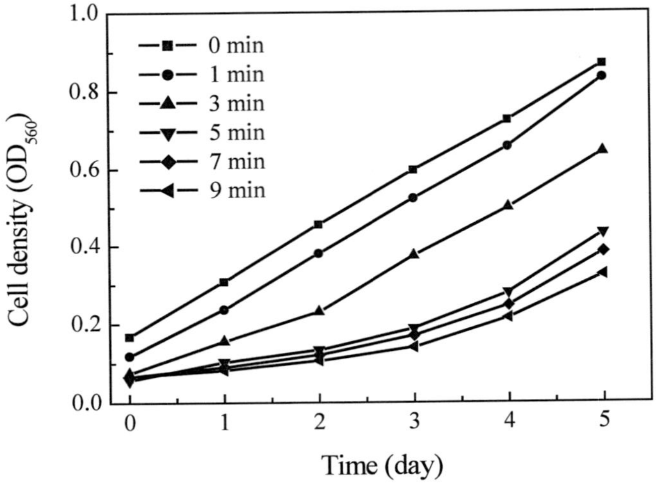

Figure 6. Influence of sonication time on ultrasonic algal growth inhibition, 1.7 MHz, 20 W.

3.1.5. Influence of Sonication Power

Four suspensions of *Spirulina platensis* were treated by 1.7 MHz ultrasound waves for 5 min and the power was 20 W, 40 W, 60 W, and 80 W respectively, and then cultured with the control sample without sonication for 8 days under the same culturing conditions. Figure 7 recorded the algal growth after sonication. Clearly, the ultrasonic power had some impacts on the effect of algal growth inhibition. When the ultrasonic power was 20 W, 40 W, 60 W, and

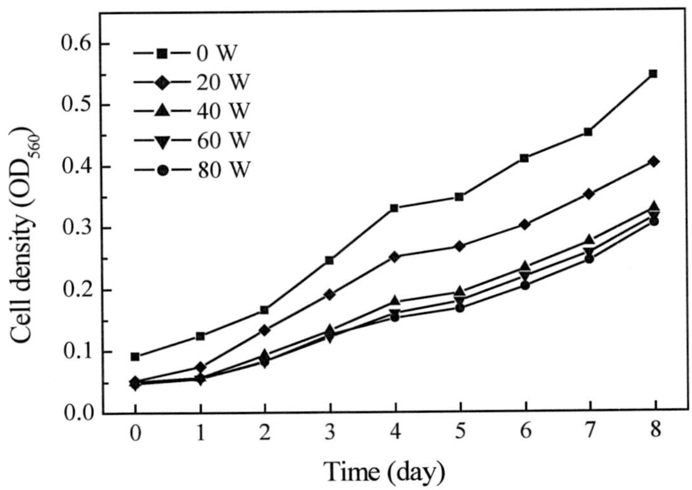

Figure 7. Influence of ultrasonic power on ultrasonic algal growth inhibition, 1.7 MHz, 5 min.

80 W, the immediate algal cell decrease was 43%, 45%, 48%, and 48% respectively, and the final algal cell decrease was 26%, 40%, 42%, and 44% respectively. Obviously, ultrasonic power higher than 40 W was unnecessary for the algal growth inhibition in this study. Similar power-saturation effects have been reported earlier for algal control and for chemical degradations (Lee et al., 2001; Hao et al., 2004a; Zhang and Hua, 2000).

3.1.6. Influence of Ultrasonic Frequency

Three suspensions of *Spirulina platensis* were treated by 40 W ultrasound waves for 5 min and the sound frequency was 20 kHz, 200 kHz, and 1.7 MHz respectively. The three sound frequencies were selected to represent typical low, middle, and high frequencies used in sonochemistry. The treated suspensions were then cultured with the control sample without sonication for 6 days under the same culturing conditions. Figure 8 recorded the algal growth after sonication. Clearly, the ultrasonic frequency had significant impacts on the effect of algal growth inhibition. The 200 kHz ultrasound waves were most efficient for *Spirulina platensis* growth control. On the other hand, Wang (Wang, 2001) examined the inhibition of *Microcystis aeruginosa* using ultrasound of the same frequencies and found that 1.7 MHz was the best for *Microcystis aeruginosa* growth. The dissimilarity may be explained by the different sizes of the algal cells. The *Microcystis aeruginosa* has an average cell size of 1-2 μm that tends to resonate in the 1.7 MHz ultrasonic field (Hao et al., 2004a), and the *Spirulina platensis* has an average cell length of 8-10 μm that tends to resonate in the 200 kHz ultrasonic field. Detailed analysis can be found in Section 4.

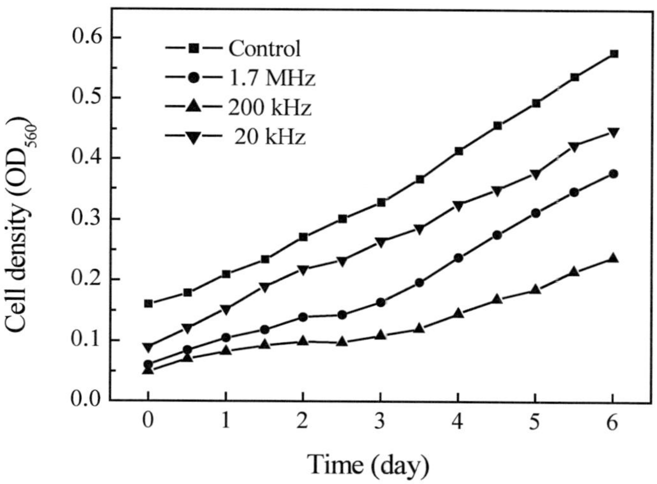

Figure 8. Influence of ultrasonic frequency on ultrasonic algal growth inhibition, 40 W, 5 min.

3.1.7. Low Dose Sonication for *Spirulina platensis* Growth

A very interesting phenomenon was found that when the sonication dose was low, the treated *Spirulina platensis* sample grew quicker than the control one. The ultrasound frequency was 200 kHz, the power was 5 W, and the time was 5 s. The power might be inaccurate since the sono-reactor could not be controlled precisely at such low power level. The results showed that the treated sample grew 5% faster than the control one but the chlorophyll *a* concentration was almost the same. The stimulation of bioactivity by low dose ultrasound has been widely observed and is believed to be a result from non-cavitation sonication (Francko, 1994; Wu and Lin, 2002).

3.2. Ultraviolet Irradiation for Cyanobacterial Growth Control

Ultraviolet light (UV) radiation is a light wave irradiation with penetrability and lethality. Ultraviolet irradiation has long been known effective to control the growth of hydrobiont, mainly on configuration, photosynthesis and nutrition, and has been widely used for disinfection. Algae and planktons dealt with UV appear slow growth and weakened energy (Simmons, 1997). This paper studies the effect of ultraviolet irradiation to control cyanobacterial growth as a more conventional method in order to compare the effects with those of ultrasonic algal growth inhibition. Suspensions of *Spirulina platensis* were used; the same algal specie was used for the sonication trials. The UV frequency of 254 nm, the most powerful biocidal frequency and commonest one in disinfection, was used.

3.2.1. Impact of UV Irradiation Time

400 ml samples of the same concentration filled in the reactor were measured for absorbency after ultraviolet light irradiation for different time at intensity of 1.03 mW/cm^2. Initial OD560 of the *Spirulina platensis* sample was 0.061 and changed little after UV irradiation for 1 min, 3 min, 5 min, 7 min and 9 min, showing that UV irradiation did not immediately decrease the cell concentration, which was different from that of ultrasonic irradiation. The main effect of ultraviolet radiation was to restrain the growth of algae, which was further proved through microscope observation that many algae lost activity but kept the shape after UV irradiation.

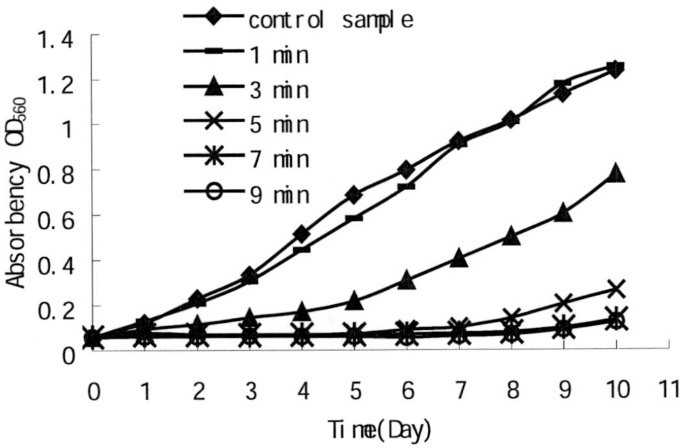

Figure 10. Impact of UV irradiation time on *Spirulina platensis* growth, 1.03 mW/cm^2.

The irradiated samples were then cultured under the same condition with the control sample without ultraviolet irradiation for 10 days. Absorbencies were measured everyday and Figure 10 reported the growth curves of *Spirulina platensis* after UV irradiation. The degree of control to *Spirulina platensis* increased with the increase of UV irradiation time from 1 min to 5 min. UV irradiation for 1 min had virtually no impact on algal growth at all. Obvious algal growth inhibition by UV irradiation was observed when the UV irradiation time was 3 min. Algae after irradiation slowly grew in 5 days and the biomass was 62% of that of the control sample after 10 days (3 min). An irradiation time of 5 min restrained the growth of *Spirulina platensis* completely. The biomass, which did not increase in 10 days, was 21% of control sample. This shows that UV can destroy some components of algae cell and restrain the photosynthesis. Further increase of irradiation time only slightly improved the inhibition effect and showed a saturation effect. The most effective and economical treat time is 5 min

3.2.2. Impact of Algal Species

To examine the effect of ultraviolet to different species of cyanobacteria, gas-vacuole free *Synechococcus* was used for contrastive experiment. Three *Synechococcus* samples of the same concentration were irradiated for different time by UV light with an intensity of 1.03 mw/cm^2, and then cultured with the control sample under the same condition. Absorbencies of OD730 were measured everyday for growth curve, and summarized in Figure 11.

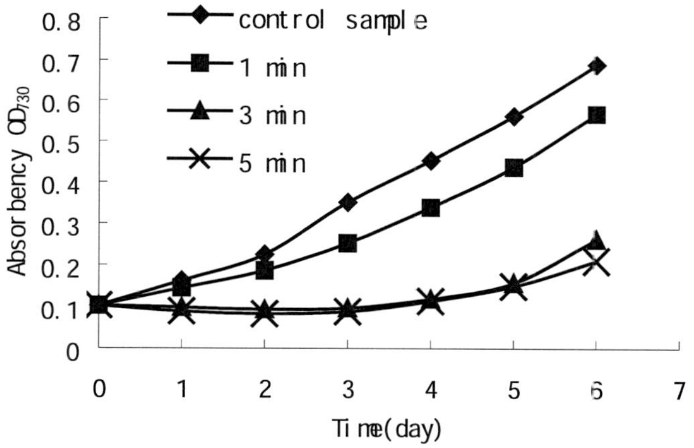

Figure 11. UV inhibition of *Synechococcus* growth, 1.03 mW/cm^2.

Figure 11 showed very similar pattern for *Synechococcus* growth control that of *Spirulina platensis* growth inhibition. UV irradiation for 1 min slightly decreased the *Synechococcus* growth and UV irradiation for 3 min or longer significantly retarded the *Synechococcus* growth. The biomass even decreased in the initial days, then increased slowly and became 45% of that of control sample after 6 days. These showed obvious control of ultraviolet light to cyanobacteria of different species and conformation, i.e., unlike sonication, UV irradiation does not depend on the existence of gas vacuoles.

3.2.3. Impact of UV Irradiation Intensity

Four *Spirulina platensis* samples of the same concentration were irradiated for 5 min with UV of different intensities separately, and then cultured under the same condition with control sample without ultraviolet irradiation. Daily measure of OD560 for growth curves was taken and summarized in Figure 12. Figure 12 showed that the growth control of *Spirulina platensis* increased with the increase of ultraviolet intensity in the UV intensity range tested. The growth of *Spirulina platensis* began to slow down when the ultraviolet density reached 0.52 mW/cm^2, which was determined as the most effective and economical UV intensity for algal control.

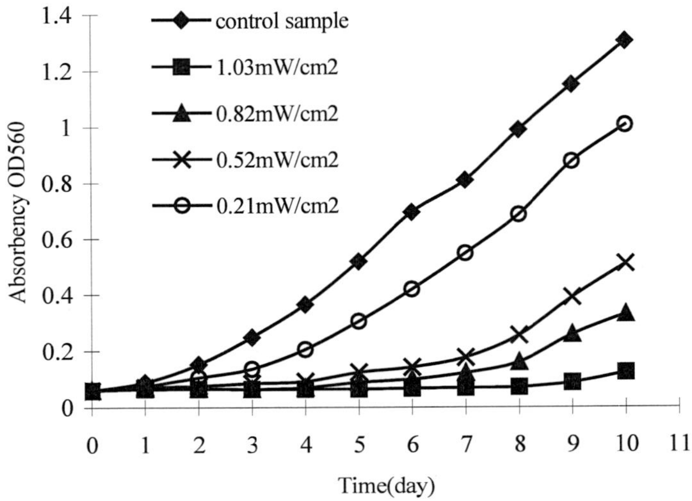

Figure 12. Impact of UV irradiation intensity on *Spirulina platensis* growth, 5 min.

3.2.4. Impact of Algal Growth Stage

Experimental trials with algal suspension in arrearage stage and exponential stage were performed to examine the impact of algal growth stage on the UV algal inhibition and to test the hypothesis of algal growth stage proposed in Section 3.1. The final concentration of the sample at the arrearage phase irradiated by ultraviolet was the lowest, and the biomass after 10 days was less than 10% of that of the control sample. Origin 7.0 was used to calculate the algae growth rates for the growth curves, and the results showed that the growth rate of the control sample increased all the time; the growth rate of the sample at arrearage phase increase little in 5 days after ultraviolet irradiation; the growth rate of the sample at exponential phase was generally the same with that of the control sample before ultraviolet irradiation, but decreased rapidly after irradiation. These observations showed that cells in the arrearage phase were more friable to damages, which agreed with the observation from the ultrasonic irradiation trials.

3.2.5. Change of Chlorophyll *a* and Cell Activity by UV Irradiation

Chlorophyll *a* is the main effectible pigment of cyanobacterial photosynthesis and its increase impacts growth of cyanobacteria mostly. The chlorophyll *a* concentration changes of *Spirulina platensis* in the control sample and the irradiated sample were measured as Figure 13. The UV irradiation did not directly decrease the algal chlorophyll *a* but significantly slowed down the formation of chlorophyll *a* during following culturing, and the final chlorophyll *a* in the treated sample was 20 % of that of the control sample after 7 days. This probably was because that ultraviolet irradiation can destroy the enzyme on thylakoid membrane, which was pivotal during the composing of chlorophyll (Jiao, 2004).

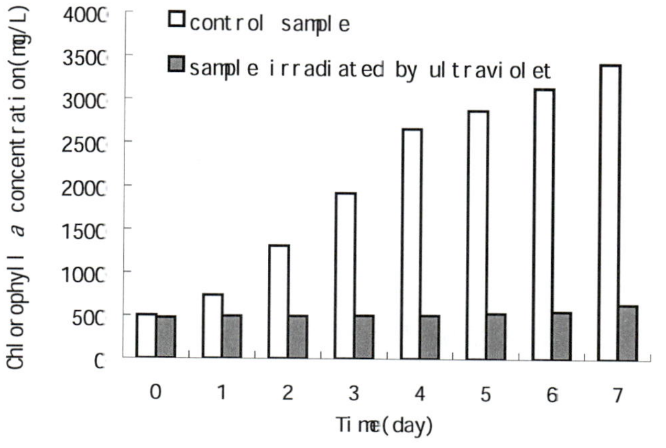

Figure 13. Chlorophyll *a* change during culturing, *Spirulina platensis*, 0.52 mW/cm^2, 5 min.

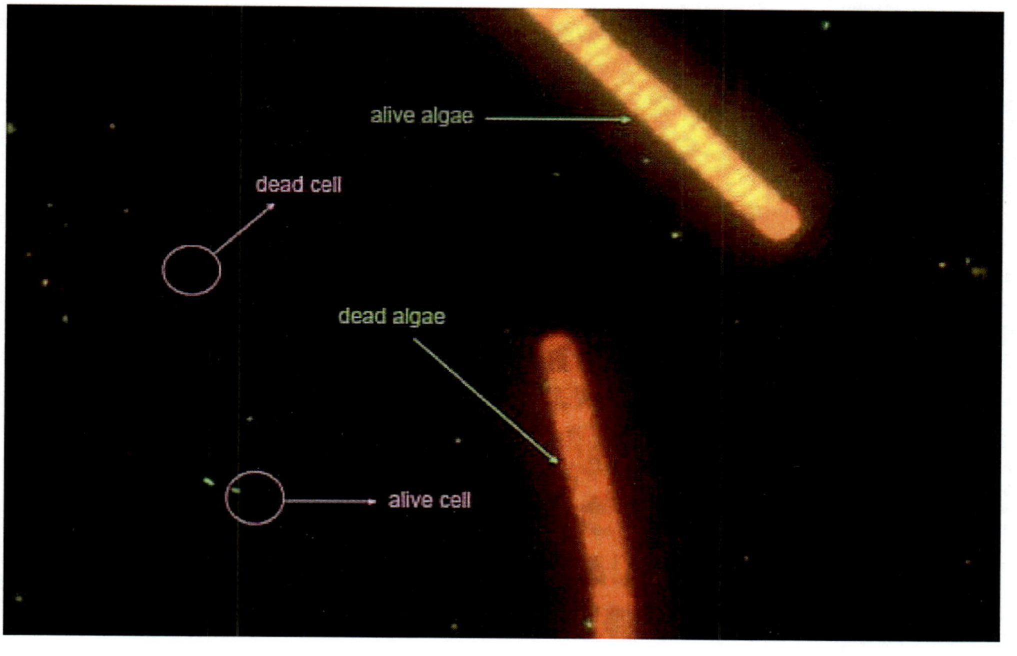

Figure 14. Fluorescent photo of *Spirulina platensis* after ultraviolet irradiation.

Figure 14 showed a fluorescent photo of *Spirulina platensis* cells before and after ultraviolet irradiation. Through fluorescent observation of the control sample and the sample irradiated by ultraviolet, it was found that cells of cyanobacteria after UV irradiation appeared

red and cells of the control sample appeared green. This showed that ultraviolet irradiation destroyed the configuration and cell membrane of algae, and most cells lost the activity. The damage of algal cells, combined with the inhibition of chlorophyll *a* formation, greatly restrain the growth of cyanobacteria. These effects had no connection with the gas vacuole, and thus UV algal inhibition did not depend on the existence of gas vacuole in the algal cells.

3.3. Ultrasonic Cyanobacterial Removal

3.3.1. Feasibility of Ultrasonic Algal Removal

The drinking water treatment process requires rapid removal of algal cells from the source water without algal toxin release. To explore the possibility of applying ultrasonic irradiation to directly remove cyanobacterial cells from the source water in water works, *Microcystis aeruginosa* suspensions were sonicated for 5 min. The ultrasonic frequency was 1320 kHz and the ultrasonic power was 20 W. The cyanobacterial cell concentration was reduced by 55% after 5 min irradiation and no significant increase was observed after 3 h storage in the dark. Clearly, sonication is an effective method to remove cyanobacteria from the source water without rebounding in pipelines or reservoirs.

3.3.2. Kinetic of Ultrasonic Algal Removal

Ultrasonic degradation of most chemicals follows pseudo-first-order reaction, and experiments of ultrasonic irradiation to cyanobacterial suspensions showed that the process is also a pseudo-first-order kinetic reaction (Figure 15). Two *Microcystis aeruginosa*

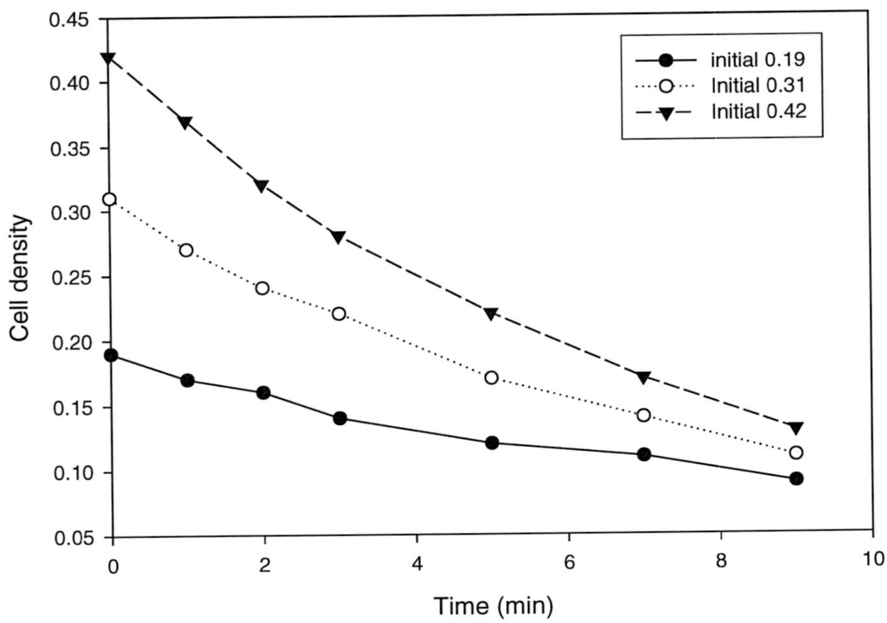

Figure 15. Kinetics in ultrasonic irradiation of *Microcystis aeruginosa*, 1320 kHz, 40W.

suspensions with different initial concentration were exposed to ultrasonic irradiation at 1320 kHz and 40W, and the plots of cyanobacterial biomass (OD684) against irradiation time were shown in Figure 15. The rate constants were calculated to be 0.13, 0.11, and 0.096 min^{-1} for the initial OD730 of 0.42, 0.31, and 0.19 respectively, so the ultrasonic degradation of cyanobacteria follows pseudo-first-order kinetic law, in the same kinetics of sonochemical degradations of organic contaminants.

3.3.3. Impact of Algal Species

Just like in the case of ultrasonic algal growth control, we found that the algal species play a key role in the effectiveness of sonication for algal cell removal (Figure 16). The existence of gas vacuoles was essential for ultrasound waves to remove the algal cells, and thus the vacuole-free *Synechococcus* was immune to sonication. The 9% cell reduction might be a result of the mechanical breakup of the *Synechococcus* cells rather than damages of the cell membrane or proteins. The configuration of algal cells also counted, and the filamous *Spirulina platensis* was more frangible to ultrasonic damage than the spherical *Microcystis aeruginosa*.

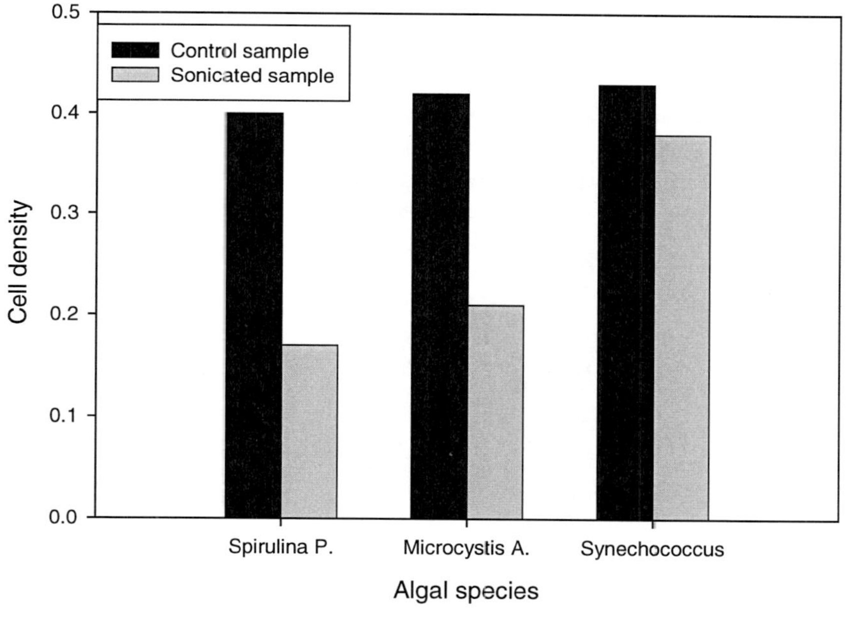

Figure 16. Algal specie influence on ultrasonic algae removal, 410 kHz, 40W, 5 min.

3.3.4. Impact of Ultrasonic Power

The ultrasound power is a very important parameter in sonochemistry. Normally, higher ultrasound power causes more violent cavitation and accelerates reactions. It was widely observed that the reaction rate for chemicals under sonication increases with the ultrasonic power employed (Kang and Hoffmann, 1998; Zhang and Hua, 2000), however it was not true in the cyanobacteria removal. Four *Microcystis aeruginosa* suspensions were irradiated by ultrasound waves at 20 W, 40 W, 60 W, and 80 W; the frequency was 1320 kHz and the sonication time was 5 min. The results showed that sonication at 20 W, 40 W, 60 W, and 80 W reduced the cell concentration by 41%, 47%, 51%, and 53% respectively. Higher power input beyond 20 W brought in little benefit and, hence, was uneconomical. With the same energy consumption, to extend the sonication time was more efficient than to increase the ultrasonic power. For example, to treat the algal suspension for 10 min with ultrasound at 20 W reduced the algal cells by 68% while to treat the algal suspension for 5 min with ultrasound at 40 W just reduced the algal cells by 47%. The saturation of power indicates that other mechanisms, immune to ultrasonic power, existed for the ultrasonic removal of cyanobacteria besides gas vesicles collapsing (Lee et al., 2000, 2001, 2002). Note that the same power saturation phenomenon existed for the ultrasonic algal growth inhibition and for the UV algal growth irradiation.

3.3.5. Impact of Ultrasonic Frequency

Ultrasound frequency is another important parameter that determines the sound field and influences the reaction kinetics (Petrier et al., 1997). Figure 17 reports the ultrasonic algae removal under different frequencies, while the ultrasonic power was 40 W. Obviously, higher ultrasound frequency was favorable for *Microcystis aeruginosa* removal. There was little difference in the algae removal rate constants among the low frequency range (20–150 kHz), but there was significant increase in the algae removal rate constant by increasing the frequency from 150 kHz to 410 kHz. The first order rate constant value was 0.022 min^{-1} when the frequency was 20 kHz and 0.11 min^{-1} when the frequency was 1320 kHz. This could be explained by the closeness of the size of algae gas vacuoles and the resonance size of

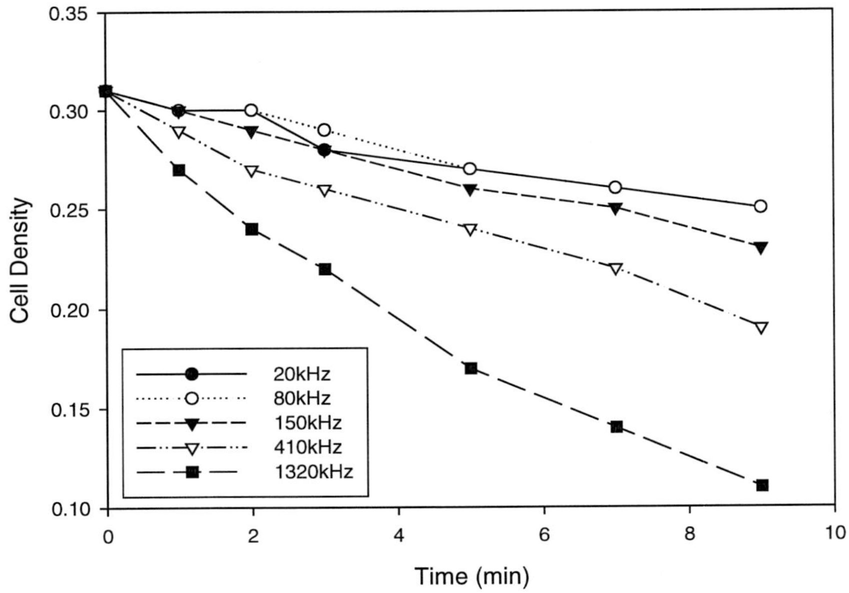

Figure 17. Frequency influence on sonication of *Microcystis aeruginosa*, 40W.

cavitation bubbles. Ultrasound can collapse gas vacuoles that control algal movement during cavitation (Lee et al., 2000, 2001, 2002; Wang, 2003; Zhang et al., 2006b). When the size of the gas vacuoles and the resonance size of cavitation bubbles are of the same order of magnitude, the gas vacuoles are more likely to resonate, undergo acoustic cavitation, and thus collapse. The resonance size of free air bubbles in water is 0.166 mm at 20 kHz and 2.47 μm at 1320 kHz (Calculation sees Section 4). Usually the gas vesicles of *microcystis aeruginosa* are up to 1 μm in length, so the *microcystis aeruginosa* gas vacuoles are more likely to resonate with the sound waves and collapse at higher frequencies than at lower frequencies. To reach 90% *microcystis aeruginosa* cell removal efficiency, 20 min was sufficient at 1320 kHz while 102 min was needed at 20 kHz. The same reason explained why middle range frequency (200 kHz) was the best for the ultrasonic removal of *Spirulina platensis* that had a length of 8-10 μm.3.3.6 Changes of other water characteristic duration sonication

Other impacts from sonication included decreases in the color, turbidity, and pH of the *Microcystis aeruginosa* samples (Figure 18). Turbidity decrease was beneficial since it reduced significantly the load for the following treatment processes in waterworks, including coagulation, sedimentation, filtration and disinfection. The pH dropped from 7.7 to 6.7. The color of the samples decreased from 150 to 12, a reduction of 92%, a combined result of settling of cyanobacterial cells, decreasing turbidity, and less chlorophyll *a*.

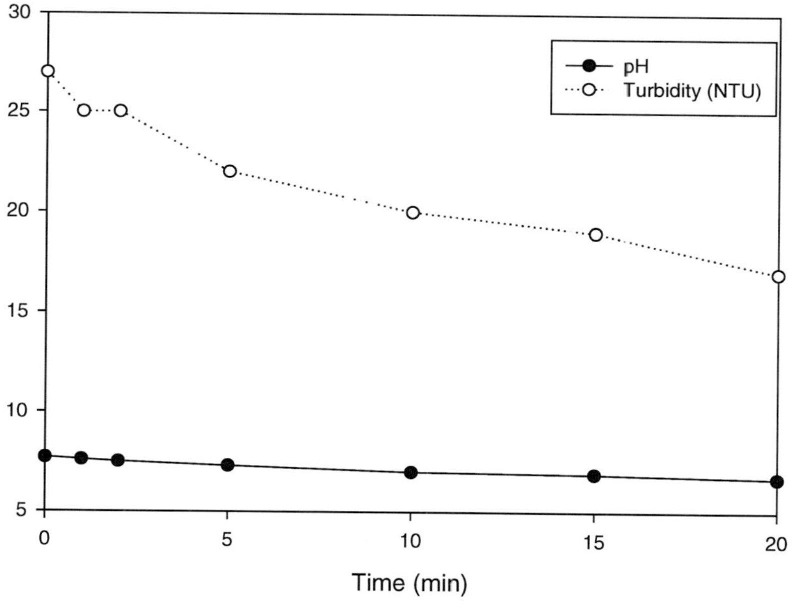

Figure 18. pH and turbidity change in ultrasonic *Microcystis aeruginosa* removal, 410 kHz, 20 W.

The mechanism of cell disruption by ultrasound is probably linked with cavitation phenomena and the resulting shock wave. Ultrasonic cell disruption generally results in very fine cell debris that is morphologically different from the coarser debris produced during other fluid shear-based processes of disrupting cells, which may be undesirable in water treatment plants since small debris may jam the filters. Besides, energy consumption is also of concern. Therefore, another, maybe better, option is to use ultrasound to assist existing drinking water treatment process to achieve rapid removal of algae cells without cell disruption. Since coagulation is one of the basic water treatment processes and is used in almost all drinking water plants, we continue this study by combining sonication with coagulation for algal reduction.

3.4. Removal of Algae by Sonication-Coagulation

For those algal cells containing gas vacuoles, acoustic cavitation effectively damages the gas vacuoles and destroys the floating of the cells, which, naturally leads to the settling of the cells. *Spirulina platensis* settling was observed in the sonication trials (Section 3.1) and the accumulation of cells was recorded as in Figure 19. Clearly, very short sonication should significantly increase the settlability of algal cells, which can be utilized for coagulation enhancement.

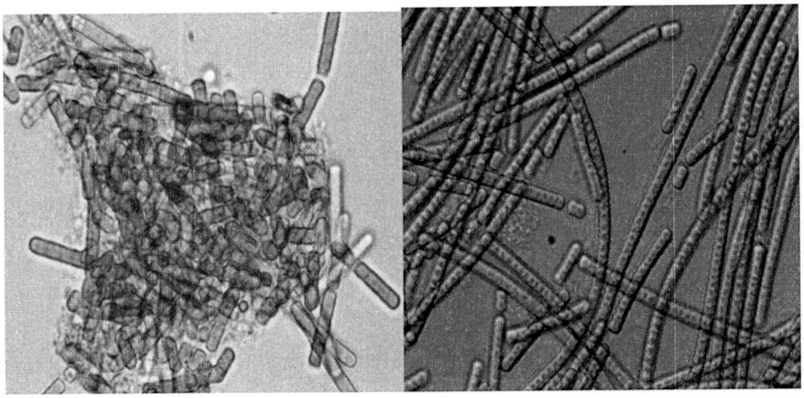

(a) (b)

Figure 19. *Spirulina platensis* cells settling before (a) and after (b) sonication, 5s, 150 kHz, 50W, photo enlarged 640 times (Wang, 2005).

The principle of the coagulation process is to use coagulant to react with colloids and small particulates suspended in water to form flocs, which grow till big enough to be removed via gravitational settling or air floatation. Coagulation is simple and dependable, and is a very important method for algal removal as well. However, when the algal concentration is high in the source water, algal cells consume a lot of coagulant and the result is usually unsatisfying. The reason is because algal cells are small and have a density closely to that of water, and thus rides freely in water without settling to the bottom or floating to the surface. Therefore, various coagulation enhancement techniques are needed to treat source water heavily loaded with algal cell, including with high coagulation dosage, multiple flocculants, pH adjustment, pre-oxidation, and particle activated carbon absorption. Figure 19 shows that simple and short sonication may effectively improve the algae settlability. Furthermore, micro-jets caused by acoustic cavitation enhance mixing (Kotyusov and Nenctsov, 1996). As a result, the coagulation process can be improved.

3.4.1. Feasibility

Sonication enhancement efficiency was examined for the coagulation removal of *Spirulina platensis* cells. The ultrasonic frequency was 150 kHz, the power was 40 W, and the duration was 5 s. The coagulant used was PAC, one of the commonest coagulants. Figure 20 reported the results. Clearly, sonication for 5 s significantly enhanced the coagulation efficiency; the lower the coagulant dosage, the bigger the improvement. When the PAC dose was 0.4 mg/L, sonication increased the algae removal efficiency from 58% to 78%. When the PAC dosage was 2 mg/L or higher, only slight improvement by sonication was observed. This phenomenon agrees with general observations that higher coagulation dose leaves less room for enhancement. For a water treatment plant that employs a daily PAC dosage of 1 mg/L, there are two options for coagulation enhancement algal cells when the source water gets polluted by algal bloom, one is to increase the PAC dosage from 1 mg/L to 4 mg/L and the other is to treated the source water with 5 s sonication before the coagulation process. Both options can achieve 90% algal cell removal, but the first option needs 4 times coagulant, and generates larger load of sludge, which in turn causes secondary pollution. Therefore, sonication for 5 s is a much better choice for coagulation enhancement.

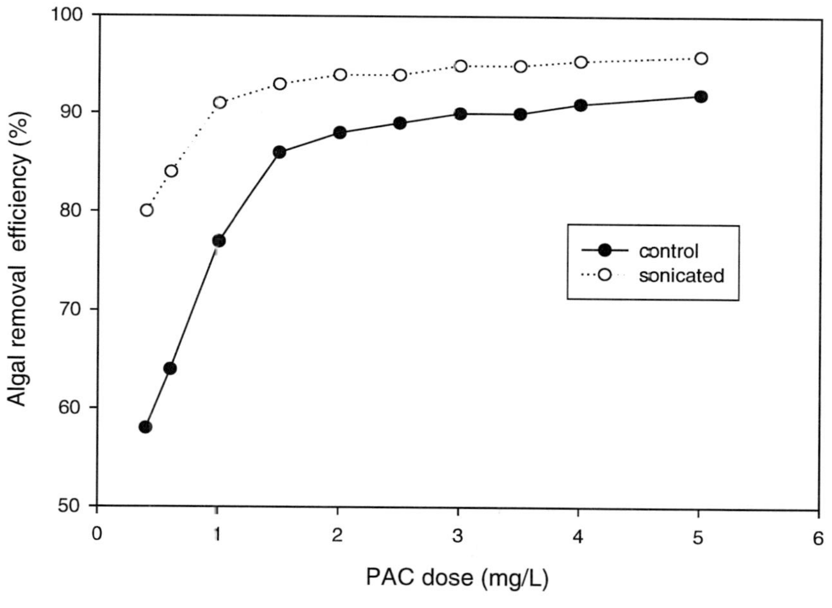

Figure 20. Ultrasonic enhancement of coagulation for algal removal, 150 kHz, 40 W, 5 s.

3.4.2. Impact of Ultrasonic Power

Ultrasonic power is an important factor in the application of ultrasound and determines, in together with sonication duration, the energy consumption of this method. Series of ultrasonic power, 20 W, 40 W, 60 W, 80 W, were compared for their influence on the sonication-coagulation process. The ultrasonic frequency was 150 kHz, the sonication time was 5 s, and the PAC dose was 1 mg/L. The results showed that little difference existed for samples treated with different ultrasonic power levels. The results agreed with those obtained from the ultrasonic algal growth inhibition (Section 3.1) and ultrasonic algal removal (Section 3.2).

3.4.3. Impact of Sonication Duration

The optimal sonication time was explored through experiments and the results were reported in Figure 21. The ultrasonic frequency was150 kHz, the power was 40 W, the PAC dose was 1.0 mg/L. Sonication for1 s, 5 s, 10 s, 30 s, 1 min, and 2 min changed the algal removal efficiency from 77% to 89%, 91%, 91%, 84%, 74%, and 67% respectively. Clearly, short sonication durations, 1 s and 5 s, were the best for coagulation enhancement. Sonication for 1 min or longer, instead of improving coagulation efficiency, decreased the coagulation process. The reason is because that sonication for 1 min or longer breached or even broke the *Spirulina platensis* cells into smaller fractures that were hard to settle down (Figure 22).

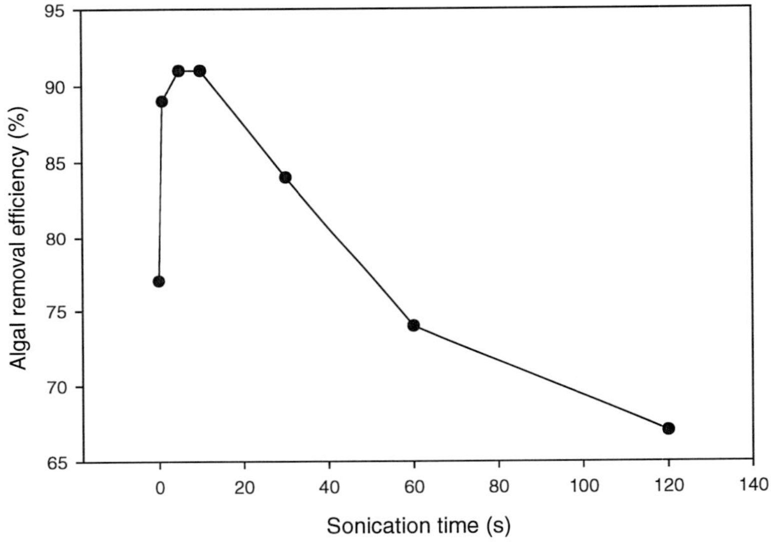

Figure 21. Impact of sonication time on sonication-coagulation, 150 kHz, 40 W.

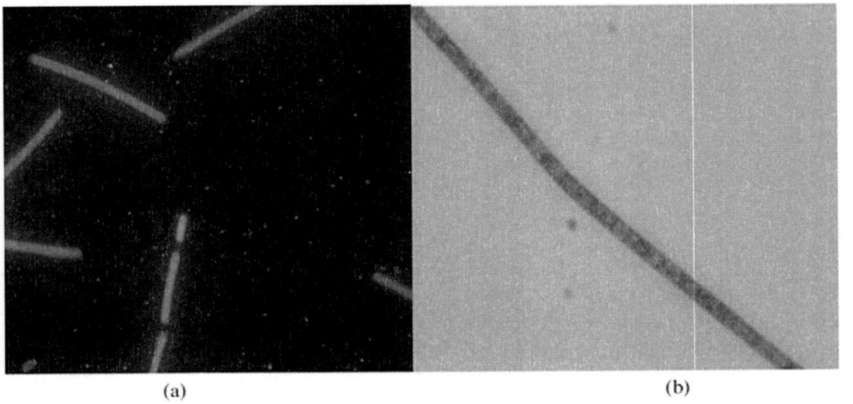

(a) (b)

Figure 22. *Spirulina platensis* cell (a) after sonication (b) control, 80 kHz, 100 W, 1 min (Hao et al., 2004b).

3.4.4. Impact of Ultrasonic Frequencies

Five ultrasonic frequencies were examined for their impact on the sonication-coagulation process, named 20, 80, 150, 410 and 1320 kHz. The ultrasonic power was 40 W, the sonication time was 5 s, and the PAC dosage was 1 mg/L. The results showed that ultrasonic frequency had little impact on the algal removal efficiency for the sonication-coagulation process, which was very different from the observations in the ultrasonic algal inhibition or removal processes.

4. Potential Mechanisms

The effects of sonication on the cyanobacterial cells can be attributed to acoustic cavitation, the most prominent phenomenon caused by ultrasound waves in water. Millions of small gas bubbles, at the size of μm, oscillate in water at the same frequency as that of the penetrating ultrasound waves under the acoustic pressure created by the transferring of the powerful acoustic pressure. When certain conditions are met, the bubbles wall contract at high speed (km/s) and collapse violently within a few μs, such vehement collapse of millions of small bubbles at one time is called acoustic cavitation. The cavitation bubbles, also called nuclei, may be gas bubbles or gas vacuoles of cyanobacteria when numerous algal cells exist. The resonance and following collapse of cyanobacteria gas vacuoles is believed to be mainly responsible for the growth inhibition and removal. Other factors of cavitation, to a less extent, also contribute to the ultrasonic algal control, including the mechanic damages, the chemical reaction with important biological molecules, and the degradation of toxins.

When the cavitation bubbles collapse, thousands of bars and degrees Kelvin can be reached in the cavities theoretically (Suslick, 1988), homolytic fission of the water molecule present in the cavities produces ·OH radicals, and high-pressure shockwave and radiation force are accompanied. The temperature and pressure are high enough to break any chemical bond in the volatile organic compounds that vaporize into the cavities. The ·OH radical, a very active and broad-spectrum oxidant, is responsible for degradation of dissolved organics in the aqueous phase (Kang and Hoffmann, 1998).

4.1. Collapse of Algal Gas Vacuoles during Sonication

In Section 3.1 and 3.2, we found that two cyanobacteria with gas vacuoles, namely the *Spirulina platensis* and *Microcystis aeruginosa,* could be effectively reduced or prohibited from reproduction by sonication. On the other hand, the gas vacuole free *Synechococcus* was virtually immune to ultrasonic treatment. The slight (less than 10%) reduction might be the result of mechanical damages and chemical damages. Since most bloom-forming cyanobacteria have gas vacuoles for buoyancy adjustment, sonication can be effective for algal blooming control or algal cell removal. Experiments confirmed the hypothesis with samples collected from water reservoirs in both southern and northern China that contained more than twenty species of algae (Section 3.1).

The gas vacuoles in cyanobacteria were made up of stacks of cylindrical vesicles cylinders, closed by conical ends, and formed by a single wall layer only 2 nm thick. Such a thin wall makes it viable to pressures. When cavitation occurred, the pressure of shock wave in the collapsing bubbles can reach 50 MPa (calculation sees section 3.4.2), so the gas vesicles are easily to collapse, resulting in the settlement and growth inhibition of gas-vacuoles cyanobacteria like *Microcystis aeruginosa*. Lee observed the detailed structure of cyanobacterial cell using Transmission Electron Microscopy (TEM), found that the gas vacuoles were disrupted after ultrasonic irradiation, as shown in Figure 23. Based on literature reports and our own experimental data, we can conclude that the dominant mechanism for ultrasonic algal treatment is the collapse of gas vacuoles subjected to the pressure of shock wave in ultrasonic cavitation.

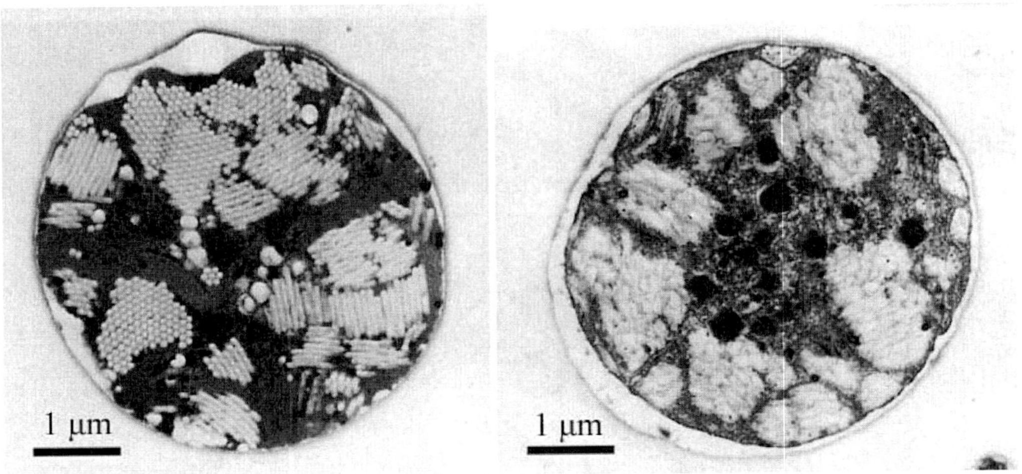

(a) Control. (b) After ultrasonic irradiation.

Figure 23. Collapse of gas vacuoles in cyanobacterial cells (Lee et al., 2000, 2001).

4.2. Mechanical Damage

The maximal pressure of shock wave in transient cavitation can be estimated as:

$$P_{\max} = P_g \left[\frac{P_m(\gamma - 1)}{P_g} \right]^{\frac{\gamma}{\gamma-1}}$$

in which P_g is the vapor pressure of liquid, P_m the ambient pressure of liquid, and γ the ratio of specific heats. Under room temperature and ambient pressure, the water P_g=2.3×10^3 Pa, P_m=10^5 Pa, and γ=1.4 (for air), the maximal pressure when cavitation collapses P_{max} was calculated to be above 5×10^7 Pa, i.e. 50 MPa. Such a high pressure is strong enough to damage the structure (Figure 22) and surface of the algal cell (Figure 23). Actually, the mechanical damages of acoustic cavitation have been well known and have been widely used for cleaning in laboratories and factories. The mechanical damages, both on the surface and on the structure, greatly slow down the growth and reproduction of the treated algal cells, and may change their behaviors during coagulation or filtration. Furthermore, the rapture of cells may release the intracellular materials such as algal toxins and proteins into water, which was observed when high intensity ultrasound was applied for long time (Section 5).

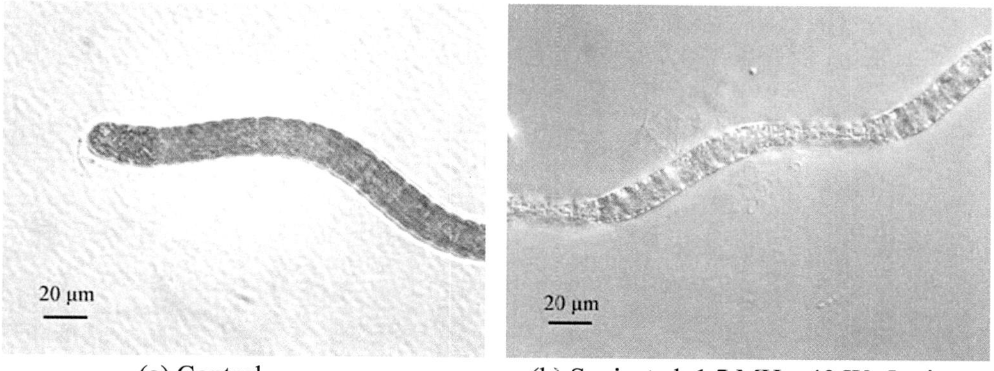

(a) Control (b) Sonicated, 1.7 MHz, 40 W, 5 min

Figure 24. Ultrasonic damage on *Spirulina platensis* cell, enlarged 1000 times (Wang, 2005).

4.3. Role of OH Free Radical

The two dominant mechanisms in sono-chemistry include free radical attack and pyrolysis inside the cavitation bubbles. The pyrolysis pathway requires the reactant to vaporize into the inside of cavitation bubbles, which is highly unlikely for the algal cells. But free radicals, especially the ·OH radical, are generated in cavitation and may react with the algal suspension and thus change the behavior of algal cells in water. Therefore, we designed a simple experimental trial to examine the potential role of ·OH on algae by adding carbonate solution into the algal suspension. It is known that CO_3^{2-} and HCO_3^- are good free radical scavengers, and the reaction rate constants are 3.9×10^8 $L{\cdot}mol^{-1}{\cdot}s^{-1}$ and 8.5×10^6 $L{\cdot}mol^{-1}{\cdot}s^{-1}$ for ·OH respectively (Zhang and Hua, 2000). When 500 mg/L carbonate was added into the solutions of 2-chlorobiphenyl and 2,4,5-chlorobiphenyl, distinctive inhibition of sonochemical degradation was observed and the sono-reaction rate constants dropped by more than 45% (Zhang and Hua, 2000). Similarly, 500 mg/L carbonate was added into the *Spirulina platensis* suspension, which was then subjected to sonication. The sonication conditions were 410 kHz, 20 W, and 1.7 MHz. The results showed that addition of 500 mg/L carbonate had little impact on the algal sonication in terms of both immediate reduction and growth inhibition. Therefore, we conclude that ·OH radicals hardly made contribution to cyanobacterial growth inhibition or cell reduction.

4.4. Ultrasonic Damages on Cyanobacterial Photosynthesis

Antenna complexes are the components in plants and algae that capture photons for photosynthesis. There are two major antenna complexes: chlorophyll–protein complex inside the cell membrane and phycobilin-peptides outside the cell membrane. PBS absorbs light in the range of 470–650 nm and chlorophyll–protein complex absorb light in the range of 430–440nm and around 670 nm, the combination of these complexes thus utilize solar lights.

There are four types of phycobilin-peptides, namely phycoerythrins, pycocyanins, phycoerythrocyanins, and allophycocyanins. For cyanobacteria, the most important chlorophyll and PBS are chlorophyll *a* and pycocyanins (Carr and Whitton, 1982), respectively, which were chosen for study in this paper. The typical structure of PC subunit is shown in Figure 25. The photo-activity was evaluated by the oxygen evolution rate since oxygen was the product of cyanobacterial photosynthesis. *Microcystis aeruginosa* was chosen as representative of cyanobacteria because it was a major bloom forming, poisonous algae specie, and was widely found in natural waters.

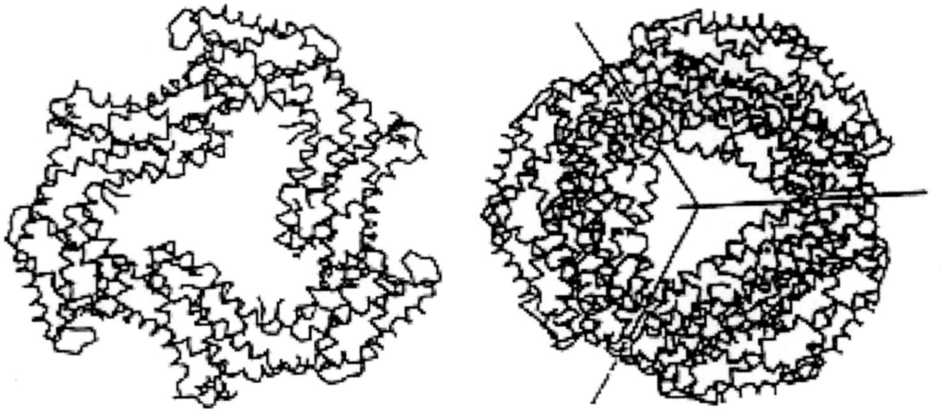

Figure 25. Typical pycocyanins subunit structure (Wang, 2003).

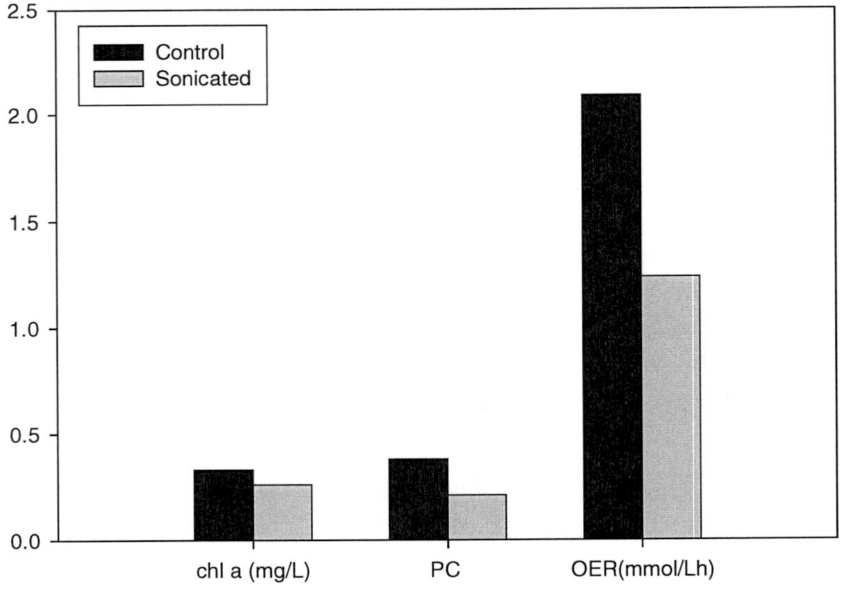

Figure 26. Ultrasonic inhibition of algal photosynthesis, chl a: chlorophyll *a,* PC: pycocyanins absorbance, OER: oxygen evolution rate, 150 kHz, 40 W, 5 min sonication.

The reductions of *Microcystis aeruginosa* chlorophyll *a* concentration, pycocyanins absorbance and oxygen evolution rate were reported in Figure 26. For *Microcystis aeruginosa*, ultrasonic treatment for 5 min (150 kHz, 40 W) reduced the chlorophyll *a* concentration by 23%, the pycocyanins absorbance by 44%, and the photo-activity by 43%, implying that sonication strongly damaged the antenna complexes and inhibited the photosynthesis of *Microcystis aeruginosa*, which significantly slowed down the cyanobacterial growth. Note that the 'empty' structure of pycocyanins made it more viable to sonication than chlorophyll *a* did, indicating that pycocyanins collapse might happen during acoustic cavitation.

4.5. Sound Frequency

Research shows that the most important operational parameters in sono-chemistry are ultrasonic power, frequency, and duration. However, previous data (Section 3) showed that among the three parameters, ultrasound frequency is by far the crucial one for algal control that markedly impacts the algal growth and algal reduction. Besides, difference algal species showed discrepancy in the selection of ultrasonic frequency. For example, the 1320 kHz or 1700 kHz are best for *Microcystis aeruginosa* control while the 410 kHz is best for *Spirulina platensis* control. The frequency discrepancy can be explained by the resonance of gas vacuoles in the ultrasonic field. When the size of the gas vacuoles and the resonance size of cavitation bubbles are of the same order of magnitude, the gas vacuoles are more likely to resonate, undergo acoustic cavitation, and thus collapse. The resonance size of free bubbles at given ultrasound frequency can be estimated by (Phillips et al., 1998):

$$f = \frac{1}{2\pi a}\sqrt{\frac{3\gamma}{\rho}\left(p_0 + \frac{2\sigma}{a}\right) - \frac{2\sigma}{a\rho}},$$

where f is the ultrasound frequency, γ is the ratio of specific heats, a is the radius of the bubble, p_0 is the pressure, σ is the surface tension, and ρ is the density of the surrounding medium. Under ambient temperature and pressure, the water surface tension can be ignored, the γ is 1.4 for air that exists in the bubble, the density of water is 1.0 g/cm3, then the resonance size of free bubble in water is 0.166 mm at 20 kHz and 2.47 μm at 1320 kHz. With an average size of 2-3 μm, the *microcystis aeruginosa* has gas vacuoles up to 1 μm in length, which is more likely to resonate with the sound waves and then collapse at higher frequencies than at lower frequencies. Thus the algae cells can be removed quicker at higher frequencies. Similarly, the size of *Spirulina platensis* is larger than that of *Microcystis aeruginosa* and tends to resonate at middle frequencies.

5. Ultrasonic Degradation of Algal Toxins

A major concern of algal blooming is that algae, especially some cyanobacteria, generate and release into water some algal toxins that may damage the health of abiotic bio-system and threaten the health of home animal or human being. To control algae toxins to acceptable level is very important in the treatment of algae. There are more than 70 algal toxins that have been identified, and among which, microcystins are of most concern. Microcystins are a group of typical algal toxins, have been found at detectable levels worldwide, and have strong toxicity. The Chinese national standard for total microcystin-LR is 1 μg/L in drinking water. Microcystins are also used for the study of ultrasonic degradation in this paper.

Microcystins are very stable due to the cyclic structure and double bonds and cannot be removed by conventional water treatment processes (Figure 27). Various methods have been studied to degrade microcystins in drinking water, including activated carbon adsorption, photic degradation, ozone oxidation, chemical oxidation, and biological degradation (Andrew et al., 1999; Jones et al., 1994; Nicholson et al, 1994; Rositano et al., 2001; Shephand et al., 2002). Sonication can effectively destroy the cyclic structure and double bonds in refractory pollutants via pyrolysis and free radical attack, which can be utilized for the degradation of microcystins. However, violent cavitation may lyse cell and release intracellular materials including microcystins into water, which is of great concerns.

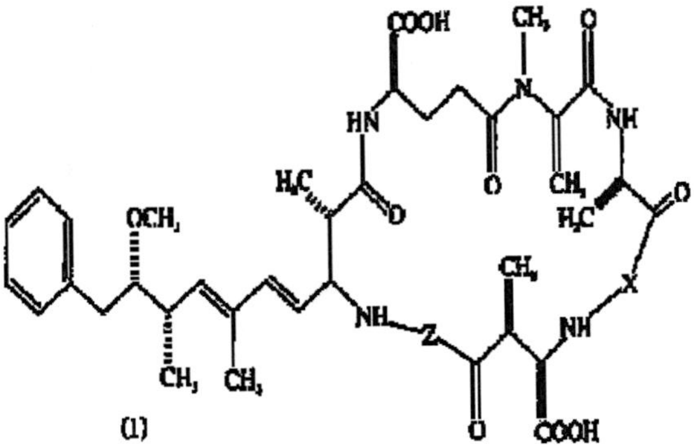

(1)

Figure 27. Structure of microcystins molecules (Zhang et al., 2003).

5.1. Release of Microcystins into Water by Sonication

Microcystis aeruginosa samples were exposed to sonication under various conditions to check the release of microcystins. No change in the microcystins concentration was observed with sonication at 80 kHz, 30 W for 5 min; little change was found with prolonged treatment of 15 min. The same conclusion could be drawn for sonication at 50 W. Therefore, sonication power below 50 W was safe in terms of microcystins release. However, higher sound power increased the extracellular microcystins concentration. Sonication (80 kHz, 80 W) increased the extracellular MCLR from 0 to 0.65 μg/L after 1 min and to 2.11 μg/L after 5 min; and increased extracellular MCRR from 0.87 to 1.0 μg/L after 5 min. In total, 5 min of sonication under 80 W, 80 kHz increased the extracellular microcystins from 0.87 to 3.1 μg/L. Three ultrasonic frequencies, namely 20, 80 and 150 kHz, were examined for their impact on microcystins release and the results showed little difference.

The increase of the concentration of extracellular microcystins might result from three pathways: the release of intracellular toxins by cell breach, the secretion of more toxins by algae cells as a defensive countermeasure to sonication, and the potential degradation of aqueous toxins by free radicals reaction caused by cavitation.

5.2. Control of Microcystins Increase during Culturing

Since ultrasonic irradiation could effectively slow down the algal growth, we suspected that the generation and release of algal toxins could also be inhibited by sonication. *Microcystis aeruginosa* sample was treated for 5 min by ultrasound at 150 kHz and 40 W, and then cultured under the same condition with the control sample without sonication. The results showed that 5 min sonication only slightly decreased the aqueous microcystins by 7.2%, but inhibited the release of microcystins from algal cells to water within the following 14 days culturing. The final extracellular microcystins concentration of the control sample was 137.7 μg/L while that of the sonicated sample was only 22.1μg/L, only 16% of that of the control sample. The cell concentration in sonicated sample was only 14.1% of that of the control after 14 days culturing. The results indicated that sonication for 5 min did not effectively degrade microcystins, but significantly control the *Microcystis aeruginosa* cell growth; and thus reproduced fewer cells and released fewer toxins.

5.3. Ultrasonic Degradation of Microcystins

The diluted microcystins solution (microcystins content was 2 μg/L) was exposed to ultrasonic irradiation at 20 kHz and 40 W, and the microcystins concentration was measured. Figure 28 showed that microcystins decomposed rapidly in the ultrasonic field and the decomposition followed a pseudo-first-order reaction. The rate constant of microcystins degradation was 0.036 min^{-1}.

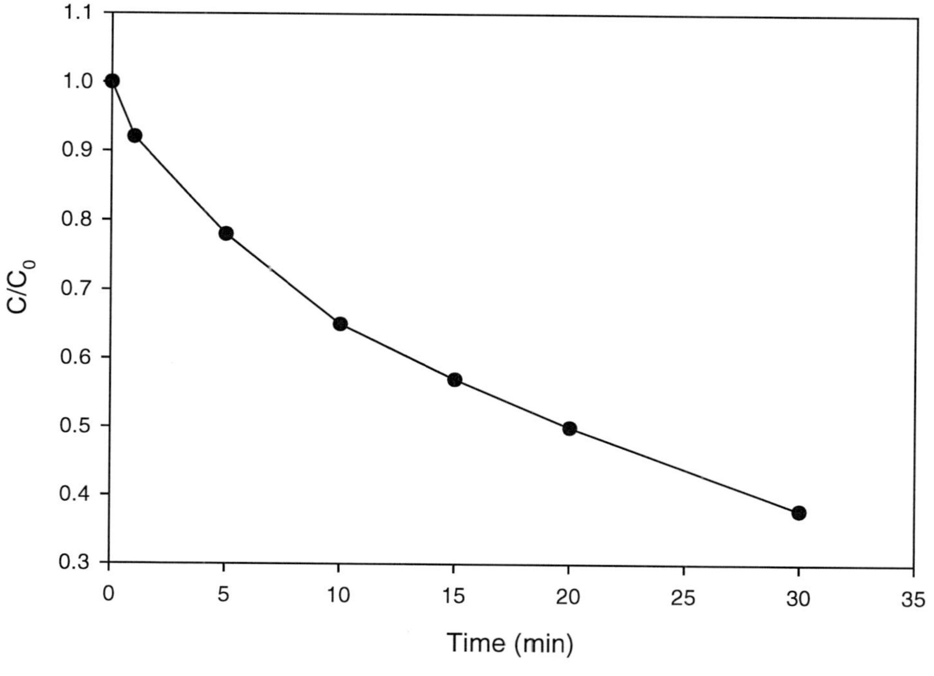

Figure 28. Kinetics of microcystins sono-degradation, 20 kHz, 40 W.

5.4. Impact of Ultrasonic Power

Three diluted microcystins solutions (2 μg/L) were exposed to ultrasonic irradiation at 20 kHz with different ultrasonic powers of 40 W, 80 W, and 120 W. The results (Figure 29) showed that within 40-120 W, the higher ultrasonic power resulted in the faster microcystins degradation speed. And the results of the samples treated with ultrasonic irradiation at 80 W and 120 W were obviously better than that at 40 W. But microcystins degradation speed of samples irradiated at 80 and 120W ultrasonic irradiation became relatively slow after 5 min, indicating a saturation of power effect.

Normally, at the same ultrasonic frequency, higher ultrasonic power leads to faster reaction within certain power ranges. It is because of the cavitation zone being extended and the associated chemical effects being enhanced. However, the sono-reaction rate decreases once the power increases beyond the certain range. Too high power slowed down the sono-degradation of CCl_4 at 20 kHz (Kang and Hoffman, 1996). That maybe because too high ultrasonic power results in too large cavitation bubbles that cannot collapse completely and thus inhibits sono-reactions. Therefore, a proper power level is needed to apply sonication in microcystins degradation.

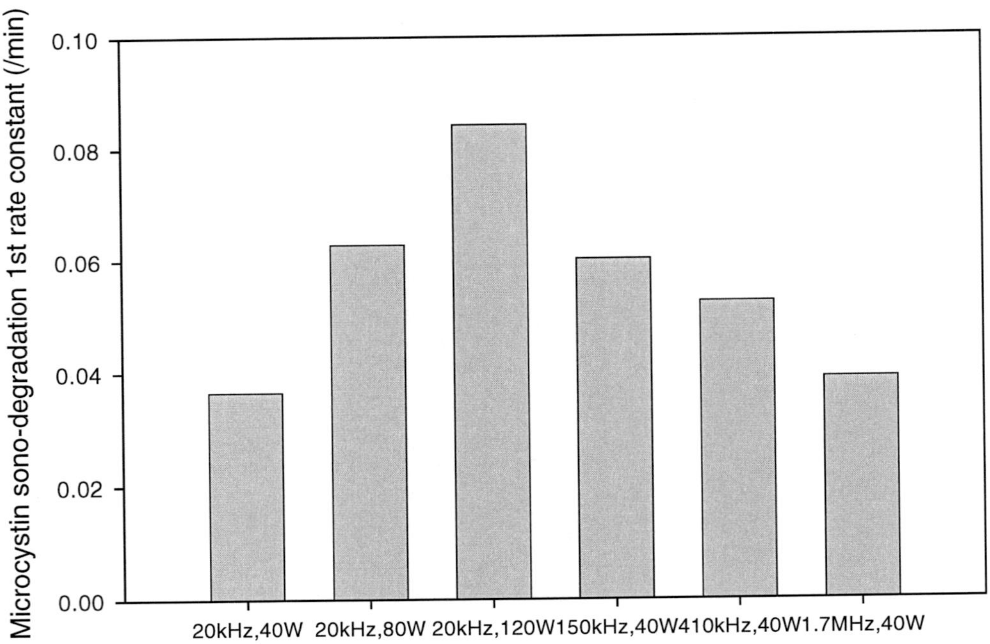

Figure 29. Impact of ultrasonic conditions on microcystins sono-degradation.

5.5. Impact of Ultrasonic Frequency on Microcystins Degradation

Four diluted microcystins solutions (2 μg/L) were exposed to ultrasonic irradiation at different frequencies of 20 kHz, 150 kHz, 410 kHz, and 1.7 MHz with the same ultrasonic power of 40 W. The results are summarized in Figure 29. After 20 min ultrasonic irradiation, the microcystins were removed by 54.7%, 70.6%, 65.2%, and 53.9%, respectively (Zhang et al., 2005). The effects of ultrasound at 150 kHz and 410 kHz were obviously better than the effects of ultrasound at 20 kHz and 1.7 MHz, and the fastest degradation was acquired at 150 kHz. Therefore, ultrasound at intermediate frequency had better effect on microcystins degradation. The main mechanism of sonochemical reaction is ultrasonic cavitation that generates hot spots for pyrolysis and releases hydroxyl free radical into water for free radical reaction. Smaller bubbles at higher frequencies limit reactions and the time required for cavitation may exceed the available compression period at frequencies beyond 1 MHz (Petrier et al., 1994). On the other hand, faster pressure cycles at higher frequencies lead to more cavitation within unit of time and faster radical ejection into the active zone for reactions (Petrier et al., 1994). Therefore, an optimum frequency usually exists in the middle.

Conclusion

Proper ultrasonic irradiation is an effective method to control algal overgrowth in water-bodies, to remove the algal cells in the water treatment plants, to decrease the release of algal toxins, and to degrade the algal toxins. Various factors influence the results of ultrasonic algal control, including but not limited to algal species, algal growth stage, water alkalinity, the ultrasound power, frequency, and duration.

For the ultrasonic algal growth control, the existence of gas vacuole in the algal cells is crucial for sonication to take effect. Filaceous algae are more prone to sonication than spherical ones. Different algal species have different optimal ultrasound frequencies and the best result occur when the algal cell size is close to the resonant nuclei size of acoustic cavitation. The method is effective for water samples collected from water reservoirs that contain more than twenty different algal species. Sonication duration and power both impact the algal growth inhibition efficiency, but longer duration or higher power do not necessarily benefit sonication, and the most economic conditions are 40 W and 5 min for this study.

Unlike sonication, UV irradiation does not decrease the algal cell concentration immediately after irradiation and does not require the existence of gas vacuole to be effective in algal growth control. Similar to the ultrasonic algal growth inhibition, certain UV light intensity is needed to retrain the growth of cyanobacteria, and the control effect increases along with the intensity or time of ultraviolet irradiation. The most economical and effective dosage of ultraviolet are intensity of 0.52 mW/cm^2 and irradiation time of 3-5 min. The algal growth stage plays the same role in UV algal control as in ultrasonic algal inhibition. As for the biologic effects of ultraviolet, ultraviolet radiation can destroy the configuration of cyanobacterial cell and possibly phycocyanin and some enzyme on thylakoid membrane in algal cells, and then destroy the composing of chlorophyll *a*, thus greatly restrain the composing of chlorophyll *a* and the growth of cyanobacteria.

Ultrasonic irradiation reduces the cell concentration in algal suspensions. The kinetics follows pseudo-first order reaction and higher initial cell concentration benefits sonication. Power saturation exists in ultrasonic algal removal as well as in the ultrasonic and UV inhibition of algal growth. With the same energy consumption, to extend the sonication time is more efficient than to increase the ultrasonic power in algal reduction. The most important parameter is the ultrasonic frequency, and the first order rate constant for ultrasonic reduction of *Microcystis aeruginosa* is 0.022 min^{-1}, 0.023 min^{-1}, 0.031 min^{-1}, 0.050 min^{-1}, and 0.11 min^{-}

[1] at 20, 80, 150, 410, and 1320 kHz respectively when the ultrasonic power is 40 W. The main cause of cell removal is the loss of buoyancy by ultrasonic collapsing of the gas vesicles. Besides cyanobacteria settling, sonication also decreases the turbidity and pH of samples, as an integrated result, the color of the *Microcystis aeruginosa* samples decreases by 92% after 17 min sonication at 20 W, 1320 kHz.

Ultrasonic irradiation improves algae settleability and changes the algae structure. Sonication-coagulation process effectively removes algal cells from the eutrophic source water, and short sonication (1 s, 5 s) significantly enhances the coagulation efficiency. To achieve 90% of algae removal, pre-sonication reduces the coagulant dose by more than half. The sonication-coagulation process is insensitive to the ultrasonic power and might be damaged by longer sonication, which breaks the cell structure and hampers the cell settling. Similarly, the ultrasonic frequency has virtually no impact on the hybrid method.

The inhibition of cyanobacterial bloom is due to the pressure and shock wave when the bubble collapses. In the ultrasonic treatment, cyanobacterial cells with gas vacuole can be easily destroyed resulting in the cyanobacterial settlement and growth inhibition. The collapse of gas vacuoles subjected to the pressure of shock wave in ultrasonic cavitation is hence the dominant mechanism for ultrasonic algal treatment. Ultrasonic irradiation also inhibits the photosynthetic process by damaging chlorophyll *a*, the antenna complex, and PC. Ultrasonic treatment for 5 min at 150 kHz and 40 W reduces the *Microcystis aeruginosa* chlorophyll *a* concentration by 23%, the pycocyanins absorbance by 44%, and the photo-activity by 43%.

High ultrasonic power and long irradiation caused microcystins release and increased the microcystins concentration. Ultrasonic irradiation effectively decreases the algal toxins concentration in water via inhibiting cell reproduction; 5 min sonication decreases 84% microcystins after 14 days culturing. Beside, sonication is also an efficient method for degradation of algal toxins via hydroxyl radical attack and/or high temperature pyrolysis generated in the acoustic cavitation. The ultrasonic microcystin decomposition follows the pseudo-first order reaction and the microcystins removal ratio reaches 60% after 5 min at 20 kHz and 120 W. The reaction rate increases with the increase of ultrasonic power input, but the power effect may saturate at high levels. The ultrasonic frequency impacts the microcystins degradation significantly, and 150 kHz is the most effective frequency.

References

Ahn, C; Park, M; Joung, S; Kim, K; Jang, K; Oh K. Growth inhibition of algae by ultrasonic radiation: laboratory and enclosure studies. *Environ. Sci. Technol.*, 2003, 37, 3031–3037.

Andrew, J; Feitz, T; Walte, D. Photo-catalytic degradation of the blue green algal toxin Microcystin-LR in a natural organic-aqueous matrix. *Environ. Sci. Technol.*, 1999, 33(2), 243–249.

Carmichael, W W. Health effects of toxin-producing cyanobacteria: the CyanoHABs. *Hum. Ecol. Risk Assess.*, 2001, 7, 1393–1407.

Carr, N G; Whitton, BA (Eds.), The Biology of Cyanobacteria. Oxford: Blackwell Scientific Publications; 1982.

Chen, Y; Liu, J; Ju, J. Flotation removal of algae from water, *Colloids Surf. B: Biointerfaces,* 1998, 12, 49–55.

Chisti, Y. and Moo-Young, M. Disruption of microbial cells for intracellular products. *Enzyme Microb. Technol.* 1986, 8, 194 – 204

Chorus, I; Bartram, J. *Toxic Cyanobacteria in Water*. New York: E & FN Spon; 1999.

Cum, G; Galli, G; Gallol, R. Role of frequency in the ultrasonic activation of chemical reactions. *Ultrasonics*, 1992, 30(4), 267–270.

Dawson, R M. The toxicology of microcystins. *Toxicon*, 1998, 26(7), 953–962.

Donati, C D; Drikas, M; Hayes, R. Microcystin-LR adsorption by powdered activated carbon. *Water Res.*, 1994, 28(8), 1735–1742.

Entezari, M. The effect of frequency on sonochemical reactions III: dissociation of carbon disulfide, *Ultrason. Sonochem.,* 1997,4, 49–54.

Francko, D.A. et al. Enhancement of nitrogen fixation in Anabaena flosaquae (Cyanobacteria) via low-dose ultrasonic treatment. *J. Appl. Phycol.* 1994, 6, 455 – 458.

Gondrexon, N; Renaudin, V; Petrier, C. Degradation or pentachlorophenol aqueous solutions using a continuous flow ultrasonic reactor: experimental performance and modeling. *Ultrason. Sonochem.*, 1999, 5(4), 125–131.

Hao, H; Wu, W; Chen, Y; Tang, J; Wu, Q. Algae bloom control by ultrasonic irradiation at 20 kHz and 1.7 MHz. *J. Environ. Sci. Health Part A*, 2004a, A39, 1435–1446.

Hao, H; Wu, W; Chen, Y; Tang, J; Wu, Q. Cavitation mechanism in cyanobacterial growth inhibition by ultrasonic irradiation. *Colloid. Surf.*, 2004b, B33, 151–156.

Hart, E J; Fischer, C H; Henglein, A. Isotopic exchange in the sonolysis of aqueous solutions containing D_2 and CH_4. *J. Phys. Chem.*, 1987, 91, 4166–4169.

Huang, J L; Feng, R; Zhu, C P. Low-MHz frequency effect on a sonochemical reaction determined by an electrical method. *Ultrason. Sonochem*, 1995, 2(2): 93–97.

Jiao, Y. The effect of ultraviolet irradiation on bacterial growth in polluted urban rivers. M.S. Thesis, 2004, Tsinghua University, Beijing

Jones, G J; Boume, D G; Robert L. Degradation of the cyanobacterial hepatotoxin microcystin by aquatic bacteria. *Natural Toxins*, 1994, 2, 228–235.

Kang, F; Hoffmann, M. Kinetics and mechanism of the sonolytic destruction of MTBE by ultrasonic irradiation in the presence of ozone. *Environ. Sci. Technol.*, 1998, 32, 3194–3199.

Kotyusov, A; Nenctsov, B. Induced coagulation of small particles under the action of sound. *Acustica,* 1996, 82, 459–463

Lian, M; Liu, Y; Yu, S. Study of microcystins in drinking water on human health. *Chinese Epidemiol. J.*, 2000, 21, 437–440.

Lee, T J; Nakano, K; Matsumura, M. A new method for the rapid evaluation of gas vacuoles regeneration and viability of cyanobacteria by flow cytometry. *Biotechnol. Lett.*, 2000, 22, 1833–1838.

Lee, T J; Nakano, K; Matsumara, M. Ultrasonic irradiation for blue-green algae bloom Control. *Environ. Technol.*, 2001, 22, 383–390.

Lee, T J; Nakano, K; Matsumara, M. A novel strategy for cyanobacterial bloom control by ultrasonic irradiation. *Water Sci. Technol.*, 2002, 46, 207–215.

Ma, J; Liu, W. Effectiveness and mechanism of potassium ferrate (VI) pre-oxidation for algae removal by coagulation. *Water Res.,* 2002, 36 , 871–878.

MacKinney, G. Absorption of light by chlorophyll solution. *J. Biol. Chem.*, 1941,140, 315–322.

Mason, T J; Lorimer, J P. *Sonochemistry: Theory, Applications and Uses of Ultrasound in Chemistry.* New York: Ellis Horwood; 1990.

Nakano, K; Lee, T F; Matsumura, M. In situ algal bloom control by the integration of ultrasonic radiation and jet circulation to flushing. *Environ. Sci. Technol.*, 2001, 35, 4941–4946.

Nicholson, B; Rositano, J; Burch, M D. Destruction of cyanobacterial peptide hepatotoxins by chlorine and chloramines. *Water Res*, 1994, 28(6), 1297–1303.

Nyborg, W L; Ziskin, M C. Biological Effects of Ultrasound. New York: Churchill Livingstone; 1985.

Peng, W. *Environmental control of algae generation in oil field.* Ph.D. Dissertation, Tsinghua University, 2001, Beijing.

Petrier, C; David, B. Ultrasonic degradation at 20 kHz and 500 kHz of atrazine and pentachlorophenol in aqueous solution: Preliminary results. *Chemosphere*, 1996, 32, 1709–1718.

Petrier, C; Jiang, Y; Lamy, M. Ultrasound and environment: sonochemical destruction of chloromatic derivatives. *Environ. Sci. Technol.,* 1998, 32, 1316–1318.

Petrier, C; Reverdy, G. Unexpected frequency effects on the rate of oxidative processes induced by ultrasound. *J. Am. Chem. Soc.,* 1992, 114, 3148–3150.

Phillips, D; Chen, X; Baggs, R; Rubens, D. Violante, M; Parker, K. Acoustic back scatter properties of the particle/bubble ultrasound contrast agent. *Ultrasonics,* 1998, 36, 883–892.

Phull, S; Newman, A; Lorimer, J; Pollet, B; Mason, T. The development and evaluation of ultrasound in the biocidal treatment of water. *Ultrason. Sonochem.,* 1997,4, 157–164.

Price, G J. Current Trends in Sonochemistry. Cambridge: Royal Society of Chemistry; 1992.

Rao, S R; Tripathi, U; Ravishankar, G A. Biotransformation of codeine to morphine in freely suspended cells and immobilized cultures of Spirulina Platensis. *World J. Microbiol. Biotechnol.*, 1999, 15, 465–469.

Repavich, WM; Sonzogni, WC; Standridge, J H; Wedepohl, R E; Meisner, LF. Cyanobacteria (blue–green algae) in wisconsin waters: Acute and chronic toxicity. *Water Res.*, 1990, 24, 225–231.

Rositano, J; Newcombe, G; Nicholson, B. Ozonation of NOM and algal toxins in four treated waters. *Water Res.*, 2001, 35, 23–32.

Rott, H D. Biological effects of ultrasound. *Nervenheilkunde*, 1998,Vol. 7, pp.16–18.

Ryding, SO; Rast W. *The Control of Eutrophication of Lakes and Reservoirs*, Paris: United Nations Educational, Scientific and Cultural Organization, 1989.

Shephand, G S; Stockenstrom, S; Villiers, D. Degradation of microcystin toxins in a falling film photocatalytic reactor with immobilized titanium dioxide catalyst. *Water Research*, 2002, 36(1), 140–146.

Shi, Y; Chen, Z; Li, G. Research on potassium permanganate composite chemicals for algae removal in labor. *J. Harbin Univ. of E. E. & Architecture*, 2003,Vol. 23, No. 4, pp.43–45.

Simmons J. Algae control and destratification at Hanning field reservoir, in: *Proceedings of IAWQ-IWSA Joint Conference on Reservoir Management and Water Supply*, Prague, Czech, 1997, 145–152.

Smith, V H; Tilman, G D; Nekola, J C. Eutrophication: impacts of excess nutrient inputs on freshwater, marine, and terrestrial ecosystems. *Environ. Pollut.*, 1999,100, 179–196.

Suslick, K S. *Ultrasound: Its Chemical, Physical, and Biological Effects*. Weinheim: VCH Publisher; 1988.

Teo, K C; Xu, Y; Yang, C. Sonochemical degradation for toxic halogenated organic compounds. *Ultrason. Sonochem.*, 2001, 8, 241–246.

Tsuji, K; Watanuki, T; Kondo, F. Stability of microcystins from cyanobacteria II. Effect of UV light on decomposition and isomerization. *Toxicon*, 1995, 33(12), 1619–1631.

Wang, B. Experimental Research on Inhibition and Removal of Algae by Ultrasound and Its Enhanced Technology. M.S.Thesis, Tsinghua University, Beijing, 2005.

Wang, H. Ultrasonic inhibition of cyanobacterial growth. M.S.Thesis, Tsinghua University, Beijing, 2003.

Wang, P; Wang, G. Study on algae removal from water by ferrous salt coagulants, *J. Wuhan University of Sci. & Tech.*, 2003, 26, 35–38.

Wu, J M. Ultrasonic destruction of chlorinated compound in aqueous solution. *Environment Progress*, 1992, 11, 195.

Wu, J.; Lin, L. Elicitor-like effects of low-energy ultrasound on plant (Panax ginseng) cells: induction of plant defense response and secondary metabolite production. *Appl. Microbiol. Biotechnol.* 2002, 59,51–57

Xu, X; Dai, X; Lu, W. The relationships between nuclei, gene and photosynthesis. *Biology Letter,* 1999, 34, 5–6.

Zhang, G; Hua, I. Cavitation chemistry of polychlorinated biphenyls: decomposition mechanisms and rates. *Environ. Sci. Technol.*, 2000, Vol. 34, pp.1529–1534.

Zhang, W; Zhang, G; Xu, X. The comparison of purification for microcystins extractant. *Acta Universitatis Sunyatseni,* 2003,42,144–146.

Zhang, G; Zhang, P; Wang, B; Liu, H. Ultrasonic frequency effects on the removal of Microcystis aeruginosa. *Ultrasonics Sonochemistry*, 2006a, 13, 446-450.

Zhang, G; Zhang, P; Wang, B; Liu, H. Algae removal by sonication-coagulation, J. *Environ. Sci. Health, Part A*, 2006b, 41, 1379-1390.

Zhang, G; Zhang, P; Liu, H; Wang, B. Ultrasonic damages on cyanobacterial photosynthesis, *Ultrasonics Sonochemistry*. 2006c, 13, 501-505.

In: Algae: Nutrition, Pollution Control and Energy Sources
Editor: Kristian N. Hagen, pp. 127-146
ISBN: 978-1-60692-008-4

Chapter 7

EFFECTS OF THE ACIDIFICATION ON PHOTOSYNTHESIS AND GROWTH OF MARINE ALGAE: A REAPPRAISAL OF THE LABORATORY DATA AND THEIR APPLICABILITY TO THE NATURAL HABITATS

Jesús M. Mercado*
Instituto Español de Oceanografía. Centro Oceanográfico de Málaga. Puerto Pesquero s/n. Apdo. 285. 29640, Fuengirola (Málaga). Spain

Abstract

The CO_2 is the substrate for the carboxylation reaction of the Rubisco that is the key enzyme implied in production of organic carbon from CO_2 by the primary producers. In the present ocean, the CO_2 concentration is not enough to saturate the Rubisco carboxylation rates. However, an increase of the CO_2 concentration by 50% with respect to the current value is expected by the year 2100, due to the acidification caused by the emission of CO_2 from the fossil fuel burning. Therefore, it could be expected that the acidification will alter the photosynthesis rates in the ocean. This hypothesis is examined by revising the results of laboratory experiments in which the effects of changes in CO_2 concentration have been researched. Most of alga taxonomic groups feature photosynthesis rates almost non-sensitive to short-term changes in the CO_2 concentration due to they have developed mechanisms for using HCO_3^- for photosynthesis (whose concentration is one magnitude order higher than CO_2) which permit to increase the CO_2 concentration around Rubisco (the so-called carbon concentrating mechanisms, CCM). Limitation of the photosynthesis rates by the CO_2 concentration (other resources being non-limited) in air-equilibrated seawater has been only described in a few alga species, including coccolithophorids (a phytoplankton group that episodically produces blooms in vast areas of the ocean) and some species of benthic brown and red algae. These latter species are shade plants whose growth at their natural habitat is primary limited by the light intensity. In spite of the photosynthesis saturation at the actual CO_2 concentration for many species, the growth at high CO_2 induces an acclimation response

* E-mail address: jesus.mercado@ma.ieo.es

that consists of down-regulation of the CCMs. Furthermore, effects of the high CO_2 on photosynthetic apparatus, growth rates and carbohydrates and protein contents have been described, although this response is un-uniform among the alga species examined. This finding suggests that acidification could induce changes in the taxonomic composition of the communities. However, the few experiments of high CO_2 performed with natural assemblages of phytoplankton do not demonstrate substantial changes in the community structure or the succession patterns during the development of alga blooms.

1. Introduction

The CO_2 in the atmosphere and ocean tend to be in equilibrium with each other, therefore the emission of CO_2 from the fossil fuel burning generates a CO_2 net flux toward the ocean. In turn, the dissolution of this CO_2 excess in the seawater produce a decrease of the pH (acidification) which deals with increases of CO_2 and HCO_3^- and decrease of $CO_3^=$. Accordingly, ocean observations indicate that the pH of the surface seawater could globally have reduced by 0.1 units with respect to the pre-industrial times (Caldeira and Wickett, 2003; Key et al., 2004). Furthermore, a doubling of the atmospheric CO_2 concentration with respect to the present level (as predicted for the year 2100; IPCC, 2003), will produce a decrease of pH by 0.5 units (Caldeira and Wickett, 2003). On average, the seawater pH in the present conditions is 8.1-8.2, with about 90% of dissolved inorganic carbon (DIC) being in the form of HCO_3^- (2.0 mM concentration) and the CO_2 being less than 1% (10-15 μM). The projected pH change by the year 2100 would imply increases of CO_2 and HCO_3^- concentrations by 50% and 6% with respect to the current values, respectively (for a more detailed analysis of the consequences of other possible CO_2 emission scenarios on the DIC chemistry in the ocean, see Royal Society, 2005).

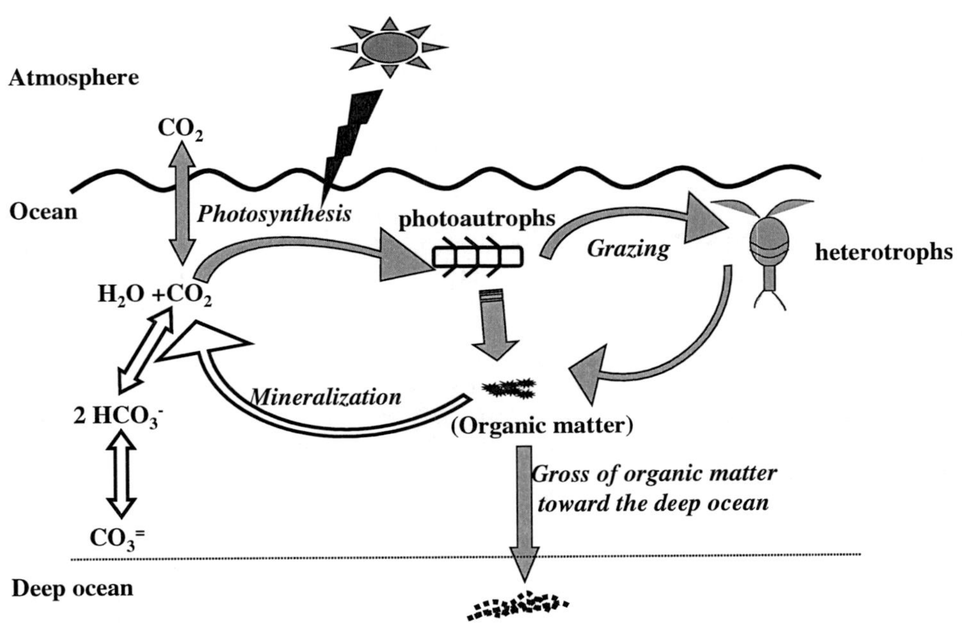

Figure 1. Functioning of the biological-pump in the ocean. The grey arrows indicate the carbon fluxes that potentially could increase due to the ocean acidification.

In contrast to our certainty about the consequences of the CO_2 emissions on the carbonate chemistry in the ocean, the effects of the acidification on the marine ecosystem functioning are almost completely unknown (Royal Society, 2005). In fact, it does not appear to be excessive to state that our understanding of the biological consequences of the ocean acidification "is in its infancy" (as asserted by Riebesell et al., 2008). Nevertheless, some documents and reviews focusing this topic have been published (see the reference list). These works highlight that a main point to be addressed is the possible effect of the acidification on the fixation rates of CO_2 by the marine photosynthetic organisms. Photosynthesis provides more than 99% of the organic matter used by the marine food webs, a part of which is exported from surface toward the deep ocean (i.e. the so-called biological-pump that slows down the accumulation of CO_2 in the atmosphere; Figure 1). The CO_2 is the substrate for the carboxylation reaction of the Rubisco that is the key enzyme implied in production of organic carbon from CO_2 by the primary producers. Therefore, it could be expected that the acidification will alter the photosynthesis rates in the ocean (which in turn would modify the ocean's capacity for absorbing atmospheric CO_2).

The aim of this review is to test this global hypothesis from an eco-physiological perspective based on the experience reported in the literature for a number of algal species and communities. Firstly, the possible role of the CO_2 as a factor limiting the algal photosynthesis will be addressed. Afterwards, the effects of the changes of CO_2 on photosynthesis and growth in algae will be revised. The analysis presented is chiefly based on results of laboratory experiments that describe the photosynthetic and growth performance of a number of algal species. In contrast to other reviews, our breakdown examine both free-living species (i.e. phytoplankton) and benthic alga. In doing so, the effects of acidification for a wide variety of marine habitats will be discussed. The marine algae are a very heterogeneous organism group from a evolutionary viewpoint, composed of at least ten phylogenetic divisions or phyla (Falkowski et al., 2004), eight out of them including free-living microscopic species (phytoplankton; Falkowski and Raven, 1997) while the benthic macroscopic forms (seaweeds or macroalgae) belong to three phyla. The main phytoplankton groups in terms of their contribution to the total algal biomass, primary production and biogeochemical cycles in the ocean, are cyanobacteria, diatoms, dinoflagellates and coccolithophorids. Globally, the contribution of the three taxonomic groups of seaweeds (Rhodophyta or red algae, Phaeophyta or brown algae and Chlorophyceae or green algae) to oceanic primary production is lower than the one by phytoplankton; however these species are dominant in rocky substrata shore habitats, which contain a variable ambient with respect to CO_2 availability and pH.

2. Is the CO_2 a Limiting Factor for the Algal Photosynthesis?

2.1. Natural Variability of the CO_2 Concentration

The diffusive transport of CO_2 in seawater is relatively slow (the diffusion coefficient at 25 °C is 1.94 10^{-9} m^2 s^{-1}, according to Kigoshi and Hashitani, 1963), therefore its concentration modifies rapidly in the immediate vicinity of the cells during the processes of photosynthesis and respiration. Normally, these changes are balanced beyond the boundary layer, where equilibrium with the bulk fluid and atmosphere is re-established (Falkowski, 1997). However, when high biological activity and calm conditions (i.e. absence of turbulence) are combined, the consumption and/or release of CO_2 are not compensated which deals with substantial changes in pH and CO_2 concentration in the bulk fluid. These two conditions are not unfrequent in coastal habitats, where the biomass density of the photosynthetic organisms (including both phytoplankton and seaweeds) is often considerably higher than in vast areas of the open ocean dominated by oligotrophic conditions. Accordingly, pH and CO_2 concentration are quite constant in open sea, while substantial variability has been reported in coastal habitats. Particularly, seasonal changes in CO_2 concentration driven by the annual production cycle have been reported in different continental shelves (Borges et al., 2006) where the CO_2 concentration could reduce by 30% in spring with respect to the other seasonal periods. The CO_2 seasonal changes become higher in coastal areas affected by hydrological fertilization mechanisms. As an example, Huertas et al. (2006) described a CO_2 annual decrease by about 70% with respect to the CO_2 concentration in air-equilibrated water, which was linked to the increase of phytoplankton biomass driven by up-welling of nutrient enriched deep water at Cádiz Bay (Atlantic Ocean). By contrast, in areas where the up-welling is very intensive and almost permanent, the CO_2 concentration keeps over its equilibrium value due to the deep water is over-saturated in CO_2 and the high hydrodynamics precludes the accumulation of high phytoplankton biomass (Cai and Dai, 2004). In spite of these seasonal cycles, more substantial changes in CO_2 concentration (lasting for a few days) are produced during episodes of coastal algal blooms. For example, Hinga (1992) documented increases of pH by 8.5 during dinoflagellate blooms in Narraganset Bay (Rhode Island). Although the values of CO_2 concentration during these algal bloom episodes were not published, it can be calculated that these pH changes dealt with reduction of the CO_2 concentration by approx. 90% with respect to pH 8.2. The pH can be high for several months in some shallow and semi-enclosed water bodies. Thus, Macedo et al. (2001) reported pH above 8.7 for part of the annual cycle in Santo André coastal Lagoon (Portugal). Similarly, pH above 9 (with peaks of 9.75) has been described from May to August in Mariager Fjord in Denmark (Hansen, 2002).

The estuaries are also water bodies in which the CO_2 concentration often differs with respect to the air-equilibrated seawater concentration, although normally the release of CO_2 by respiration exceeds its consumption by photosynthesis (Gattuso et al., 1998), and consequently the annually-averaged CO_2 concentration is over its equilibrium value. However, in these systems (and other shallow water bodies), where the benthic macrophytes are the dominant algal species, the diurnal variability could become more relevant than the seasonal one. Thus, Yates et al (2007) documented that the range of diurnal variation for pH linked to photosynthetic activity in Tampa Bay (a shallow tidal estuary on the west coast of Florida) was 93% of the seasonal range of variability. Similarly, Middelboe and Hansen (2007) described peaks of pH above 9 at midday in shallow water sites located at the Danish shore. Values of pH above 10 and CO_2 concentration lower than 1 μM, have been described during the diurnal cycle in the rockpools that occupies the upper inter-tidal fringe (Larsson et al., 1997). In these systems, the diurnal pH changes are coupled with the tidal cycle as the pH peaks are produced when low tide coincides with high solar radiation. In turn, the coastal systems also arrange habitats where some seaweeds find a CO_2 enriched medium. It occurs for a number of species inhabiting crevices at the upper inter-tidal fringe. These species spend part of their dial cycle emersed and receive continuous splashes of water. Photosynthesis under these conditions favours the use of air-CO_2 because of the pH of the water film surrounding the thallus arises rapidly and produces a CO_2 gradient between the atmosphere and the plant surface several times higher than when submerged (Mercado and Niell, 2000). Similarly, the CO_2 concentration keeps over its equilibrium value around stems and roots of vascular plants or in the proximity of heterotrophic organism communities, where some alga species grow strongly attached (Raven, 1997). From these data, it is clear that pH and CO_2 concentration could significantly differ with respect to the CO_2 global environment, which must be taken into account in studying the possible limitation of the photosynthesis by CO_2 in each particular habitat.

2.2. Kinetic of the Rubisco and Use of HCO_3^-

The key enzyme implied in the CO_2 fixation is the ribulose-biphosphate carboxylase oxygenase (Rubisco), which uses CO_2 as substrate for its carboxylation reaction. Besides, the enzyme combines with O_2 to initiate photorespiration. Classically, the photorespiration is viewed as a process that decreases the energetic efficiency of the photosynthesis (in terms of fixed carbon per absorbed light quantum mol). Two forms of Rubisco have been described in alga (forms I and II) which differ in their affinity to CO_2 and O_2 (see Badger et al., 1998 for a review about the molecular and kinetic properties of the different algal Rubisco families).

Table 1. Values of photosynthetic semi-saturation constant of the photosynthesis for CO_2 ($K_{0.5(CO2)}$) by different species and algal groups.

Taxonomic group	$K_{0.5(CO2)}$ (μM)	References
Cyanobacteria		
Synechocystis	0.2	Benschop et al., 2003
Diatoms		
Phaeodactylum tricornutum	0.3 – 4.7	Matsuda et al., 2001; Burkhardt et al., 2001
Thalassiosira weissflogii	2.5	Burkhardt et al., 2001
Skeletonema costatum	0.3 – 2.7	Rost et al., 2003
Coccolithophorids		
Phaeocystis globosa	1.5	Rost et al., 2003
Pleurochrysis	2.5 – 5.0	Israel and González, 1996
Emiliania huxleyi	27.3-12.5	Nimer et al., 1994; Rost et al., 2003
Dinoflagellates		
Prorocentrum minimum	4	Rost et al., 2006
Prorocentrum micans	2.5	Nimer et al., 1999
Heterocapsa triquetra	12	Rost et al., 2006
Seaweeds		
Green-macroalgae	1.61 – 3.19	Mercado et al., 1998a
Brown-macroalgae	2.3 - >13.5	Mercado et al., 1998a
Red-macroalgae	3.1 - >13.5	Mercado et al., 1998a

However, the carboxylation reaction by all the Rubisco forms is relatively un-efficient (probably due to its early evolutionary origin, Raven, 1991), since its semi-saturation constant for CO_2 ranges from 30 to 250 µM, i.e. it is higher than the CO_2 concentration in air-equilibrated seawater (in addition to note that the biological activity in some habitats can deal with lower CO_2 concentrations and relatively high O_2 concentration which favours photorrespiration). Consequently, on average it could be expected that the CO_2 concentration is not enough to saturate the photosynthesis rates in the present ocean. However, the kinetic properties of the purified Rubisco could hardly explain the high photosynthetic affinity to CO_2 described in terms of CO_2 semi-saturation constants for the photosynthesis ($K_{0.5(CO2)}$) in many alga species (Table 1). The values of $K_{0.5(CO2)}$ shown in Table 1 were obtained in the laboratory by determining the response of the photosynthesis to increasing CO_2 doses at pH 8.0-8.2 and under light saturation. There are values of $K_{0.5(CO2)}$ published for species of the main groups of marine phytoplankton (cyanobacteria, diatoms, dinoflagellates and coccolithophorids) and for the three main taxonomic groups of seaweeds (i.e. red, green and brown macroalgae). However, it has to be noted that the number of marine phytoplankton species analysed is relatively scarce in comparison to the amount of available data for macroalgae, possibly because the macrophytes are more easily isolated from their natural habitat. The analysed marine cyanobacteria feature values of $K_{0.5(CO2)}$ for photosynthesis around 0.5 µM, that are remarkably lower than $K_{0.5(CO2)}$ of the cyanobacterial Rubisco (about 250 µM according to Andrews and Abel, 1981). This high affinity to CO_2 means that photosynthesis is saturated at the CO_2 concentration in seawater at pH 8.0-8.2 (i.e. at about 10-15 µM CO_2). The same thing can be said for the few marine diatom species analysed since their $K_{0.5(CO2)}$ for photosynthesis varies from 0.3 to 5.0 µM (Table 1). The values of $K_{0.5(CO2)}$ reported for dinoflagellates are some more variable as they range from 2.0 to 12 µM. This latter value would imply that some dinoflagellates species photosynthesise under sub-saturating CO_2 concentration (Colman et al., 2002; Dason et al., 2004), however this statement has been recently questioned by Rost et al. (2006) who reported $K_{0.5(CO2)}$ substantially lower than 10-12 µM for some species of this group. The available data for coccolithophorids indicate a photosynthetic affinity to CO_2 lower than for cyanobacteria and diatoms (Table 1), although the values of $K_{0.5(CO2)}$ for photosynthesis reported are still lower than the semi-saturation constant for Rubisco form I (the one presents in this alga group). According to the data collected in Table 1, photosynthesis by *E. huxleyi* (the most important coccolithophorid in vast ocean areas) at pH 8.0-8.2 under un-limited light could be lower by 50-80% than the photosynthesis rate at CO_2 saturation. The photosynthetic affinity to CO_2 is quite variable within the seaweeds. Thus, all the researched green macroalgae have $K_{0.5(CO2)}$ substantially lower than the semi-saturation constant of the Rubisco form II. However, some red macroalgae exhibit values of $K_{0.5(CO2)}$ for photosynthesis which do not differ substantially with respect to the value expected from the Rubisco kinetic. On overall, about 70% of the analysed seaweeds feature photosynthesis rates saturated at 10-15 µM of CO_2 (Maberly, 1990; Surif and Raven, 1990; Uusitalo et al., 1990; Haglund et al., 1992; Börjk et al., 1992; Mercado et al. 1998a). The degree of photosynthesis saturation for the rest of species (which includes both brown and red macroalgae) is very variable, particularly among the red macroalgae. For this group, low photosynthetic affinity to CO_2 appears to be more frequent among some families of the class Florideophyceae (Raven et al., 2002). This finding indicates that the photosynthetic affinity to CO_2 can vary substantially among families (or even genera) belonging to the same phylum. This variability probably also occurs within the main phytoplankton groups, although the low amount of analysed species prevents to support this hypothesis.

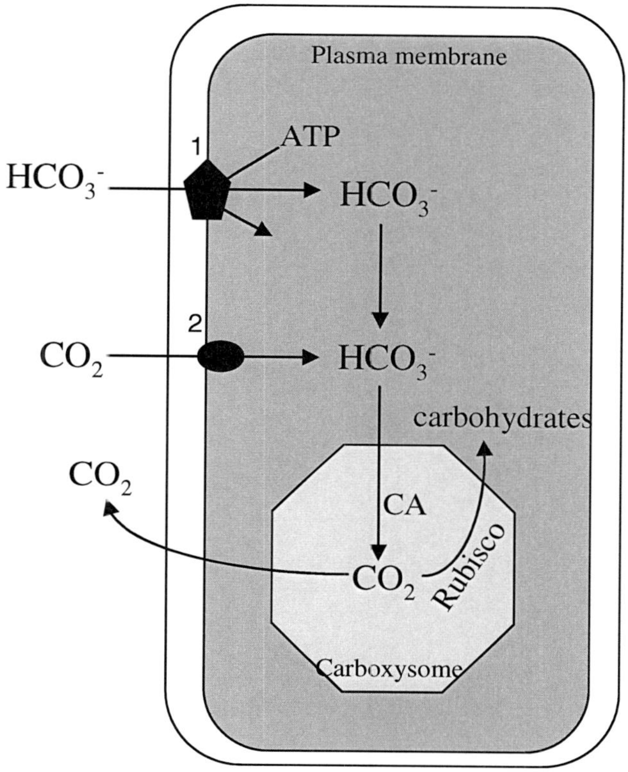

Figure 2. A model for the inorganic carbon transport and CO_2 accumulation in cyanobacteria. The model represents two types of inorganic carbon transport, HCO_3^- (1) and CO_2 (2) transport (note that different transporters for both inorganic carbon sources have been described in cyanoacteria). Besides, the presence of internal carbonic anhydrase in the carboxysime is assumed. Modified from Giordano et al., (2005).

The physiological reason for this variability in photosynthetic affinity to CO_2 is that many algal species depict a mechanism that permits to increase the CO_2 concentration around the Rubisco (the so-called carbon concentrating mechanism, CCM). The functioning of the CCMs has been extensively studied in cyanobacteria and microalgae (see Colman et al., 2002, and Badger, 2003, for recent reviews) and poorly in macroalgae. The molecular components of the CCMs differ among the main algal taxonomic groups, likely due to their diverse polyphyletic origin throughout evolution. However, all the CCMs are based on the use of HCO_3^- for photosynthesis as a source of DIC* (note that the HCO_3^- concentration in seawater at pH 8.2 is about 100 times higher than the CO_2 concentration). An internal form of the enzyme carbonic anhydrase (CA_{int}) is other component that appears to be present in all the

* The presence of CCMs based on the accumulation of a C_4 intermediate that is decarboxylate in the chloroplast to provide CO_2 for Rubisco (C_4-like metabolism) has been suggested in different algae, although the evidence is weak (see Giordano et al. 2005, for a complete discussion).

CCMs. CA_{int} catalyses the reversible conversion of HCO_3^- to CO_2 inside the cells (Figures 2 and 3) and then contributes to arrange a CO_2 enriched ambient around the Rubisco. The use of HCO_3^- can occur directly by means of an active transport mechanism located at the plasma-membrane outside and/or indirectly after its external transformation into CO_2 that afterwards penetrates the cell by diffusion or active transport. The conversion of HCO_3^- into CO_2 is performed by a external CA isoenzyme (CA_{ext}) attached to the plasma-membrane or cell wall. The figures 2 and 3 schematise the functioning of the different inorganic carbon transport modes and CO_2 accumulation mechanisms in cyanobacteria and eukaryotic algae

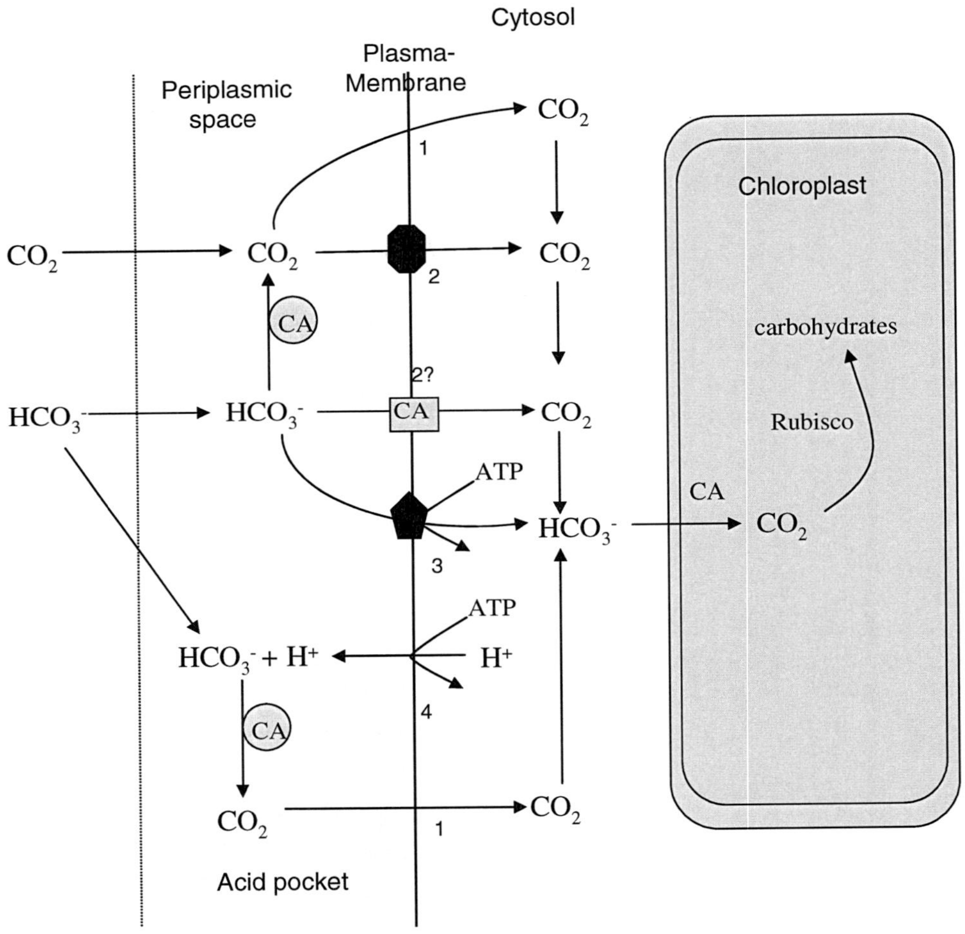

Figure 3. Models for the inorganic carbon transport and CO_2 accumulation in eukaryotic algae. Four different transport systems are represented: 1. Diffusive entry of CO_2; 2. CO_2 active transport (including the hypothetical possibility of CO_2 transport mediated by CA-like activity, 2?); 3. HCO_3^- active transport; 4. H^+ pumping that generates acid pockets where transformation of HCO_3^- into CO_2 is favoured (Mercado et al. 2006). The model assumes that the inorganic carbon is accumulated in the form of HCO_3^- inside the cell, and that HCO_3^- is transformed into CO_2 by an internal carbonic anhydrase (CA) within the chloroplast.

The extension at which HCO_3^- uptake and/or CO_2 entry (via diffusive or active transport) contributes to the DIC flux toward inside the cells is variable. Thus, CA_{ext} appears to be absent in cyanobacteria (Figure 2) that use HCO_3^- and CO_2 by means of active transport (Price et al., 2002; Shibata et al., 2002; Badger, 2003). In contrast, CA_{ext} is implied in the HCO_3^- use in some diatom species (Colman and Rotatore, 1995; Rotatore et al., 1995; Morel et al., 2002) although the most important mechanism within this group appears to be the direct uptake of HCO_3^- (Figure 3). The HCO_3^- use capacity has been questioned in dinoflagellates, however recent evidence indicates that some species use HCO_3^- by means of CA_{ext}. Besides, Rost et al. (2006) demonstrated that HCO_3^- uptake was predominant in three dinoflagellate species. The uptake of HCO_3^- has been inferred in some green macroalgae by determining the effects on the photosynthesis of an inhibitor for a plasma-membrane anion exchanger (reviewed by Larsson and Axelsson, 1999). However, the HCO_3^- use is mediated by CA_{ext} in the most of seaweeds studied for which high photosynthetic affinity to CO_2 has been described (Cook et al., 1986; Axelsson and Uusitalo, 1988; Smith and Bidwell, 1989; Surif and Raven, 1989; Maberly, 1990; Björk et al. 1992; Haglund et al., 1992; Johnston et al., 1992; Mercado et al., 1997a, 1998a). It is interesting note that CA_{ext} activity has been also reported in some red seaweeds altough they exhibit photosynthesis rates non-saturated at normal concentration (Mercado and Niell, 1999). This result involves that the only presence of CA_{ext} does not guarantee photosynthesis saturation at the normal pH in seawater. In other words, the external CA must be operating in association to some kind of *additional mechanism* to provide a high photosynthetic affinity to CO_2 (Mercado et al., 1997b). In fact, different reports indicate that the CA_{ext} operates jointly with H^+ pumps that contribute to create acid regions outside the cells in some macroalgae (Mercado et al., 2006). Figure 3 illustrates the possible functioning of this mechanism.

In summary, the available data indicate that the photosynthetic affinity to CO_2 (and therefore the saturation degree of the photosynthesis at pH 8.0-8.2, is primary determined by the capacity of each species for using HCO_3^-, which is a feature determined evolutionary (although it does not preclude that the species are able to modulate their CCMs in response to changes in CO_2 concentration).

2.3. CO_2 Limitation into an Ecological Context

The data shown in Table 1 could be extrapolated to the natural strains in order to research the *in situ* saturation degree of the photosynthesis, provided the conditions under which photosynthesis occurs in the habitat are comparable with the laboratory conditions under which the $K_{0.5(CO_2)}$ values were obtained (i.e. other resources like the light being non-limited). Furthermore, it could expected that the photosynthetic affinity to CO_2 in mixed communities reflects the DIC acquisition kinetic of the dominant species or taxonomic group. In fact, a few reports performed with natural phytoplankton assemblages demonstrate that the importance of the direct HCO_3^- uptake relative to HCO_3^- use via CA_{ext} is related with the dominance of diatoms relative to other taxonomic groups (Tortell et al., 2000; Tortell and Morel, 2002)

On average, the phytoplankton community composition in the open ocean and the coastal areas differs both functionally and taxonomically. Thus, in stratified oligotrophic regions, the phytoplankton is dominated by pico-plankton (i.e. cells of 0.2 to 2 μm in size) which is mainly composed of cyanobacteria and a diverse eukaryotic cell group characterised by high abundance of Chlorophytes of the class Prasinophycea. The CO_2 concentration in these oceanic regions is relatively constant; besides the cyanobacteria have capacity for HCO_3^- efficient use. Therefore, CO_2 could hardly become a factor limiting the photosynthesis in these communities. However, semi-oligotrophic conditions deal with the formation of coccolithophorid blooms in open sea The blooms occur mainly during summer at high latitudes (Iglesias-Rodríguez et al., 2002), and are normally dominated by *Emiliania huxleyi* (although *Gephyrocapsa oceanica* also can form blooms). Although substantial changes in the CO_2 concentration are not produced during the different bloom phases (Robertson et al., 1994), the growth of the coccolithophorid dominated communities could be limited by the CO_2 concentration since the photosynthetic affinity to CO_2 by this group is relatively low (Table 1).

In the coastal areas, high regimes of mixing and eutrophic conditions favour the dominance of diatoms (Yallop, 2001). These conditions are fulfilled when deep-water upwelling events normally fuelled by wind stress are produced. The fertilization of the surface layer deals with blooms of diatoms that are usually replaced by dinoflagellates when calm conditions are re-established. Although the phytoplankton growth stimulation produces a measurable decrease of the CO_2 concentration, the high photosynthetic affinity to CO_2 of diatoms and dinoflagellates assures that CO_2 is not a nutrient limiting the photosynthesis in these communities. By contrast, in shallow and semi-closed water bodies, the CO_2 concentration reduction is more substantial since this could drop below the $K_{0.5(CO_2)}$ values reported for diatoms and dinoflagellates. The CO_2 depletion is particularly extreme during red tide episodes which are produced by dinoflagellates. Therefore, CO_2 could become a factor limiting photosynthesis in these communities. However, both diatoms and dinoflagellate have capacity to increase their photosynthetic affinity to CO_2 in response to decreasing CO_2 concentration. Thus, acclimation experiments at low CO_2 (or high pH) performed with dinoflagellates demonstrate a reduction of $K_{0.5(CO_2)}$ for photosynthesis after 5 day acclimation (Ros et al., 2006). Similar results have been described for different diatom species (Goldman, 1999; Burkhardt et al., 2001) although in this case acclimation was produced within 48 h. These results are interpreted as the growth at low CO_2 (or at high pH) induces over-expression of the CCMs. Normally, the CA_{ext} is one of the CCM elements implied in the acclimation process (Nimer et al., 1997, 1998; Hobson et al., 2001; Harada et al., 2005). The few researched macroalgae exhibit also capacity to increase their CO_2 photosynthetic affinity at low CO_2 concentration. However, Larsson et al. (1997) demonstrated that photosynthesis was severely hampered at midday in the rockpools subject to CO_2 extreme conditions in spite of they are dominated by green macroalgae with relatively low values of $K_{0.5(CO_2)}$.

As obtained in the laboratory, it should be expected that the acclimation mechanisms at low CO_2 are operating in the natural habitats. Concordantly, Berman-Frank and Zohary (1994) demonstrated that CA activity increased in natural communities dominated by the dinoflagellate *Peridinium gatunense* when the DIC concentration declined during the development of a bloom. Tortell et al. (2006) and Martin and Tortell (2006) also obtained a correlation between CA_{ext} activity and CO_2 partial pressure in different assemblages of phytoplankton, although it was not cleared if the inter-community variability in carbon acquisition resulted from a direct response to the changing CO_2 concentration or from differences in taxonomic composition, as both variable usually co-varies in the natural habitat.

On the other hand, it could be expected that CO_2 is a limiting growth factor in the macroalgae with low (or absent) capacity to use HCO_3^-. However, Raven et al. (2002) questioned this assessment since most of these species inhabit crevices in the upper inter-tidal fringe or sub-tidal habitats. In these shaded mediums, the light is much reduced and probably limits the algal growth. Several reports demonstrate that the CO_2 concentration that is necessary to reach the maximal photosynthesis rates at a given low light intensity is lower than at high light intensity (Mercado et al., 2000a). Therefore, the CO_2 could play a minor role in limiting the photosynthesis by these species. Besides, some other macroalga species apparently lacking a CCM grow attached to stems and/or roots of vascular plants or in association with heterotrophic organisms, where the respiratory CO_2 produces CO_2 enriched micro-habitats. There are not available data about the CO_2 concentration actually available for photosynthesis in these habitats; however Mercado et al. (1998b) calculated the impact of the respiratory CO_2 released by a bryozoan (*Electra pilosa*) on the photosynthesis of the red macroalga *Gelidium sesquipedale*. The bryozoan grows tightly attached to the alga thallus that features a light-saturated photosynthesis rate limited at the CO_2 concentration in pH 8.0-8.2 seawater. According to Mercado et al. (1998b), the use of the bryozoan respiratory CO_2 increases the photosynthesis rate of the alga by 40% with respect to the rate reached after removing the bryozoan.

In summary, the photosynthesis rates for many alga species could be non-sensitive to decrease of the free-CO_2 concentration because they have capacity to up-regulate their CCMs. For the species with low (or absent) capacity for using HCO_3^-, other habitat-dependent factors should be taken into account to asses if or not CO_2 could be limiting photosynthesis. The photosynthesis limitation by CO_2 is only feasible in some particular phytoplankton communities dominated by coccolithophorids or in some macroalgae communities subject to extreme CO_2 conditions. However, these conclusions are based on very short-term experiments while acidification would induce a long-term response, which is not necessary similar. In other words, saturation of the photosynthesis at the actual CO_2 concentration in many alga species does not preclude they are sensitive to CO_2 concentration increases since the effects of high CO_2 on photosynthesis are not necessary coupled with the growth response (Beardall and Raven, 2004).

3. Testing the Effects of the Acidification on Photosynthesis and Growth

3.1. Laboratory Experience

There are a few experiments describing the effects of high CO_2 on growth in marine algae. Most of these experiments have been performed in bottles where the CO_2 concentration is modified by bumbling CO_2-enriched air or by diminishing the medium pH. These techniques have several problems which are not easy to overcome. For example, CO_2 enrichment normally deals with changes in pH and HCO_3^- concentration whose magnitude depends on the biological activity within the experimentation bottles and the aeration rate. Furthermore, sometimes the CO_2 level used in the experiments is un-naturally high (although normally the assays are performed at CO_2 concentration doubled with respect to the present value, i.e. the conditions expected by the year 2100 are tested). For these reasons, the extrapolation of the laboratory results to the field must be done with caution.

Theoretically, the high CO_2 should stimulate the growth in the alga species that are not able to use HCO_3^- for photosynthesis while the effects should be more limited in algae featuring a CCM. The effects of high CO_2 have been studied only in one species dependent on CO_2 diffusive entry, the brown macroalga *Lomentaria articulata*, while most of the researched species are HCO_3^- users. As expected, the growth rates were stimulated in *Lomentaria articulata* cultured at high CO_2 (Kübler et al., 1999). In contrast, the growth response among the HCO_3^- user algae is variable. The only common response for all these species consists of down-regulation of the CCM components, particularly a decline of the CA activity (see Giordano et al., 2005, for a review). Normally, these re-adjustments of the CCM do not deal with changes in the growth rates as could be also expected from the short-term photosynthetic response. Thus, non-stimulation of the growth by high CO_2 has been demonstrated for the cyanobacteria *Prochlorococcus* (Fu et al., 2007), the diatoms *Phaeodactylum tricornutum* (Laws et al., 1997), *Skeletonema costatum* and *Thalassiosira weissflogii* (Burkhardt et al., 1999) and the dinoflagellate *Prorocentrum minimum* (Fu et al., 2008). Similarly, the culture in high CO_2 did not affect the growth rates by different red, green and brown macroalgae (Israel and Hophy, 2002; Mercado et al., 2000b). On the contrary, increases of the growth rates from 33% to 100% in high CO_2 have been described in some other HCO_3^- users. Thus, Yu et al. (2006), Hutchins et al. (2007) and Fu et al (2007, 2008) reported increased growth in the cyanobacteria *Trichodesmium* and *Synechococcus* and the diatom *Nitzschia closterium* (although the CO_2 concentration assayed for this last species was fairly high, ca. 5% CO_2). The growth at high CO_2 was also stimulated in the macroalgae *Ulva lactuca* (Gordillo et al., 2003), *Hizikia fusiforme* (Zou et al., 2003) and *Gracilaria* sp (Andría et al., 1998). By contrast, high CO_2 affected negatively the growth of the red macroalga *Porphyra leucosticta* (Mercado et al., 2000b). Negative growth response to high CO_2 has been also described for the coccolithophorid *Emiliania huxleyi* (Nimer, 1994).

The reason for this diversity of growth responses to high CO_2 is not clear although that should be related with the effects of the CO_2 on other cellular components in addition to CCM. Particularly, it has been demonstrated that high CO_2 affects negatively to the calcification rates in coccolithophorids (Riebesell et al., 2000; Engel et al., 2005), which explains the adverse effects on growth for this group. For other species with active transport systems of CO_2 and/or HCO_3^-, it has been speculated that high CO_2 should stimulate the growth since less energy is invested in DIC acquisition (Beardall and Raven, 2004), provided CO_2 diffusive entry vs. DIC active uptake is favoured. As demonstrated by Kaplan and Reinhold (1999), Badger and Spalding (2000) and Colman et al. (2002), the extra ATP that is necessary for functioning of the HCO_3^- and/or CO_2 transport systems (beyond that required for functioning of the Calvin Cycle) is provided from photosynthesis in most microalgae (a dependence of DIC transport on the mitochondrial ATP has been demonstrated only in a few species; Huertas et al., 2002). In fact, a positive relationship between the activation state of the CCM and the growth light intensity has been reported (Shiraiwa and Miyachi, 1985; Berman-Frank et al., 1998; Kübler and Raven, 1994; Rost et al., 2006), which has been ascribed to the ATP necessary for DIC uptake only can provide under conditions of light non-limitation. Therefore, it could be expected that acclimation to high CO_2 produces re-adjustment of the photosynthetic apparatus to fit the change in energy demand. The symptoms of re-adjustments in the photosynthetic apparatus consist of changes in the pigment cell content, steochiometry of photosystems and/or quantum efficiency of the photosynthesis (i.e. the number of photons necessary to fix a mol of DIC). There are scarce published data about the effects of high CO_2 on the photosynthetic performance (other than the effects on the CCM components). Nevertheless, variations of accessory pigment relative to chlorophyll a content have been described in some species grown at high CO_2, but the effects are non-uniform. Thus, the accessory pigment content increased in *Synechoccoccus* (Fu et al., 2007) while that was un-affected in *Prochlorococcus*. On contrary, accessory pigments decreased in the red macroalgae *Gracilaria* and *Porphyra* at high CO_2 (García-Sánchez et al., 1994; Mercado et al., 2000b). Photosystem re-adjustment at high CO_2 has been described in *Synechococcus* (Rost et al., 2006). On the other hand, Fu et al. (2008) reported increase of the chlorophyll a photosynthetic efficiency at high CO_2 in *Heterosigma* and *Prorocentrum*. Therefore, the available data suggest that high CO_2 will affect the photosynthetic performance of many alga species, although a common pattern of acclimation can be not depicted at present.

High CO_2 also induces changes in some other cellular components. For example, high CO_2 stimulated the accumulation of carbohydrates in some macroalgae (Mercado et al., 2000b; Gordillo et al., 2001) as well the synthesis of lipids in some phytoplankton species (Pratt and Johnson, 1964). On the contrary, proteins increased and carbohydrates decreased in *Chaetoceros* sp. growing at high CO_2 (Araujo et al., 2005). According to Burkhardt and Riebesell (1997), the change in allocation of the DIC assimilated to intracellular carbon pool should produce alterations in the carbon relative nitrogen cellular content (i.e. in the C:N ratio of the organic matter). Concordantly, these authors observed a decrease of the C:N ratio at high CO_2 in different phytoplankton species. However, the C:N ratio increased in the macroalgae *Ulva rigida* and *Porphyra leucostica* at high CO_2 (Mercado et al., 2000a; Gordillo et al., 2001).

3.2. From Laboratory to Field. Effects of the CO_2 Changes on Photosynthesis in Natural Communities

The presented data indicate that the effects of high CO_2 could be very variable among different species. Besides, it has been demonstrated that the acclimation response to high CO_2 for a particular species could differ depending on the growth light intensity and the nutritional status. For example, Sciandra et al. (2003) demonstrated that different CO_2 treatments had a limited effect on the ratio of calcification to organic carbon production in *Emiliania huxleyi* growing at low light intensity (in contrast to that occurred at high light intensity). Gordillo et al. (2003) also demonstrated that the effect of high CO_2 on the growth rate in *Ulva* was much reduced in N-starvation conditions compared to N-repletion. The growth conditions (light intensity and nutrient concentration) in the high CO_2 experiments discussed above are often not clearly defined with respect to the optimal conditions for each studied species. Therefore, it can be not discarded that the apparent diversity of responses to high CO_2 among the examined algae results, at least partially, from the fact of growth conditions used in different laboratories being not fully comparable.

In spite of this, it is clear that acidification will favour the growth of same species and consequently shifts in the taxonomic structure of the communities would be expected. The only available experiments examining the effects of high CO_2 at community level have been performed with natural assemblages of phytoplankton, while there are not available data for benthic alga communities. The early experiments performed by Tortell et al. (1997, 2000, 2002a) consisted of small-scale incubations at high CO_2 lasting a few days (2-4 d). On general, the treatments of high CO_2 did not affected the growth or taxonomic composition of the communities, in spite of significant changes in the CCMs were produced. Tortell et al. (2000) argued that non-modification of the phytoplankton composition in their experiments could be due to the rather short duration of the incubations. Concordantly, experiments of longer term (11 d) revealed important changes in the taxonomic composition of the community (Tortell et al., 2002b), although primary productivity and phytoplankton net growth were also not affected. In more recent experiments, Engel et al. (2005) tested the effects of CO_2 enrichment on the development of a bloom of coccolithophorids that was induced in mesocosm experiments performed in a Fjord in Norway. The CO_2 did not affect the succession dynamics of the phytoplankton communities in the mesocoms, and finally a community dominated by coccolithophorids was obtained irrespective of the CO_2 level. Similar results were reported by Riebesell et al. (2007) who also investigated the effects of high CO_2 on the phytoplankton succession in mesocosm experiments. Although from these experiments it can be concluded that neither phytoplankton composition nor succession were significantly affected, the CO_2 treatment affected other relevant processes at ecosystem level Particularly, a higher loss of organic carbon from the upper layer was produced at high CO_2 that means CO_2 stimulated the organic carbon production and its transfer to other trophic levels, (i.e. the so-call biological pump, Figure 1).

4. Conclusion

Most of the algae have developed carbon concentrating mechanisms that increase the CO_2 concentration around the Rubisco. Consequently, the photosynthesis by these species is not sensitive to changes in CO_2 concentration as that due to the ocean acidification. Only a reduced group of species will potentially increase their photosynthesis rates at CO_2 concentration higher than the present one. This species group includes the coccolithophorids that dominate vast areas of the ocean during part of the seasonal cycle. Photosynthesis could be also limited in some benthic macroalga communities that are sporadically subject to pH and CO_2 extreme conditions. For some other brown and red macroalgae lacking a CCM, *in situ* photosynthesis is limited by the light intensity instead of the CO_2 concentration. In spite of the presence of a CCM, the acclimation at high CO_2 affects the growth rates, the photosynthetic performance and the protein and carbohydrate contents in different species. This finding suggests that acidification could favour the growth of some species and consequently the acidification would produce changes in the succession patterns. However, experiments performed with natural phytoplankton communities indicate that neither phytoplankton composition nor the succession patterns were significantly affected by the high CO_2. In any case, the available experimental evidence is very scarce and consequently the effects of the ocean acidification at community and ecosystem level are basically unknown.

Acknowledgments

This work has been supported from the Project NITROALBORAN (CTM2006-00426) funded from Education and Science Ministry of Spain and co-funded from FEDER, EU.

References

Andrews, T. J., & Abel, K.M. (1981). Kinetics and subunit composition of RUBISCO from the cyanobacterium, *Synechococcus* sp. *Journal of Biological Chemistry,* **256**, 8445-8451.

Axelsson, L., & Uusitalo, J. (1988). Carbon acquisition strategies for marine macroalgae. I. Utilization of proton exchanges visualized during photosynthesis in a closed system. *Marine Biolology,* **97**, 295-300.

Badger, M. R. (2003). The roles of carbonic anhydrases in photosynthetic CO_2 concentrating mechanisms. *Photosynthesis Research* **77**, 83-94.

Badger, M. R., & Spalding, M. H. (2000). CO_2 acquisition, concentration and fixation in cyanobacteria and algae. In R. C. Leegood, T. D. Sharkey T. D. & S. von Caemmerer S. (Eds.), *Photosynthesis: physiology and metabolism.* (367-397). Dordrecht, The Netherlands: Kluwer.

Badger, M. R., Andrews, T. J., Whitney, S. M., Ludwig, M., Yellowlees, D. C., Leggat, W., & Price, P. G. (1998). The diversity and co-evolution of Rubisco, plastids, pyrenoids and chloroplast-based CCMs in the algae. *Canadian Journal of Botany,* **76**, 1052-1071.

Beardall, J., & Raven, J. A. (2004) The potential effects of global climate change on microalgal photosynthesis, growth and ecology. *Phycologia,* **43**, 26-40.

Berman-Frank, I., & Zohary, T. (1994). CO_2 availability, carbonic anhydrase, and the annual dinoflagellate bloom in Lake Kinneret. *Limnolology & Oceanography* **39**(8), 1822-1834.

Berman-Frank, I., Erez, J., & Kaplan, A. (1998). Changes in inorganic carbon uptake during the progression of a dinoflagellate bloom in a lake ecosystem. *Canadian Journal of Botany* **76**(6), 1043-1051.

Björk, M., Haglund, K., Ramazanov, Z., García-Reina, G., & Pedersen, M. (1992). Inorganic-carbon assimilation in the green seaweed *Ulva rigida* C. Ag. (Chlorophyta). *Planta,* **187,** 15-46.

Borges, A.V., Schiettecatte, L.-S., Abril, G., Delille, B., & Gazeau, F. (2006). Carbon dioxide in European coastal waters. *Estuarine Coastal and Shelf Science,* **70**, 375-387

Burkhardt, S., & Riebesell, U. (1997). CO_2 availability affects elemental composition (C:N:P) of the marine diatom Skeletonema costatum. *Marine Ecology Progress Series,* **155**, 67-76.

Burkhardt, S., Zondervan, I., & Riebesell, U. (1999). Effect of CO_2 concentration on C:N:P ratio in marine phytoplankton: A species comparison. *Limnology and Oceanography,* **44**, 683-690.

Burkhardt, S., Amoroso, G., Riebesell, U., & Sültemeyer, D. (2001). CO_2 and HCO_3^- uptake in marine diatoms acclimated to different CO_2 concentrations. *Limnology and Oceanography,* **46**, 1378-1391.

Cai, W. J., & Dai, M. (2004). Comment on "enhanced open ocean storage of CO_2 from shelf sea pumping". *Science* **306**, 1477 doi: 10.1126/science.1102132.

Caldeira. K., & Wickett, M. E. (2003). Anthropogenic carbon and ocean pH. *Nature,* **425**, 365.

Colman, B., & Rotatore, C. (1995). Photosynthetic inorganic carbon uptake and accumulation in two marine diatoms. *Plant Cell Environment,* **18**, 919-924.

Colman, B., Huertas, I. E., Bhatti, S. & Dason, J. (2002). The diversity of inorganic carbon acquisition mechanisms in eukaryotic microalgae. *Functional Plant Biology,* **29**, 261-270.

Cook, C. M., Lanaras, T., & Colman, B. (1986). Evidence for bicarbonate transport in species of red and brown macrophytic algae. *Journal of Experimental Botany,* **39**, 1185-1198.

Dason, J. S., Huertas, I. E. & Colman, B. (2004). Sources of inorganic carbon for photosynthesis in two marine dinoflagellates. *Journal of Phycology,* **40**, 285-292.

Engel, A., Zondervan, I., Aerts, K., Beufort, L., Benthien, A., Chou, L., Delille, B., Gattuso, J.-P., Harlay, J., Heeman, C., Hoffman, L., Jacquet, S., Nejsgaard, J., Pizay, M.-D., Rochelle-Newall, E., Schneider, U., Terbruggen, A., & Riebesell, U. (2005). Testing the direct effect of CO_2 concentration on a bloom of the coccolithophorid *Emiliania huxleyi* in mesocosm experiments. *Limnology & Oceanogrphy* **50**, 493-507.

Falkowski, P. G. (1997). The paradox of carbon dioxide efflux. *Current Biology,* **7,** R637-R639.

Falkowski, P. G., & Raven, J. A. (1997). Aquatic photosynthesis. Oxford, Blackwell Scientific Publishers.

Falkowski, P. G., Katz, M. E., Knoll, A. H., Quigg, A., Raven, J. A., Schofield, O., & Taylor, F. J. R. (2004). The Evolution of Modern Eukaryotic Phytoplankton. *Science,* **305**, 354-360.

Fu, F.-X., Warner, M. E., Zhang, Y. H., Feng, Y. Y., & Hutchins, D. A. (2007). Effects of increased temperature and CO_2 on photosynthesis, growth, and elemental ratios in marine *Synechococcus* and *Prochlorococcus* (Cyanobacteria). *Journal of Phycology,* **43**, 485-496.

Fu, F.-X., Zhang, Y., Warner, M. E., Feng, Y., Sun, J., & Hutchins, D. A. (2008). A comparison of future increased CO_2 and temperature effects on sympatric *Heterosigma akashiwo* and *Prorocentrum minimum. Harmful algae,* **7,** 76-90.

García-Sánchez, M. J., Fernández, J. A., & Niell, F. X. (1994). Effect of inorganic carbon supply on the photosynthetic physiology of Gracilaria tenuistipitata. *Planta,* **194**, 55-61.

Gattuso, J.-P., Frankignoulle, M., & Wollast, R. (1998). Carbon and carbonate metabolism in coastal aquatic ecosystems. *Annual Review of Ecological Systematics,* **29**, 405-433.

Giordano, M., Beardall, J., & Raven J. A. (2005). CO_2 concentrating mechanisms in algae: mechanisms, environmental modulation and evolution. *Annual Review of Biology,* **56**, 99-131.

Goldman, J. C. (1999). Inorganic carbon availability and the growth of large marine diatoms. *Marine Ecology Progress Series,* **180**, 81-91.

Gordillo, F. J. L., Jiménez C., Goutx, M., & Niell, X. (2001). Effects of CO_2 and nitrogen supply on the biochemical composition of *Ulva rigida* with special emphasis on lipid class analysis. *Journal of Plant Physiology,* **158**, 367-373.

Gordillo, F. J. L., Figueroa, F. L., & Niell, F. X. (2003). Photon- and carbon-use efficiency in Ulva rigida at different CO_2 and N levels. *Planta,* **218**, 315-322.

Haglund, K., Björk, M., Ramazanov, Z., García-Reina, G., & Pedersen, M. (1992). Role of carbonic anhydrase in photosynthesis and inorganic-carbon assimilation in the red alga *Gracilaria tenuistipitata*. *Planta,* **187**, 275-281.

Hansen, P. J. (2002). Effect of high pH on the growth and survival of marine phytoplankton: implications for species succession. *Aquatic Microbial Ecology,* **28**, 279-288.

Harada, H., Nakatsuma, D., Ishida, M., & Matsuda, Y. (2005). Regulation of the expresión of intracellular beta-carbonic anhydrase in response to CO_2 and light in the marine diatom *Phaedodactylum tricornutum. Plant Physiology,* **139**, 1041-1050.

Hinga, K. R. (1992). Co-occurrence of dinoflagellate blooms and high pH in marine enclosures. *Marine Ecology Progress Series,* **86**, 181-187.

Hobson, L.A., Hanson, C. E., & Holeton, C. (2001). An ecological basisfor extracellular carbonic anyydrase in marine unicellular algae. *Journal of Phycology,* **37**, 717-723.

Huertas, I. E., Espie, G. S., & Colman, B. (2002). The nergy source for CO_2 transport in the marine microalga *Nannochloris atomus*. *Planta* **214**: 947-953.

Huertas, I. E., Navarro, G., Rodríguez-Gálvez, S., & Lubián, L. M. (2006). Temporal patterns of carbon dioxide in relation to hydrological conditions and primary production in the northeastern shelf of the Gulf of Cadiz (SW Spain). *Deep-Sea Research Part II*, **53**, 1344-1362.

Hutchins, D. A., Fu, F. X., Zhang, Y., Warner, M. E., Feng, Y., Portune, K., Bernhardt, P. W., & Mulholland, M. R. (2007). CO_2 control of *Trichodesmium* N_2 fixation, photosynthesis, growth rates, and elemental ratios: implications for past, present, and future ocean biogeochemistry. *Limnology & Oceanography,* **52**, 1293-1304.

Iglesias-Rodríguez, M. D., Brown, C. W., Doney, S. C., Kleypas, J., Kolber, D., Kolber, Z., Hayer, P. K., & Falkowski, P. G. (2002). Representing key phytoplankton functional groups in ocean carbon cycle models: Coccolithophorids. *Global Biogeochemical Cycles,* **16**, doi: 10.1029/2001GB001454.

Israel, A., & González, E. L. (1996). Photosynthesis and inorganic carbon utilization in *Pleurochrysis* sp. (Haptophyta), a coccolithophorid alga. *Marine Ecology Progress Series*, **137**, 243-250.

Israel, A., & Hophy, M. (2002). Growth, photosynthetic properties and amounts of marine macroalgae grown under current and elevated seaweater CO_2 concentrations. *Global Change Biology*, **8**, 831-840.

Kaplan , A., & Reinhold, L. (1999). CO_2 concentrating mechanisms in photosynthetic microorganisms. *Annual Review of Plant Physiology Plant Molecular Biology,* **50**, 539-570.

Key, R. M., Kozyr, A., Sabine, C. L., Lee, K., Wannikof, R., Bullister, J., Feely, R. A., Millero, F., Mordy, C, & Pen, T.-H. (2004). A global ocean carbon climatology: results from GLODAP. *Global Biogeochemical Cycles,* **18**, GB4031.

Kigoshi, K., & Hashitani, T. (1963). The shelf-diffusion coefficients of carbon dioxide, hydrogen carbonate and carbonate ions in aqueous solution. *Bulletin of the chemical Society of Japan* **36**, 1372.

Kübler, J. E., & Raven, J. A. (1994). Consequences of light limitation for carbon acquisition in 3 Rhodophytes. *Marine Ecology Progress Series,* **110**, 203-209.

Kübler, J. E., Johnston, A. M., & Raven, J. A. (1999). The effects of reduced and elevated CO_2 and O_2 on the seweed *Lomentaria artuculata. Plant Cell Environment,* **22**, 1303-1310.

Larsson, C., & Axelsson, L. (1999). Bicarbonate uptake and utilization in marine macroalgae. *European Journal of Phycology,* **34**, 79-86.

Larsson, C. Axelsson L., Ryberg, H., & Beer, S. (1997). Photosynthetic carbon utilization by *Enteromorpha intestinalis* (Chloroplyta) from a Swedish rockpool. *European Journal of Phycology,* **32**, 49-54.

Laws, E. A., Bidigare, R. R., & Popp, B. N. (1997). Effect of growth rate and CO_2 concentration on carbon isotopic fractionation by the marine diatom *Phaedodactylum tricournutum. Limnology & Oceanography,* **42**, 1552-1560.

Maberly, S. C. (1990). Exogenous sources of inorganic carbon for photosynthesis by marine macroalgae. *Journal of Phycology,* **26**, 439-449.

Macedo, M. F., Duarte, P., Mendes, P., & Ferreira, J. G. (2001). Annual variation of enviromental variables, phytoplankton species composition and photosynthetic parameters in a coastal lagoon. *Journal of Plankton Research,* **23**, 719-732.

Martin, C. L, & Tortell, P. D. (2006). Bicarbonate transport and extracellular carbonic anhydrase in Bering Sea phytoplankton assemblages: Results from isotope disequilibrium experiments. *Limnology & Oceanography,* **51**, 2111-2121

Mercado, J. M., Figueroa, F. L., Niell, F. X., & Axelsson, L. (1997a). A new method for estimating external carbonic anhydrase activity in macroalgae. *Journal of Phycology,* **33**, 999-1006

Mercado, J. M., Niell, F. X., & Figueroa, F. L. (1997b). Regulation of the mechanism for HCO3- use by the inorganic carbon level in *Porphyra leucosticta* Thur. in Le Jollis (Rhodophyta). *Planta,* **201**, 319-325.

Mercado, J. M., Gordillo, F. J. L., Figueroa, F. L., & Niell, F. X. (1998a). External carbonic anhydrase and affinity for inorganic carbon in intertidal macroalgae. *Journal of Experimental Marine Biology and Ecology,* **221**, 209-220.

Mercado, J. M., Carmona R., & Niell, F. X. (1998b). Epiphytic bryozoans enhance the inorganic carbon acquisition in *Gelidium sesquipedale. Journal of Phycology,* **34**, 925-927.

Mercado, J. M., & Niell, F. X. (1999). Carbonic anhydrase and use of HCO_3^- in *Bostrychia scorpioides* (Ceramiales, Rhodophyceae). *European Journal of Phycology,* **34**, 13-19.

Mercado, J. M., & Niell, F. X. (2000). Uptake of CO_2 by *Bostrychia scorpioides* (Rhodophyceae) under merged conditions. *European Journal of Phycology,* **35**, 53-59.

Mercado, J. M., Carmona, R., & Niell, F. X. (2000a). Affinity for inorganic carbon of *Gracilaria tenuistipitata* cultured at low and high irradiance. *Planta* **210**, 758-764.

Mercado, J. M., Gordillo, F. J. L., Niell, F. X., & Figueroa, F. L. (2000b). Effects of diferent levels of CO_2 on photosynthesis and cell component of the red alga Porphyra leucostica. *Journal of Applied Phycology,* **11**, 455-461.

Mercado, J. M., Andrea, J., Pérez-Llorens, J. L., Vergara, J. J., & Axelsson, L. (2006). CO_2 concentrating mechanism in the plasmamembrane of *Laminaria sacharina. Photosynthesis Research,* **88**, 259-268.

Middelboe, A. L., & Hansen, P. J. (2007). High pH in shallow-water macroalgal habiyays. *Marine Ecology Progress Series,* **338**, 107-117.

Morel, F. M., Cox, E. H., Kraepiel, A. M. L., Lane, T. W., Milligan, A. J., Schaperdoth, I., Reinfelder, J. R., & Tortell, P. D. (2002). Acquisition of inorganic carbon by the marine diatom *Thalassiosira weissflogii. Functional Plant Biology,* **29**, 301-308.

Nimer, N. A., Brwnlee, C., & Merrett, M. J. (1994). Carbon dioxide availability, intracellular pH and growth rate of the coccolithophore *Emiliania huxleyi. Marine Ecology Progress Series,* **109**, 257-262.

Nimer, N. A., Iglesias-Rodríguez, M. D., & Merrett, M. J. (1997). Bicarbonate utilization by marine phytoplankton species. *Journal of Phycology,* **33**, 625-631.

Nimer, N. A., Warren, M., & Merrett, M. J. (1998). The regulation of photosynthetic rate and activation of extracellular carbonic anhydrase under CO_2-limiting conditions in the marine diatom *Skeletonema costatum. Plant Cell Environment,* **21**, 805-812.

Nimer, N., Brownlee, C., & Merrett, M. (1999). Extracellular carboni anhydrase facilities carbon diaxide availability for photosynthesis in the marine dinoflagellate *Prorocentrum micans. Plant Physiology,* **120**, 105-111.

Price, G. D., Maeda, S.-I., Omata, T., & Badger, M. R. (2002). Modes of inorganic carbon uptake in the cyanobacterium *Synechococcus* sp. PCC7942. *Functional Plant Biology,* **29**, 131-149.

Raven, J.A. (1991). Implication of inorganic carbon utilization: ecology, evolution and geochemistry. *Canadian Journal of Botany,* **69**, 908-924.

Raven, J. A. (1997) Inorganic carbon acquisition by Marine Autotrophs. *Advances in Botanical Research,* **27**, 85-209.

Raven, J. A., Johnston, A. M., Kübler, J. E., Korb, R., McInroy, S. G., Handley, L. L., Scrimgeour, C. M., Walker, D. I., Beardall, J., Clayton, M.N., Vanderklift, M., Fredriksen, S., & Dunton, K. H. (2002). Seaweeds in Cold Seas: Evolution and Carbon Acquisition. *Annals of Botany Company,* **90**, 525-536.

Riebesell, U., Zondervan, I., Rost, B., Tortell, P. D., Zeebe, R. E., & Morel, F. M. M. (2000). Reduced calcification of marine plankton in response to increased atmospheric CO_2. *Nature,* **407**(6802), 364-367.

Riebesell, U., Shulz, K. G., Bellerby, R. G. J., Botros, M., Fritsche, P., Meyerhofer, M., Neill, C., Nondal, G., Oschlies, A., Wohlers, J., & Zollner, E. (2007). Enhanced biological carbon consumption in a high CO_2 ocean. *Nature,* **450**, doi:10.1038.

Riebesell, U., Bellerby, R. G. J., Grossart, H.-P., & Thingstad, F. (2008). Mesocosm CO_2 perturbation studies: from organisms to community level. *Biogeosciences Discussions,* **5,** 641-659.

Robertson, J. E., Robinson. C., Turner, D. R., Holligan, P., Watson, A. J., Boyd, P., Fernández, E., & Finch, M. (1994). The impact of a coccolithophore bloom on oceanic carbon uptake in the Northeast Atlantic during summer 1991. *Deep-Sea Research Part I,* **41**, 297-.

Rost, B., Riebesell, U., & Burkhardt, S. (2003). Carbon acquisition of bloom-forming marine phytoplankton. *Limnology & Oceanography,* **48**, 55-67.

Rost, B., Richter, K.-U., Riebesell, U., & Hansen, P. J. (2006). Inorganic carbon acquisition in red tide dinoflagellates. *Plant Cell Environment,* **29**, 810-822.

Rotatore, C. B., Colman, B., & Kuzma, M. (1995). The active uptake carbon dioxide by the marine diatoms *Phaedodactylum tricornutum* and *Cyclotella* sp. *Plant Cell Environment,* **18**, 913-918.

Royal Society (2005). *Ocean acidification due to increasing atmospheric carbon dioxide.* The Royal Society, Policy document 12/05, London.

Sciandra, A., Harlay, J., Lefevre, D., Lemee, R., Rimmelin, P., Denis, M., & Gattuso, J.-P. (2003). Response of coccolithophorid Emiliania huxleyi to elevated partial pressure of CO_2 under nitrogen limitation. *Marine Ecology Progress Series,* **261**, 111-122.

Shibata, M., Katoh, H., Sonoda, M., Ohkawa, H., Shimoyama, M., Fukuzawa, H., Kaplan, A. M., & Ogawa, T. (2002). Genes essential to sodium-dependent bicarbonate transport in cyanobacteria: function and phylogenetic analysis. *Journal of Biological Chemistry,* **277**, 18658-18664.

Shiraiwa, Y., & Miyachi, S. (1985). Factors controlling induction of carbonic anhydrase and efficiency of photosynthesis in Chlorella-vulgaris 11H-cells. *Plant Cell Physiology,* **24**, 919-923.

Smith, R. G., & Bidwell, R. G. S. (1989). Mechanism of photosynthetic carbon dioxide uptake by the red macroalga *Chondrus crispus*. *Plant Physiology,* **89**, 93-99.

Surif, M. B., & Raven, J. A. (1989). Exogenous inorganic carbon sources for photosynthesis in seawater by members of the Fucales and the Laminariales (Phaeophyta): Ecological and taxonomic implications. *Oecologia,* **78**, 97-105.

Surif, M.B., & Raven, J.A. (1990). Photosynthetic gas exchange under emersed conditions in eulittoral and normally submersed members of the Fucales and the Laminariales: interpretation in relation to C isotope ratio and N and water use efficiency. *Oecologia,* **82**, 68-80.

Tortell, P. D. (2000). Evolutionary and ecological perspectives on carbon acquisition in phytoplankton. *Limnology & Oceanography,* **45**, 744-750.

Tortell, P. D., Rau, G. H., & Morel, F. M. M. (2000). Inorganic carbon acquistion in coastal Pacific phytoplankton communities. *Limnology & Oceanography,* **45**, 1485-1500.

Tortell, P. D., DiTullio, G. R., Sigman, D. M., & Morel, F. M. M. (2002) CO_2 effects on taxonomic composition and nutrient utilization in an Equatorial Pacific phytoplankton assemblage. *Marine Ecology Progress Series,* **236**, 37-43.

Tortell, P. D., Martin, C. L., & Corkum, M. E. (2006). Inorganic carbon uptake and intracellular assimilation by subartic Pacific phytoplankton assemblages. *Limnology & Oceanography,* **51**, 2102-2110.

Tortell, P. D., & Morel, F. M. M. (2002). Sources of inorganic carbon for phytoplankton in te eastern Subtropical and Equatorial Pacific Ocean. *Limnolgy & Oceanography,* **47**, 1012-1022.

Tortell, P. D., Reinfelder, J. R., & Morel, F. M. M. (1997). Active uptake of bicarbonate by diatoms. *Nature* **390**, 243-244.

Yates, K. K., Dufore, C., Smiley, N., Jackson, C., & Halley, R. B. (2007). Diurnal variation of oxygen and carbonate system in Tampa Bay and Florida Bay. *Marine Chemistry,* **104**, 110-124.

Yallop, M. L. (2001). Distribution patterns and biomass estimates of diatoms and autotrophic dinoflagelattes in the NE Atlantic during june and July 1996. *Deep-sea Research Part II,* **48**, 825-844.

Yu, J., Tang, J., Zhang, P., & Dong, S. (2006). Effects of elevated CO_2 on sensitivity of six species of algae and interspecific competition of three species of algae. *Journal of Experimental Science,* **18**, 353-358.

Zou, D., Gao, K., & Xia, J. (2003). Photosynthetic utilization of inorganic carbon in the economic brown alga, *Hizikia fusiforme* (Sargassaceae) from the South China Sea. *Journal of Phycology,* **39**, 1095-1100.

In: Algae: Nutrition, Pollution Control and Energy Sources
Editor: Kristian N. Hagen, pp. 147-175
ISBN: 978-1-60692-008-4

Chapter 8

SULFATED POLYSACCHARIDES FROM ALGAE: CHARACTERISTIC STRUCTURES AND THEIR MEDICINAL APPLICATIONS

Jung-Bum Lee [*] ***and Toshimitsu Hayashi***
Graduate School of Medicine and Pharmaceutical Sciences for Research, University of Toyama, 2630 Sugitani, Toyama, Toyama 930-0194, Japan

Abstract

Sulfated polysaccharides, in which at least some of the hydroxyl groups of the sugar residues are substituted by sulfate groups, are universal and characteristic ingredients of algae. So far, various types of polysaccharides have been isolated from algae and studied their chemical structures. Sulfated galactans including carrageenans and agaroids and sulfated fucans (fucoidans) are well known sulfated polysaccharides isolated from red algae and brown algae, respectively. Various types of sulfated polysaccharides have been isolated from green and blue-green algae. On the other hand, sulfated polysaccharides are well known to possess multiple biological activities such as anticoagulant, antiviral, antitumor, antioxidant and immunomodulating effects. Therefore, algal sulfated polysaccharides are promising candidates for developing novel pharmaceuticals. In addition, the third function of food has been attracted a great deal of attention to maintain human health, recently, therefore, it is possible to develop functional foods based on the biological potencies of sulfated polysaccharides. In this chapter, we will summarize the chemical characteristics and biological activities of sulfated polysaccharides from algae.

Introduction

Algae include a wide variety of plants that range from unicellular organisms to seaweeds extending over 30 m. The majority of the algae are found in fresh or salt water and growing

[*] E-mail address: lee@pha.u-toyama.ac.jp. Tel: +81-76-434-7580, Fax: +81-76-434-5170. (To whom author correspondence should be addressed)

most abundantly at the intertidal level. In particular, marine algae (seaweeds) are regarded to be important sources of food, fodder, fertilizer, pharmaceuticals, and chemicals.

Algae produce various metabolites and have been recognized as promising targets in the search for biologically active compounds [1,2]. It is well known that algal polysaccharides are one of the most abundant resources in nature, and their structures are different from those of higher land plants. Among them, phycocolloids such as alginates, carrageenans, and agars are unique algal polysaccharides, which have been used in the food, biomedical, pharmaceutical and bioindustrial areas. Except for these biopolymers, large numbers of polysaccharides, in particular sulfated polysaccharides which some of hydroxyl groups of the sugar residues are substituted by sulfate groups, are obtained from various algae. Sulfated polysaccharides might be promising candidates of pharmaceutics because of their multiple biological activities.

Here, we summarize the structures and biological activities of sulfated polysaccharides from algae, and discussed on possibilities of their medicinal applications.

Sulfated Polysaccharides from Blue-Green Algae

Blue-green algae (Cyanophyta) are photosynthetic prokaryotes, and are distributed in various environments from the Arctic Circle to desert. Some of them have been used as foods for a long time. In Japan, the following four species have been utilized from ancient times: *Aphanothece sacrum*, *Nostoc verrucosum*, *N. commune*, and *Blachytrichia quoyi* [3]. *Spirulina* (*Arthrospira*) is a microscopic and filamentous blue-green alga that also has a long history of use as food [4]. Furthermore, *S. platensis* and *S. maxima* have been cultivated and sold as healthy food and feed for the last 20 years. Thus, *Spirulina* is one of the most important industrially cultivated microalgae.

They are well known to produce unique metabolites including polysaccharides owing to their ecological and biochemical diversity. Furthermore, they produce copious amount of polysaccharides in the form of sheaths, slim capsules. However, a little knowledge of characterization and/or structural elucidation of sulfated polysaccharides have been reported. It is noteworthy that hot water extract from *S. platensis* have been found to possess antiviral effects [5,6]. It has been proved that the sulfated polysaccharide, sodium spirulan (Na-SP) or calcium spirulan (Ca-SP), could contribute to the effects. Na-SP is a unique sulfated polysaccharide mainly consisted of 3-linked α-L-rhamnose and 2-linked α-L-3-*O*-methylrhamnose residues with D-glucuronic acid and D-galacturonic acid [7]. Sulfate groups were suggested to be partially substituted at C-2 or C-4 of rhamnosyl and C-4 of 3-*O*-methylrhamnose residues. Furthermore, both glucuronic and galacturonic acids were found to be 4- and 3,4-linked residues. Therefore, these hexuronosyl residues might be a branching point in the polysaccharide. On the other hand, ESI-MS/MS analyses of Na-SP-derived oligosaccharides revealed that the polysaccharide consisted of two types of repeating units, →3)-rhamnose-(1→2)-3-*O*-methylrhamnose-(1→ and →3)-rhamnose-(1→4)-glucuronic acid or galacturonic acid-(1→ [8]. Therefore, it was suggested that Na-SP is a highly branched sulfated polysaccharides with core structure consisted of aldobiouronic acids, and unique rhamnan-type chain is attached to the core structure (Figure 1).

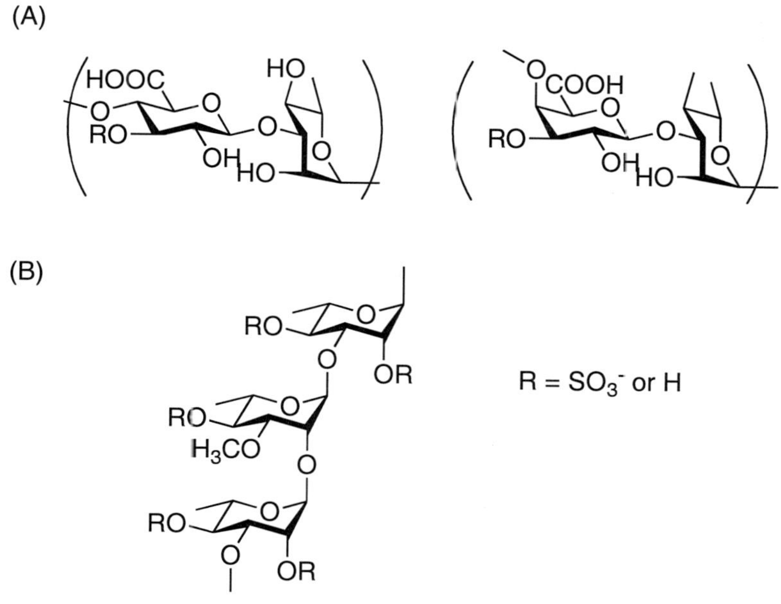

Figure 1. Proposed structure of sodium spirulan from *Spirulina platensis*.

Chemical characteristics of sulfated polysaccharides isolated from blue-green algae are listed in Table 1. *Aphanocapsa halophytia* MN-11 is a halophilic blue-green alga, and it produces sulfated exopolysaccharide in cultured media [9]. The polysaccharide constituted of fucose, glucose and mannose with a ratio of 10:5:3. Sulfate and protein contents were 11.9% and 10.3%, respectively. *Gloeothece* sp. PCC6909 has also been reported to produce sulfated polysaccharide composed of galactose, glucose, mannose, rhamnose, xylose, 2-*O*-methylxylose and sulfate [10]. De Philippes *et al.* surveyed fifteen *Cyanothece* strains isolated from saline environments whether they produce or not sulfated polysaccharides, and showed six strains at least produced sulfated polysaccharides [11]. *Synechocystis aquatilis* was found to release a sulfated polysaccharide composed mainly of arabinose (45%) and fucose (47%) [12].

Sulfated Polysaccharides from Red Algae

Several red algae such as *Porphyra, Ceramium, Gracilaria, Gelidium,* and *Gloiopeltis* have been eaten as food, particularly in Japan, and they are also used as a drug in traditional Chinese herbal medicine for over 1,000 years. Red algae (Rhodophyta) produce considerable amount of polysaccharides, and the major polysaccharides are sulfated galactans, carrageenans and agars, which are the main components of their cell walls. These polymers are also called phycocolloids and are used in a variety of laboratory and industrial applications [13,14]. However, their chemical structures are very heterogeneous and are correlated to the algal sources, the lifestages, and the extraction procedures of the polysaccharides.

Table 1. Characteristics of sulfated polysaccharides isolated from blue-green algae.

Species	Neutral sugars (molar ratio)								UA (%)*	SO_4^- (%)	Other substitutes	MW (kDa)	Ref.
	Rha	Fuc	Ara	Xyl	Man	Glc	Gal	Other					
Spirulina platensis	1.0	tr	tr	tr	tr	tr	–	a-d	16.5	17.3		245	7
Aphanocapsa halophytia MN-11	1.0	26.5	–	1.5	7.5	12.5	1.5	–	–	11.9		nd	9
Gloeothece sp. PCC6909	1.0	–	–	0.6	1.8	3.3	2.7	e	5.0	13.8		nd	10
Cyanothece sp. TP 5	1.0	1.2	0.4	0.3	–	0.9	–	–	40.4	+	Pyr	nd	11
Cyanothece sp. TP 10	1.0	0.8	1.8	–	0.1	0.6	–	–	31.3	+	Pyr	nd	11
Cyanothece sp. 16Som2	1.0	1.0	–	1.2	0.4	1.8	0.1	–	20.6	+		nd	11
Cyanothece sp. VI 13	1.0	1.5	–	1.5	0.5	2.0	0.1	–	32.1	+	Pyr	nd	11
Cyanothece sp. VI 22	1.0	1.8	–	1.8	1.0	2.8	0.3	–	40.8	+	Ac, Pyr	nd	11
Cyanothece sp. IR 20	1.0	0.2	–	–	0.2	tr	tr	–	9.8	+	Ac, Pyr	nd	11

Abbreviations: Rha = rhamnose; Fuc = fucose; Ara = arabinose; Xyl = xylose; Man = mannose; Glc = glucose; Gal = galactose; UA = uronic acids; + = present; – = absent; tr = traces; nd = not determined; Ac = acetyl; Pyr = pyruvate.

a: 3-*O*-methylrhamnose (0.6). b: 2,3-di-*O*-methylrhamnose (tr). c: 3-*O*-methylfucose (tr). d; 3-*O*-methylxylose (tr). e: 2-*O*-methylxylose (0.5).

Carrageenans and Related Polysaccharides

Carrageenans are apparently named after the coastal town of Carragheen in Ireland where for hundreds of years dried and bleached red seaweed referred to as 'Irish moss'. These polysaccharides are found particularly in members belonging Gigartinales, *Gigartina, Chondrus, Iridaea, Hypnea, and Eucheuma*, and they are widely used in the food and industries as thickening and stabilizing agents. Most of the carrageenans are produced from cultivated species of *Eucheuma* and *Kapaphycus* [15].

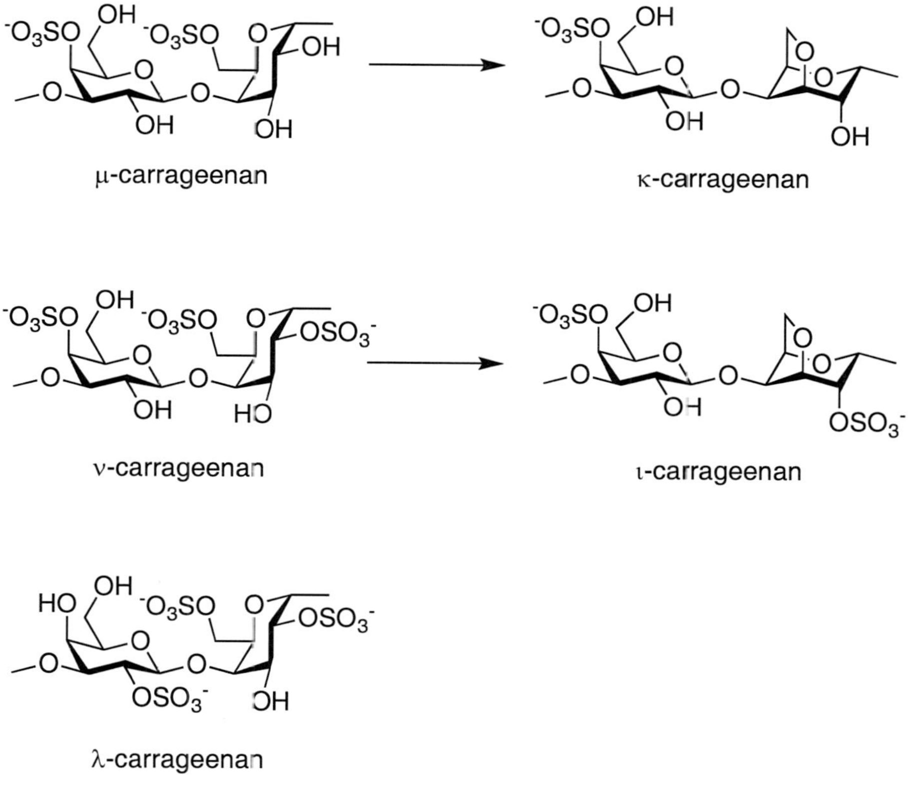

Figure 2. Representative disaccharide structures of carrageenans.

Carrageenans are made up of linear chains of D-galactose with alternating α-(1→3) and β-(1→4) linkages [13]. Furthermore, carrageenans are classified according to the number and the position of sulfated ester and by the occurrence of α-linked 3,6-anhydro-galactosyl residues (Table 2). For example, κ (kappa)-carrageenan has one sulfate group at C-4 of galactose, whereas ι (iota)- has two sulfates at C-4 of galactose and C-2 of 3,6-anhydro-galactose. On the other hand, μ (mu)- and ν (nu)-carrageenans are not contained 3,6-anhydro-galactose, and they are suggested to be precursors of κ- and ι-carrageenan, respectively (Figure 2). On the other hand, λ (lambda)-carrageenan is also one of important biopolymer, which contained three sulfate groups in disaccharide unit. Characteristic differences between carrageenans are gelation properties and solubilities. κ- and ι-carrageenans form hard and soft thermoreversible gels, respectively, in the presence of salt. The gel strength depends on the polysaccharide concentration and type of salt with $K^+ > Ca^{2+} > Na^+$. On the other hand, λ-carrageenan do not form gel in the presence of salt. Solubilities of κ- and ι-carrageenans also depends on their counter cation, *e.g.*, sodium salts can dissolve at room temperature, whereas potassium salt can dissolve at more than 60 °C. On the other hand, λ-carrageenan is soluble at all temperature.

Table 2. Classification of carrageenans

	3-linked β-D-galactose (G)	4-linked α-D-galactose (D)
κ-family		
κ (kappa)	4-sulfate	3,6-anhydro
ι (iota)	4-sulfate	3,6-anhydro 2-sulfate
μ (mu)	4-sulfate	6-sulfate
ν (nu)	4-sulfate	2,6-disulfate
λ-family		
λ (lambda)	2-sulfate	2,6-disulfate
ξ (xi)	2-sulfate	2-sulfate
π (pi)	2-sulfate, 4,6-(1-carboxyethyliden)	2-sulfate
θ (theta)	2-sulfate	3,6-anhydro 2 sulfate
β-family		
β (beta)	–	3,6-anhydro
α (alpha)	–	3,6-anhydro 2-sulfate
γ (gamma)	–	6-sulfate
δ (delta)	–	2,6-disulfate
ω-family		
ω (omega)	6-sulfate	3,6-anhydro
ψ (psi)	6-sulfate	6-sulfate

So far, various species of red algae have been examined to characterize their carrageenans. Usually, gametophytes and sporophytes contain κ- and λ-family of carrageenans, respectively [15]. However, some variants with substituents have been reported from several algae. θ-carrageenan, which composed of 3-linked galactose 2-sulfate and 4-linked 3,6-anhydrogalactose 2-sulfate, has been isolated from *Callophyllis homobroniana* [16]. Rodríguez *et al* reported that three galactan sulfates from *C. variegata* have been characterized that F1 consists mainly of 3-linked galactose 2,4-di-sulfate and 4-linked 3,6-anhydrogalactose 2-sulfate or galactose 2,3,6-tri-sulfate and contribution of θ-carrageenan was observed in F2 and F3 [17]. Liao *et al* reported that sulfated galactan from *Phacelocarpus peperocarpos* are mainly composed of disaccharide units with 3-linked galactose 4,6-di-sulfate and 4-linked 3,6-anhydrogalactose [18]. Sulfated galactan consisted of repeating disaccharide, 4-linked α-D-galactose 2,3-di-sulfate or 2-sulfate and 3-linked β-D-galactose was isolated from *Botryocladia occidentalis* [19].

Agars and Related Sulfated Polysaccharides

Agar consists of agarose and agaropectin, and the former (also called agaran) is a linear polysaccharide isolated from red seaweed and is commercially extracted mainly from *Gelidium, Gracilaria,* and *Pterocladia*. Agarose self-assembles into stiff and brittle gels, thus it has been used as gelling and icing agents in food industry. The polysaccharide consisted of 3-linked β-D- and 4-linked α-L-galactose residues, with the later predominantly in the form of a 3,6-anhydride and lack of sulfate groups (Figure 3). On the other hand, structure of agaropectin is similar to agarose but slightly branched and sulfated, and they may have methyl and pyruvic acid ketal substituents. So far, there are several investigations of the related polysaccharide to agar. Porphyran is a representative one which is isolated from *Porphyra* species [20-22]. The polysaccharide is highly substituted by 6-*O*-sulfation of the L-galactose and 6-*O*-methylation of the D-galactose residues (Figure 3). Furthermore, it has been reported that sulfated polysaccharides isolated from *Gloiopeltis complanata* contained at least two components, 6-sulfated agarose (funoran) and the precursor disaccharide structure consisted of 3-linked β-D-galactose 6-sulfate and 4-linked α-L-galactose 6-sulfate [23].

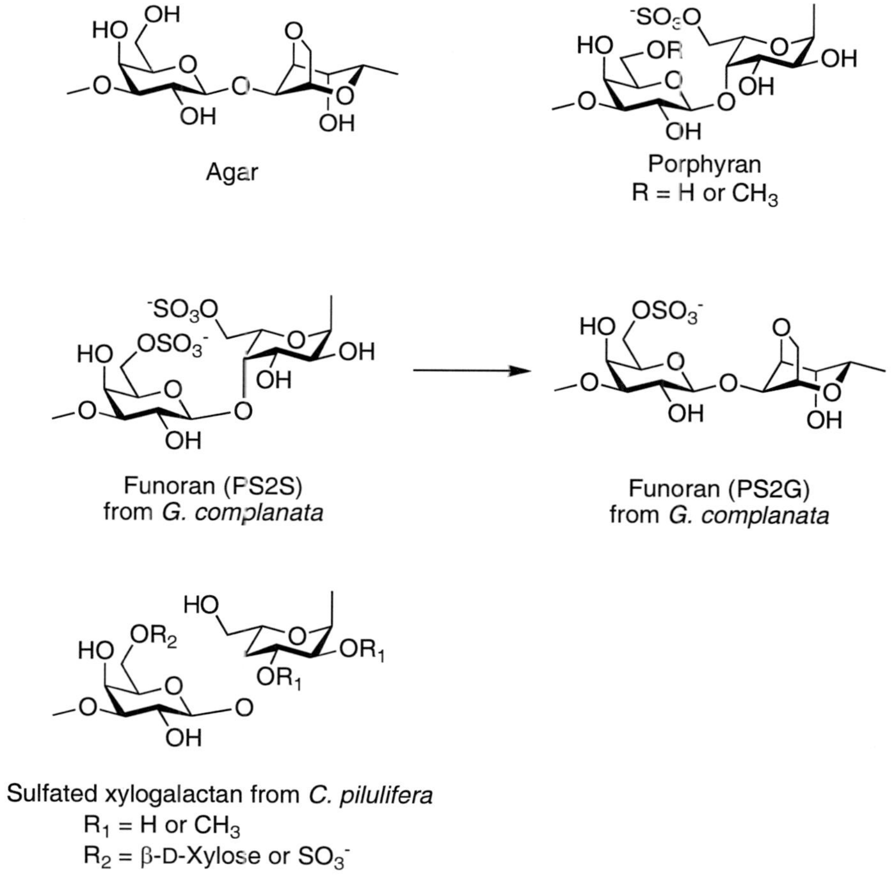

Figure 3. Structures of agar and related sulfated polysaccharides from red algae.

DL-Hybrid Sulfated Galactans

As described in the above, two diasteromeric polysaccharide groups are defined as carrageenan and agaran, and the red algae that biosynthesize these polysaccharides are called carrageenophytes and agarophytes, respectively. Nevertheless, it has been found there are small but significant amounts of polysaccharides containing α-D- and α-L-galactose and 3,6-anhydro-α-D- and L-galactose units in seaweeds belonging to the Gigartinaceae [24,25]. On the other hand, it has also been found that some agarophytes contain small amounts of 4-linked α-D-galactose residues [26,27]. Although it is not clear whether they are present in one molecule or not, DL-Hybrid galactans have been proposed to contain in *Gymnogongrus torulosus* [28], *Kapaphycus alvarezii* [29], and *Cryptonemia crenulata* [30].

Sulfated Xylogalactans

Frequently, sulfated galactans are found to accept various structural modifications such as *O*-sulfation, *O*-methylation, and glycosylation with β-D-xylosyl residues. Some of them, significant amounts of xylosyl residues are present and regarded to be sulfated xylogalactans. Structural investigation of sulfated xylogalactans named corallinans revealed that the backbone of the polysaccharides had an alternating disaccharide units, 3-linked β-D-galactose and 4-linked α-L-galactose. C-6 position of 3-linked residues is substituted mainly by β-D-xylosyl stubs or sulfate ester groups [31]. Similar sulfated xylogalactan has also been isolated from *C. pilulifera*, but sulfate substitution patterns were different from those of *C. officinalis* [32]. Other sulfated xylogalactans have been isolated from *Georgiella confluens* [33], and *Pterocladiella capillacea* [34].

Sulfated Xylomannans

As described above section, the major water soluble polysaccharides of the red algae are galactan sulfates. Nevertheless, a less red seaweed, *Nothogenia fastigiata*, produces a mixture of sulfated polysaccharides from which sulfated α-D-(1→3)-mannans with single stubs of β-D-xylosyl residue at C-2 position [35,36]. The polysaccharides were substituted with sulfate groups at C-2 and C-6 of 3-linked mannose residues.

Sulfated Polysaccharides from Brown Algae

Many brown algae have been used as food in East Asia, especially Japan, Korea and China, where seaweed cultivation has become a major industry. *Laminaria, Undaria*, *Ecklonia*, *Hizikia, Nemacystus,* and *Sargassum* are not only representative edible seaweeds but also have been used as traditional Chinese herbal medicines.

So far, laminarans, alginates, and fucoidans have been obtained from brown algae as unique polysaccharides. Among them, fucoidans are one of the most famous sulfated polysaccharides in nature. They have extensively been isolated from brown algae and are well studied their structure and biological activities. These polysaccharides mainly constituted of sulfated L-fucose with or without xylose, galactose, mannose, and/or uronic acid. The composition was observed to change according to the algal species.

Sulfated Fucans (Fucoidans)

Commercially available fucoidan has been prepared from *Fucus vesiculosus*, and its structure was proposed to be composed of 2-linked (major) and 3-linked α-L-fucose (minor) residues with single stubs of α-L-fucose residue at C-2 position and 4-*O*-sulfation [37]. On the other hand, revised structure has been proposed that commercially available fucoidan was mainly composed of 3-linked α-L-fucose with stubs of α-L-fucose residue at C-2 or C-4 positions [38]. It was also found that the sulfate groups were mainly at C-4. On the other hand, it has been reported that fucoidan from *F. evanescens* contained partially acetylated fucoidan which contained a linear backbone of alternating 3- and 4-linked α-L-fucose 2-sulfate residues [39]. Furthermore, a fucoidan from *F. distichus* also reported to possess an alternating disaccharide unit consisted of 3-linked α-L-fucose 2,4-disulfated and 4-linked α-L-fucose 2-sulfate residues [40]. On the other hand, a fucoidan from *F. serratus* seemed to be different from fucoidans from *F. evanescens* and *distichus* [41]. The fucoidan from *F. serratus* was suggested to build up with a backbone of alternating 3- and 4-linked α-L-fucose residues. However, about half of the 3-linked residues being substituted at C-4 by trifucoside units, α-L-Fuc*p*-(1→4)-α-L-Fuc*p*-(1→3)-α-L-Fuc*p*-(1→. From these observations, fucoidans from the algae belonging to the order Fucales may be consisted of alternating 3- and 4-linked α-L-fucose residues.

A highly branched sulfated fucan has been isolated from *Sargassum horneri*, which consists of 3-linked, 4-linked and 3,4-linked α-L-fucose residues. It was suggested that 4-linked and 3-linked fucose residues were sulfated at *O*-2 and *O*-4, respectively [42]. Furthermore, sulfated fucan whom structure is different from sodium hornan has also been isolated from *S. horneri* [43]. It contained 2-linked, 3-linked, and 4-linked α-L-fucose residues. Similar fucoidan has also been isolated from *Ascophyllum nodosum* [44].

On the other hand, sulfated fucans having a linear backbone composed with 3-linked α-L-fucose residues were isolated from *Cladosiphon okamuranus* and *Chorda filum* [45,46]. *C. okamuranus* is a cultivated brown alga in Okinawa, Japan as food resource, and structural analyses revealed that fucoidan from *C. okamuranus* has a linear backbone of 3-linked α-L-fucose residues with single stub of α-D-glucuronic acid residues at C-2 of fucosyl residues. Furthermore, it has also been suggested that half of each fucose residue was sulfated and one *O*-acetyl ester was present in every six fucose residues. On the other hand, *C. filum* also contained 3-linked α-L-fucose backbone with single stubs of α-L-fucose residue. Some fucose residues are sulfated at *O*-4 (mainly) and *O*-2 positions and 2-*O*-acetylated. Fucoidans from *Laminaria cichorioides* and *Analipus japonicus* have been reported to consist mainly of 3-linked α-L-fucose residue but they seemed to be highly branched [47,48].

Sulfated Heterofucans

Some of fucoidans have been reported to consist of large amount of fucose residues with other sugars like galactose, xylose, mannose and uronic acid residues. Fucoidans from *Sargassum thunbergii* and *S. stenophyllum* contain those type of polymer [49,50]. The later has been suggested to be composed of a linear core, 6-linked β-D-galactose and/or 2-linked β-D-mannose residues, with branched chains of fucans consisted of 3- and/or 4-linked α-L-fucose, 4-linked β-D-glucuronic acid, terminal β-D-xylose, and, sometimes, 4-linked β-D-glucose residues. Similarly, Li *et al* also reported the structure of a fucoidan containing a fucose-free core from *Hizikia fusiforme* [157]. The fucose-free core was mainly composed of alternating units of 2-linked α-D-mannose and 4-linked β-D-glucuronic acid residues with a minor portion of 4-linked β-D-galactose residues. Fucose residues were present at the nonreducing ends (two-thirds) and the remainders were 2-, 3- and 4-linked. About two-thirds of xylose residues were also present at the nonreducing ends.

Sulfated Galactofucans

In contrast to fucoidans as described above, some sulfated polysaccharides isolated from brown algae contained considerable amounts of galactose residue. Thus, this class of sulfated polysaccharides can be regarded to be sulfated galactofucans, and isolated from many brown algae. Two highly branched sulfated galactofucans were isolated from *Ecklonia kurome*, they were mainly composed of 3-linked α-L-fucose, 3-linked α-L-fucose 4-sulfate, and galactose residues with various glycosidic linkages [51,52]. Similar results have also been reported in the case of *Adenocystis uticularis* [53]. Fucoidans from *Undaria pinnatifida* contained almost same amounts of fucose and galactose residues [54]. Structural analysis revealed the fucoidan to possess highly branched structure. On the other hand, a highly sulfated galactofucan has been isolated from Tasmanian *U. pinnatifida*, and it has been shown to consist predominantly of 3-linked α-L-fucose 2,4-disulfate and 3- and 4-linked galactose residues [55]. On the other hand, unique sulfated polysaccharide has been isolated from *Spatoglossum schroederi* [56]. The isolated polysaccharide contained fucose, xylose, galactose and sulfate in a molar ratio of 1.0:0.5:2.0:2.0, and consisted of 4-linked β-D-galactose, some residues possess 3-sulfate, core units with branches of oligosaccharide chains composed of α-L-fucose 3-sulfate and β-D-xylose residues.

Sulfated Polysaccharides from Green Algae

Diversity of sugar composition of their cell wall polysaccharides including sulfated polysaccharides from green algae has been observed and it has been seemed to be dependent on their taxa. For example, microfibril polysaccharides of many green algae are cellulose, whereas β-D-(1→4)-mannan and β-D-(1→3)-xylans have been isolated from *Codium* sp and *Caulerpa* sp, respectively. Therefore, polysaccharide composition of green algae might reflect the difference of evolution process in each taxa. However, research for sulfated polysaccharides from green algae is behind when compared with those of red and brown algae.

Rhamnose-Containing Sulfated Polysaccharides

Monostroma sp. is belonging to Ulotrichales, and *M. nitidum* and *latissimum* are cultivated as food resources in Japan. Sulfated polysaccharides from these algae are rhamnan sulfates, which consisted of mainly α-L-rhamnose [57-60]. They are composed of much amounts of α-L-rhamnose with xylose, glucose and glucuronic acid. Rhamnan sulfate from *M. nitidum* has been shown to consist of 2-linked and 3-linked α-L-rhamnose 2-sulfate residues, whereas the polysaccharide from *M. latissimum* are composed of 2-linked α-L-rhamnose 3- or 4-sulfate and 3-linked α-L-rhamnose residues. On the other hand, sulfated glucuronoxylorhamnan were isolated from *Urospora wormskioldii*, which is also belonging to Ulotrichales [61].

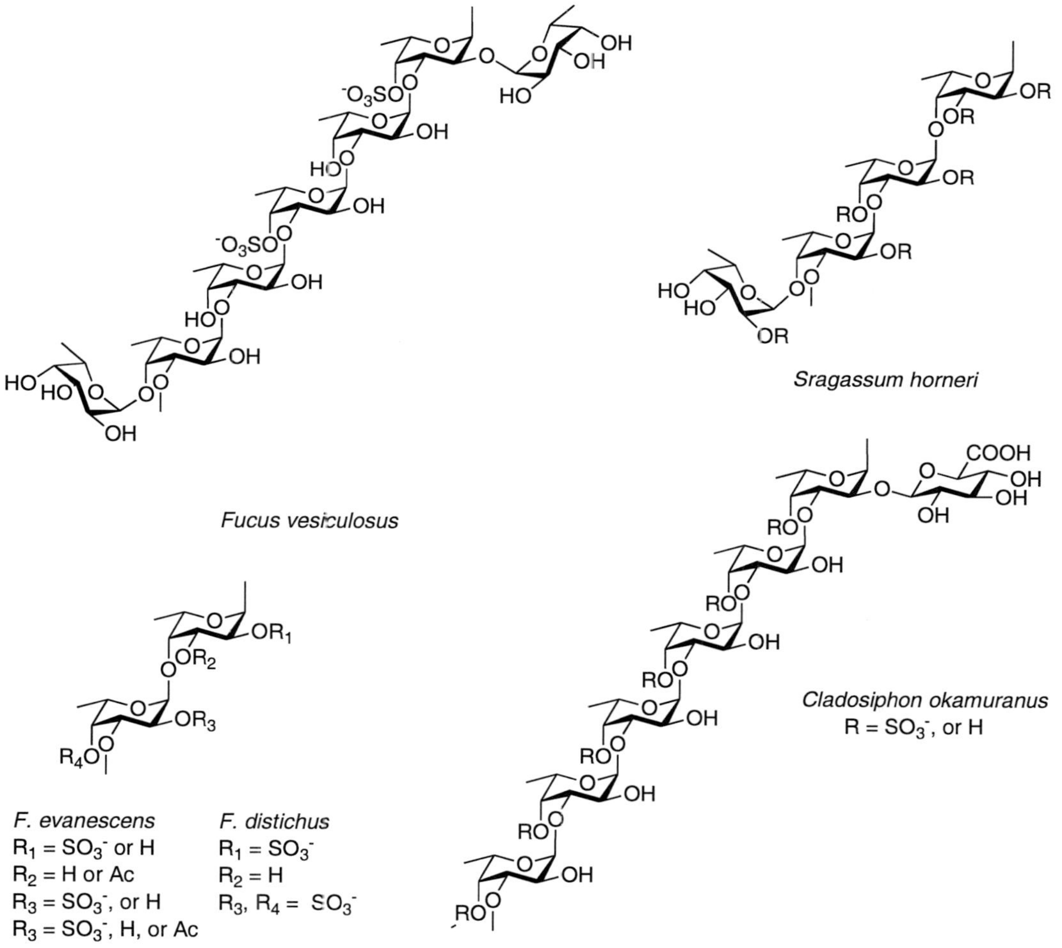

Figure 4. Structures of sulfated fucans from brown algae.

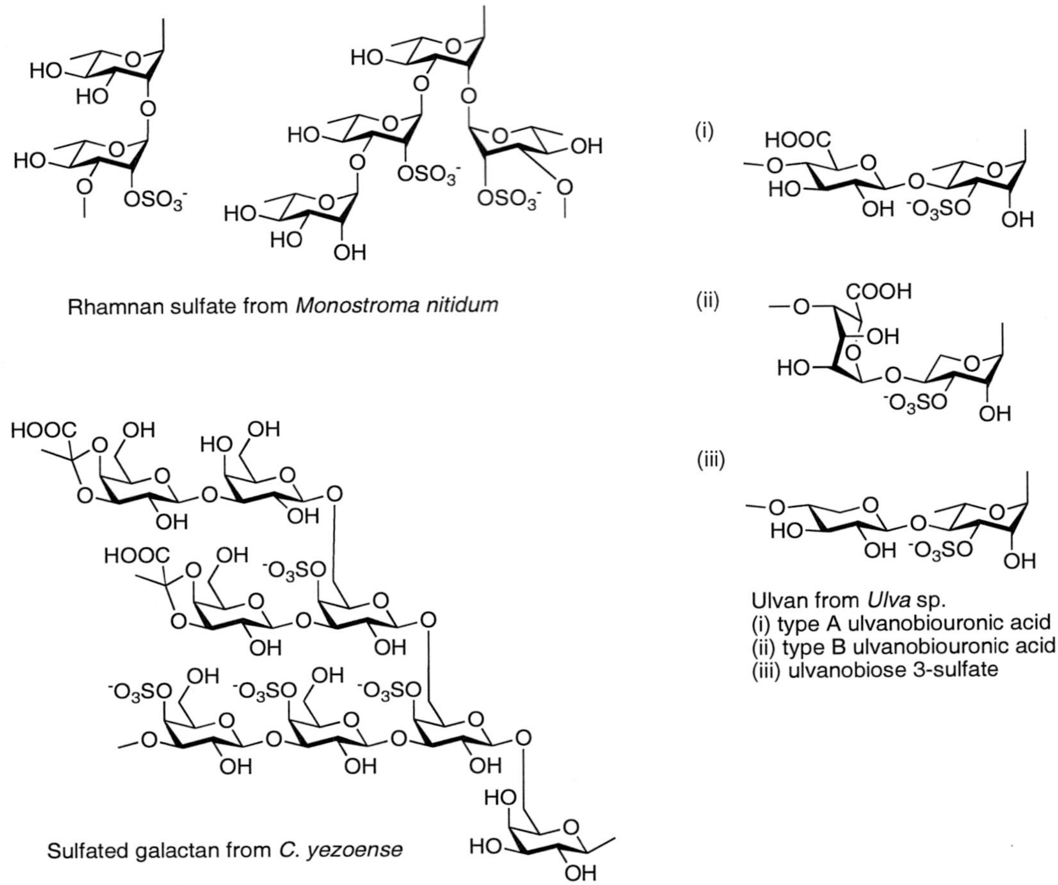

Figure 5. Structure of sulfated polysaccharides from green algae.

Ulva and *Enteromorpha* are belonging to Ulvales, and they distribute in worldwide, and ulvan is used to refer their sulfated polysaccharides. Sugar composition of ulvan is reported to be rhamnose, xylose, glucose and uronic acid such as glucuronic and iduronic acids. Structure of ulvans is well reviewed by Lahaye and Rovic [62]. Briefly, main repeating disaccharide in ulvan is type A ulvanobiouronic acid, →4)-β-D-glucuronic acid-(1→4)-α-L-rhamnose-(1→ (Figure 5). In addition to the characteristic disaccharide, many oligosaccharides composed of α-L-rhamnose, β-D-xylose, β-D-glucuronic acid and α-L-iduronic acid with or without single stubs of glucuronosyl residue have been characterized in ulvans from *Ulva* sp. Some α-L-rhamnose and β-D-xylose residues have been suggested to possess sulfate esters at C-3 and C-2 positions, respectively. Recently, sulfated heteroglycan has been isolated from *Enteromorpha compressa*, and the polysaccharide was composed of 4- and 2,4-linked L-rhamnose 3-sulfate, 4-linked D-glucose, 3- and 6-linked D-galactose, 4- and terminally linked D-glucuronic acid, and 4-linked D-xylose partially sulfated at C-2 [63].

Arabinose and/or Galactose-Containing Sulfated Polysaccharides

There are several reports concerning isolation and characterization of sulfated polysaccharides composed of arabinose and/or galactose. Sulfated polysaccharide from *Chaetomorpha anteninna* contained L-arabinose, D-galactose, and L-rhamnose [64]. The polysaccharide was suggested to possess a branched structure with 4-linked L-arabinose, 4- and 3-linked D-galactose, and 4-linked L-rhamnose. Sulfate groups were suggested to locate partially at C-2 of 4-linked arabinose, C-4 of terminal arabinose, and C-3 of galactose residues. It is noteworthy that identification of oligosaccharide fragments composed of arabinose and galactose revealed the presence of sulfated heteropolysaccharide. Same researchers also reported that sulfated heteropolysaccharide from *Cladophora socialis* were branching polymer and were composed of 3-linked D-galactose, 4-linked L-arabinose, and terminal D-xylose residues [65]. Sulfated polysaccharides from *Caulerpa racemosa* were reported to consist of 3-linked galactose 6-sulfate, terminal and 4-linked xylose, 4-linked arabinose 3-sulfate and 3,4-linked arabinose residues [66].

The genus *Codium* is very diverse group distributed from the temperate and tropical zones, and some of them have been used as food and traditional Chinese herbal medicine. Love and Percival studied on the polysaccharide from *C. fragile*, and they found that it was composed of 3-linked galactose and arabinose [67]. However, there is a question whether such sugar residues are present in the same polymer or not. Uehara *et al* reported that sulfated arabinan was isolated from *C. latum* as an anticoagulant principle [68]. Moreover, sulfated arabinan and sulfated arabionogalactan were also isolated from *Codium dwarkense* [69], however, their structures have not well been resolved. On the other hand, the presence of sulfated galactans have been reported from *C. cylindricum* [70]. Furthermore, structural elucidation of a highly pyruvylated sulfated galactan from *C. yezoense* has been performed by Bilan *et al* [71]. The polysaccharide contained a highly ramified structure, which contained 3-linked β-D-galactose backbone, and about 40% of the residues were additionally substituted at C-6 (Figure 5). Sulfate and pyruvate groups were found to be mainly present at C-4 and at *O*-3 and *O*-4 as five-membered ketals, respectively. In the same way, sulfated galactan consisted preponderantly of 3-linked β-D-galactose 4-sulfate with sulfation and glycosylation at C-6 as minor amounts has also been isolated from *Codium isthmocladum* [72]. In the polysaccharide, pyruvate groups are also found to form five-membered cyclic ketal as 3,4-*O*-(1-carboxy)ethylidene-β-D-galactose residue. On the other hand, sulfated proteoglycan consisted of glucose with trace amounts of arabinose and galactose was isolated from *C. pugniformis* as anticoagulant, however, the proteoglycan was a minor constituent [73].

Biological Activities of Sulfated Polysaccharides from Algae

Anticoagulant and Antithrombin Activities

Anticoagulant effect of sulfated polysaccharides occurs predominantly by inhibiting the key coagulation serine proteases, thrombin and factor Xa. The anticoagulation is facilitated by accelerating the activity of major serine protease inhibitors (serpin), antithrombin III (AT III) and heparin cofactor II (HC II) [74]. The former inhibits all of the proteinases involved in the intrinsic coagulation cascade along with thrombin and factor Xa, whereas the later exclusively inhibits the thrombin. So far, numerous studies of the evaluation of anticoagulation or antithrombin effects of algal sulfated polysaccharides have been performed, and their elucidated antithrombin mechanism(s) are listed in Table 3.

Table 3. Anticoagulant Sulfated Polysaccharides from Algae and Inhibitory Mechanism(s)

Species	Polysaccharides	Mechanisms*	Ref
Blue-green algae			
Spirulina platnesis	Rhamnan	HC	75,76,77
Red algae			
Botryocladia occidentalis	Galactan	AT, HC	20
Gelidium crinale	Galactan	AT	78
Brown algae			
Ascophyllum nodosum	Fucoidan	HC	79
	Fucoidan	Direct	80
Fucus evanescens	Fucoidan	HC	81
F. vesiculosus	Fucoidan	AT, HC, Direct	80,82
Ecklonia cava	Fucoidan	AT	83
E. kurome	Fucoidan	Direct	84
Laminaria brasiliensis	Fucoidan	Direct	80
L. cichorioides	Fucoidan	HC	85
Green algae			
Monostroma nitidum	Rhamnan	HC	86
M. latissimum	Rhamnan	HC	86
Ulva conglobata	Ulvan	HC, Direct	87
Caulerpa brachypus	Galactan	HC	86
Codium cylindricum	Galactan	Direct	70
C. fragile	n/a	AT,HC	88
C. latum	Arabinan	HC	86
C. pugniformis	Galactoarabinoglucan	AT, Direct	73

* HC: Heparin cofactor II mediated, AT: Antithrombin III mediated, Direct: Direct inhibition of thrombin.

On the other hand, potency of antithrombin effect has been generally suggested to be dependent on sulfate contents and molecular weight [57,59,89,90]. However, several sulfated polysaccharides such as sulfated mannan and carrageenans show no or low anticoagulant activities whereas they are high molecular weight polymers and possess enough amounts of sulfate groups [36,91,92]. Thus, it is suggested that other factors such as sugar composition and linkage patterns might be contributed the anticoagulant effects. Chevolot *et al* was prepared low molecular weight fucans by chemical method to compare their anticoagulant activities, and authors concluded that anticoagulant activity was apparently related not only to molecular weight and sulfate content, but also to sulfation level at specific position [93]. Similarly, Pereira *et al* revealed that the different structural feature determines not only the anticoagulant potency of the sulfated fucans but also the mechanism by which they exert the activity [80].

To consider clinical applications of sulfated polysaccharides for the treatment of thrombosis, fibrinolytic effects and/or enhancement of local anticoagulation effects are also important effects. In the fibrinolytic system, plasmin is the principle enzyme formed from plasminogen by action of plasminogen activators. Fucoidans have been reported to show the activation of plasminogen tissue type and urokinase type plasminogen activators (t-PA and u-PA) [94-96]. Calcium spirulan also showed the induction of t-PA in human fetal lung fibroblasts [97]. However, the polysaccharide induced u-PA and suppressed plasminogen activator inhibitor type 1 (PAI-1) secretion on vascular endothelial cells [98]. In addition, vascular endothelial cells show antithrombogenic properties by synthesizing and secreting anticoagulant substances like prostacyclin and t-PA. Furthermore, the cells synthesize heparan sulfate proteoglycans (HSPG) and chondoroitin/dermatan sulfate proteoglycans (CS/DS PG), and HSPG and CS/DS PG exhibit an antithrombin activity by activation of AT III and HC II, respectively. Thus, it is also suggested that stimulation of these proteoglycan releases may contribute to prevent thrombosis. Indeed, sulfated galactofucan and xylofucoglucuronan from *Spatoglossum schröederi* showed stimulatory effects on heparan sulfate release from vascular endothelial cells, whereas they had no or low anticoagulant activities [56,99]. Sodium spirulan from *Spirulina platensis* also stimulates the release of HSPG and CS/DS PG from vascular endothelial cells [100].

Antiviral Activities

Human immunodeficiency virus type 1 (HIV-1) is a retrovirus that causes the acquired immunodeficiency syndrome (AIDS). An increasing number of patients with HIV infection and/or AIDS cannot use the currently approved drugs including the reverse transcriptase and protease inhibitors, due to the adverse effects and the emergence of drug resistance. Thus, the urgent need for new anti-HIV/AIDS drugs is a global concern. On the other hand, herpes simplex virus type 1 and 2 (HSV-1 and 2) are major opportunistic agents involved in the pathogenesis of AIDS. They can act as co-factors by interacting with HIV-1, thereby influencing disease progression. Therefore, to control HIV infection and AIDS, developments of anti-HSV drugs are also demanded.

Many algal sulfated polysaccharides possess antiviral activities against enveloped viruses as shown in Table 4. Comparative evaluation of sulfated polysaccharide by Baba *et al* showed that fucoidan and κ- and λ-carrageenans inhibited enveloped virus replication *in vitro* [101]. The mechanism by which sulfated polysaccharides inhibit virus replication has been generally explained by inhibition of virus adsorption to host cells and subsequent virus-cell fusion or virus penetration [119,120]. However, some sulfated polysaccharides showed significant antiviral effects when added any time after virus penetration [42,109,115,121,122]. Calcium spirulan and sodium hornan showed antiviral effects when added after 2 h after virus adsorption, whereas dextran sulfate showed no inhibitory effect. Similar effects were also observed in some sulfated polysaccharides from green algae. In the case of carrageenan, λ-carrageenan showed antiherpetic effects when added after 8 h postinfection, whereas κ/ι- and μ/ν-carrageenan did not show any effects on virus replication [122]. From these observations, these effects seemed to be dependent on the structures of sulfated polysaccharides. However, antiviral target(s) of sulfated polysaccharides at this stage are still unclear although it was observed that fluorescein-labeled spirulan could internalize into HSV-1 infected cells [123]. On the other hand, λ-carrageenan, DL-hybrid galactan and fucoidan have been reported to show the virucidal effect against HSV [105,106,122]. Such effect is defined as a direct inactivation of virions, and it might be useful property for using as one of microbicide.

Table 4. Antiviral Sulfated Polysaccharides from Algae

Species	Polysaccharide types	Effective viruses*	Ref
Blue-green algae			
Spirulina platnesis	Spirulan	HSV-1, HCMV, Measles, Mumps, IFV, HIV	102
	Spirulan?	HSV-1, HHV-6, HCMV, HIV-1	103
Red algae			
Bostrychia montagnei	Agaran□	HSV-1 and -2	104
Callophyllis variegata	Carrageenan	HSV-1 and -2, DENV-2	17
Cryptonemia crenulata	DL-Galactan	HSV-1 and -2, DENV-2	108,110
Gigartiana skottsbergii	κ/ι-Carrageenan	HSV-1 and -2	91
	μ/ν-Carrageenan	HSV-1 and -2	91
	λ-Carrageenan	HSV-1 and -2	91
Grateloupia filicina	Galactan	HIV-1	118
G. indica	Galactan	HSV-1 and -2	117
G. longifolia	Galactan	HIV-1	118
Gymnogongrus torulosus	DL-Galactan	HSV-2, DENV-2	105
	κ/ι/ν-Carrageenan	HSV-1 and -2, DENV-2	108,110
Meristiella gelidium	ι/κ/ν-Carrageenan	HSV-2, DENV-2	111
Nothogenia fastigiata	Mannan	HSV-1	36
Schizymenia binderi	DL-Galactan	HSV-1 and -2	112
Stenogramme interrupta	Carrageenan	HSV-1 and -2	116
Brown algae			
Adenocystis utricularis	Fucoidan	HSV-1 and -2	53
Fucus vesiculosus	Fucoidan	HSV-1 and -2, HCMV, VSV, Sindbis, HIV-1	101
Sargassum horneri	Fucan	HIV, HCMV	42
	Fucan	HSV-1	43
S. patens	Fucoidan	HSV-2	106
	Fucoidan	HSV-1	113
Stoechospermum marginatum	Fucan	HSV-1 and -2	114
Undaria pinnatifida	Galactofucan	HSV-1 and -2, HCMV	54
	Galactofucan	HSV-1 and -2, HCMV	55
Green algae			
Caulerpa brachypus	Galactan	HSV-1	109
C. okamurai	Galactan	HSV-1	109
C. racemosa	Arabinogalactan	HSV-1 and -2	107
C. scapelliformis	Galactan	HSV-1	109
Chaetomorpha crassa	Galactoarabinan	HSV-1	109
C. spiralis	Galactoarabinan	HSV-1	109
Codium fragile	Arabinan	HSV-1	109
C. latum	Arabinan	HSV-1	109
Enteromorpha compressa	Ulvan	HSV-1	109
Monostroma latissimum	Rhamnan	HSV-1, HCMV, HIV-1	115
M. nitidum	Rhamnan	HSV-1	109

Antitumor Activities of Sulfated Polysaccharides

Screening study on antitumor activity of marine algae in Japan revealed that fucoidan, carrageenans, and porphyran were active principles for the inhibition of Ehrlich carcinoma [124]. In addition, there are several reports that fucoidans and carrageenans possess antitumor effects [49,125-129]. The mechanism by which polysaccharides, in one part, inhibit tumor growth is suggested to be due to the enhancement of immune systems. Indeed, it has been reported that carrageenan oligosaccharides stimulated humoral and cell-mediated immunity [128,129]. On the other hand, fucoidans were also reported to show anti-angiogenic activities, and it might contribute to their antitumor effects [125,126]. Angiogenesis is an indispensable biological event to form new blood vessel. Tumor growth requires it to supply nutrients and oxygen, and utilizes the newly formed blood vessels as conduits to disseminate invasive tumor cells. It has been reviewed that heparins can interfere with tumor angiogenesis, immunity system, cancer cell motility and adhesion [130]. Beside angiogenesis, prevention of tumor metastasis is also an important promising target, and some polysaccharides such as calcium spirulan from *S. platensis* and fucoidan from *F. evanescens* showed antimetastatic activity [131-132].

Immunomodulating Activities against Macrophages

So far, polysaccharides from algae have been shown to concern with immune response, and beneficial therapeutic properties are thought to be due to the modulation of innate immunity and macrophage function [133,134]. Indeed, fucoidan probably worked as immunomodulator for macrophages. For instance, commercially available fucoidan from *F. vesiculosus* reported to stimulate nitric oxide, which is an important regulatory molecule in physiological functions like host defense, by inducing the inducible nitric oxide synthase (iNOS) [135-138]. Furthermore, the fucoidan stimulated the tumorcidal and phagocytic activities of macrophages and the production of tumor necrosis factor-α (TNF-α) and interleukin 6 (IL-6) [136]. Similarly, the polysaccharide showed the immunomodulatory effects, such as increment of cell viability and stimulation of the production of IL-12 and TNF-α on dendritic cells [139]. On the other hand, water-soluble acidic polysaccharides, probably ulvan, from *U. rigida* showed the activation of macrophages, including expression of inflammatory cytokines and receptors, nitric oxide and prostaglandin E2 production, and nitric oxide synthase 2 and cyclooxygenase-2 gene expression [140].

Antiproliferative Effects on Vascular Smooth Muscle Cells

Like glycosaminoglycans, algal sulfated polysaccharides have antiproliferative effects on vascular smooth muscle cells (SMC). Inhibition of SMC hyperproliferation is one of the key pharmacological strategies for prevention of atherosclerosis because it is initiated by vascular endothelial cell damage followed by an intimal hyperplasia of SMCs. Fucoidans were found to be potent inhibitors of SMC growth *in vitro* and *in vivo* [141-145]. The mechanism by which fucoidan inhibits SMC proliferation was suggested to be mediated through interaction with growth factors, alteration of the expression of fibronectin and thrombospondin, and inhibition of the mitogen-activated protein kinase pathway. Sodium spirulan also showed antiproliferative effects on not only SMC but also aortic endothelial cells [146,147].

Antioxidant Activities

In general, marine algae are being considered to be a rich source of antioxidants and active substances are suggested to be polyphenols [149]. Various algal sulfated polysaccharides have also been shown to possess antioxidant activities [22,150-154]. By the comparative study, homofucan from *F. vesiculosus* and λ-carrageen showed stronger antioxidant effect than those of heterofucan from *Padina gymnospora* and κ- and ι-carrageenans. [154]. Additionally, antioxidant effect of sulfated polysaccharides seemed to be dependent on their sulfated content and molecular weight [152,155].

Future Perspective

During the past three decades, many sulfated polysaccharides have been isolated from algae, and their biological activities have attracted more attention recently in biomedical and medical fields. A large amount of experimental and clinical data corroborates the usefulness of algal sulfated polysaccharides.

Anticoagulant and antithrombotic effects of algal sulfated polysaccharides might contribute the development of the heparin alternatives and novel drugs. Heparin, which is a glycosaminoglycan, has been widely used as a therapeutic anticoagulant to prevent and treat with thromboembolic disorders. However, there are some problems related to its clinical application, e.g., it is ineffective in treating antithrombin-deficient patients because of bleeding complications and leading to heparin-induced thrombocytopenia. On the other hand, one of the sources of heparin is bovine tissue, thus, there is a risk of contamination of prions which cause bovine spongiform encephalitis (BSE). Taking these problems into account, it is significant to obtain anticoagulant polysaccharides from non-animal sources. From the point of view, algal sulfated polysaccharides might be important sources since most of sulfated polysaccharides from algae possess anticoagulant effects.

Antitumor and antiviral effects of sulfated polysaccharides, in part, may be affiliated with immunomodulating activities. Recent advances have been made toward understanding host immune responses to infectious diseases and cancer. Thus, one of the most promising recent alternatives to classical antibiotic treatment is the use of immunomodulators (formally biological response modifier) for enhancing host defense responses [133]. In other words, the development of preventative strategies to resist disease could certainly be a more efficient and possibly more effective long-term healthcare strategy.

Teas *et al* showed an interesting hypothesis that regular consumption of dietary algae might help to prevent HIV infection and suppress viral load among those infected since dramatic difference in HIV/AIDS prevalence rates exist between algae-eating populations in Eastern Asia and most of Africa [156]. Although the reason why ingestion of algae exerts to the HIV/AIDS prevalence is not clear, it is suggested to be mediated by sulfated polysaccharides contained in the algae. Hayashi *et al* reported that oral intake of fucoidan might take the protective effects through direct inhibition of viral replication and stimulation of both innate and adaptive immune systems [138]. Similarly, GFS which was a capsule prepared from Tasmanian *U. pinnatifida* associated with increased healing rates in patients with active infections and inhibition of reactivation of latent infection [158]. Furthermore, *Helicobacter pylori*-induced gastritis and the prevalence of *H. pylori* were markedly reduced by administration of fucoidan from *Cladosiphon in vivo* [159]. On the other hand, orally administered λ-carrageenans also showed *in vivo* antitumor and immunomodulatory effects [128]. These results strongly revealed that the algal sulfated polysaccharides are potent candidates to enhance host immune responses against infectious organisms and cancer. On the other hand, fucoidan from *C. okamuranus* showed prophylactic effect against enteral prion infection by orally administration [160]. Although its prophylactic mechanism of fucoidan is still unknown, it might be an important finding for future application of sulfated polysaccharides. In addition, many algae have a long history as important food resources and crude drugs. Therefore, there are no problems to develop functional foods based on algal sulfated polysaccharides.

In the application of algal sulfated polysaccharides as functional foods, their antioxidant activities are also considered to be valuable effects for human health. The free radical theory of aging suggests that damage produced by the interaction of oxygen-derived free radicals with cellular macromolecules results in cellular senescence and aging [148]. In addition, lipid peroxidation is a major cause of many pathological effects such as cardiovascular disease, cancer, and brain dysfunction. Thus, it is essential to develop and utilize effective and natural antioxidants so that they can protect the human body from free radicals and retard the progress of many chronic diseases.

References

[1] Mayer AM, Rodríguez AD, Berlinck RG, Hamann MT. Marine pharmacology in 2003–4: marine compounds with anthelmintic, antibacterial, anticoagulant, antifungal, anti-inflammatory, antimalarial, antiplatelet, antiprotozoal, antituberculosis, and antiviral activities; affecting the cardiovascular, immune and nervous systems, and otehr miscellaneous mechanisms of action. *Comp Biochem Physiol.* 2007;145:553-581.

[2] Blunt JW, Copp BR, Hu WP, Munro MH, Northcote PT, Prinsep MR. Marine natural products. *Nat Prod Rep.* 2007;24:31-86.

[3] Watanabe A. Studies on the application of Cyanophyta in Japan. *Aqua Sci.* 1970;32:566-569.

[4] Abdulqader G, Barsanti L, Tredlici M. Harvest of *Arthrospira platensis* from Lake Kossorom (Chad) and its household usage among the Kanembu. *J Appl Phycol.* 2000;12:493-498.

[5] Hayashi K, Hayashi T, Morita N. An extract from *Spirulina platensis* is a selective inhibitor of Herpes simplex virus type 1 penetration into HeLa cells. *Phytother Res.* 1993;7:76-80.

[6] Ayehunie S, Belay A, Baba TW, Reprecht RM. Inhibition of HIV-1 replication by an aqueous extract of *Spirulina platensis* (*Arthrospira platensis*). *J Acquired Immune Deficiency Syndromes Hum Retroviruses.* 1998;18:7-12.

[7] Lee JB, Hayashi T, Hayashi K, Sankawa U, Maeda M, Nemoto T, Nakanishi H. Further purification and structural analysis of calcium spirulan from *Spirulina platnesis*. *J Nat Prod.* 1998;61:1101-1104.

[8] Lee JB, Hayashi T, Hayashi K, Sankawa U. Structural analysis of calcium spirulan (Ca-SP)-derived oligosaccharides using electrospray ionization mass spectrometry. *J Nat Prod.* 2000;63:136-138.

[9] Sudo H, Burgess JG, Takemasa H, Nakamura N, Matsunaga T. Sulfated exopolysaccharide production by the halophilic cyanobacterium *Aphanocapsa halophytia*. *Curr Microbiol.* 1995;30:219-222.

[10] Tease B, Jürgens UJ, Golecki JR, Heinrich UR, Rippka R, Weckesser J. Fine-structural and chemical analyses on inner and outer shealth of the cyanobacterium *Gloeothece* sp. PCC 6909. *Antonie von Leeuwenhoek.* 1991;59:27-34.

[11] De Phillippis R, Margheri MC, Materassi R, Vincenzini M. Potential of unicellular cyanobacteria from saline environments as exopolysaccharide producers. *Appl Environ Microbiol.* 1998;64:1130-1132.

[12] Volk RB, Venzke K, Blaschek W. Alban S. Complement modulating and anticoagulant effects of a sulfated exopolysaccharide released by the cyanobacterium *Synechocystis aquatilis. Planta Med.* 2006;72:1424-1427.

[13] Rees D. Structure, conformation, and mechanism in the formation of polysaccharide gels and networks. *Adv Carbohydr Chem Biochem.* 1969;24:267-332.

[14] Layaye M. Developments on gelling algal galactans, their structure and physico-chemistry. *J Appl Phycol.* 2001;13:173-184.

[15] Chipin T, Kerin BF, Mazerolle R. Phycocolloid chemistry as a taxonomic indicator of phylogeny in the Gigartinales, Rhodphyceae: a review and current developments using Fourier transformation infrared diffuse reflectance spectrometry. *Phycol Res.* 1999;47:167-188.

[16] Falshaw R, Frneaux RH, Stevenson DE. Structural analysis of carrageenans from the red alga, *Callophyllis homobroniana* Mont. Kütz (Kallymeniaceae, Rhodophyta). *Carbohydr Res.* 2005;340:1149-1158.

[17] Rodríguez MC, Merino ER, Pujol CA, Damonte EB, Cerezo AS, Matulewicz MC. Galactans from cystocarpic plants of the red seaweed *Callophyllis variegata* (Kallymeniaceae, Gigartinales). *Carbohydr Res.* 2005;340:2742-2751.

[18] Liao ML, Chiovitti A, Munro SLA, Craik DJ, Kraft GT, Bacic A. Sulfated galactans from Australian specimens of the red alga *Phacelocarpus peperocarpos* (Gigartinales, Rhodophyta). *Carbohydr Res.* 1996;296:237-247.

[19] Farias WRL, Valente A, Pereira M, Mourão PAS. Structure and anticoagulant activity of sulfated galactans. *J Biol Chem.* 2000;275:29299-29307.

[20] Rees DA, Conway E. The structure and biosynthesis of porphyran: a comparison of some samples. *Biochem J.* 1962;84:411-416.

[21] Morrice LM, McLEan MW, Long WF, Willamson FB. Porphyran primary structure. *Eur J Biochem.* 1983;133:673-684.

[22] Zhang Q, Li N, Liu X, Zhao Z, Li Z, Xu Z. The structure of a sulfated galactan from *Porphyra haitanensis* and its in vivo antioxidant activity. *Carbohydr Res.* 2004;339:105-111.

[23] Takano R, Hayashi K, Hara S, Hirase S. Funoran from the red seaweed, *Gloiopertis complanata*: polysaccharides with sulphated agarose structure and their precursor structure. *Carbohydr Polym.* 1995;27:305-311.
[24] Ciancia M, Matulewicz MC, Cerezo AS. A L-galactose-containing carrageenan from cystocarpic *Gigartina skottsbergii*. *Phytochemistry.* 1997;45:1009-1013.
[25] Shortz CA, Cases MR, Cerezo AS. The system of agaroids and carrageenans from the soluble fraction of the tetrasporic stage of the red seaweed *Iridaea undulosa*. *Carbohydr Polym.* 1997;34:61-65.
[26] Navarro DA, Stortz CA. Determination of the configuration of 3,6-anhydrogalactose and cyclized a-galactose 6-sulfate units in red seaweed galactans. *Carbohydr Res.* 2003;338:2111-2118.
[27] Estevez JM, Ciancia M, Cerezo AS. DL-Galactan hybrids and agarans from gametophytes of the red seaweed *Gymnogongrus torulosus*. *Carbohydr Res.* 2001;331:27-41.
[28] Estevez JM, Ciancia M, Cerezo AS. DL-Galactan hybrids and agarans from gametophytes of the red seaweed *Gymnogongrus torulosus*. *Carbohydr Res.* 2001;331:27-41.
[29] Estevez JM, Ciancia M, Cerezo AS. The system of galactans of the red seaweed, *Kappaphycus alvarezii*, with emphasis on its minor constituents. *Carbohydr Res.* 2004;339:2575-2592.
[30] Zibetti, RGM, Noseda MD, Cerezo AS, Duarte MER. The system of galactans from *Cryptonemia crenulata* (Halymeniaceae, Halaymeniales) and the structure of two major fractions. Kinetic studies on the alkaline cyclization of the unusual diad G2S→D(L)6S. *Carbohydr Res.* 2005; 340:711-722.
[31] Cases MR, Stortz CA, Cerezo AS. Structure of the 'corallinans' – sulfated xylogalactans from *Corallina officinalis*. *Int J Biol Macromol.* 1994;16:93-97.
[32] Usov AI, Bilan MI, Shashkov AS. Structure of a sulfated xylogalactan from the calcareous red alga *Corallina pilulifera* P. *et* R. (Rhodophyta, Corallinaceae). *Carbohydr Res.* 1997;303:93-102.
[33] Kolender AA, Matulewicz MC. Sulfated polysaccharides from the red seaweed *Georgiella confluens*. *Carbohydr Res.* 2002;337:57-68.
[34] Errea MI, Matulewicz MC. Unusual structures in the polysaccharides from the red seaweed *Pterocladiella capillacea* (Gelidiaceae, Gelidiales). *Carbohydr Res.* 2003;338:943-953.
[35] Kolender AA, Matulewicz MC, Cerezo AS. Structural analysis of antiviral sulfated α-D-(1→3)-mannans. *Carbohydr Res.* 1995;273:179-185.
[36] Kolender AA, Pujol CA, Damonte EB, Matulewicz MC, Cerezo AS. The system of sulfated α-(1→3)-linked D-mannans from the red seaweed *Nothogenia fastigiata*: structures, antiherpetic and anticoagulant properties. *Carbohydr Res.* 1997;304:53-60.
[37] Percival E, Ross AG. The isolation and purification of fucoidin from brown seaweeds. *J Chem Soc.* 1950:717-720.
[38] Patankar MS, Oehninger S, Barnett T, Wiiliams RL, Clark GF. A revised structure for fucoidan may explain some of its biological activities. *J Biol Chem.* 1993;268:21770-21776.

[39] Bilan MI, Grachev AA, Ustuzhanina NE, Shashkov AS, Nifantiev NE, Usov AI. Structure of a fucoidan from the brown seaweed *Fucus evanescens* C.Ag. *Carbohydr Res.* 2002;337:719-730.

[40] Bilan MI, Grachev AA, Ustuzhanina NE, Shashkov AS, Nifantiev NE, Usov AI. A highly regular fraction of a fucoidan from the brown seaweed *Fucus distichus* L. *Carbohydr Res.* 2004;339:511-517.

[41] Bilan MI, Grachev AA, Shashkov AS, Nifantiev NE, Usov AI. Structure of a fucoidan from the brown seaweed *Fucus serratus* L. *Carbohydr Res.* 2006;341:238-245.

[42] Hoshino T, Hayash T, Hayashi K, Hamada J, Lee JB, Sankawa U. An antivirally active sulfated polysaccharide from *Sargassum horneri*. *Biol Pharm Bull.* 1998;21:730-734.

[43] Preeprame S, Hayashi K, Lee JB, Sankawa U, Hayashi T. A novel antivirally active fucan sulfate derived from an edible brown alga, *Sargassum horneri*. *Chem Pharm Bull.* 2001;49:484-485.

[44] Marais MF, Joseleau JP. A fucoidan from *Ascophyllum nodosum*. *Carbohydr Res.* 2001;336:155-159.

[45] Nagaoka M, Shibata H, Kimura-Takagi I, Hashimoto S, Kimura K, Makino T, Aiyama R, Ueyama S, Yokokura T. Structural study of fucoidan from *Cladisiphon okamuranus* TOKIDA. *Glycoconj J.* 1999;16:19-26.

[46] Chizhov AO, Dell A. Morris HR, Haslam SM, McDowell RA, Shashkov AS, Nifant'ev NE, Khatuntseva EA, Usov AI. A study of fucoidan from the brown seaweed *Chorda filum*. *Carbohydr Res.* 1999;320:108-119.

[47] Zvyagintseva TN, Shevchenko NM, Chizhov AO, Krupnova TN, Sundukova EV, Isakov VV. Water-soluble polysaccharides of some far-eastern brown seaweeds. Distribution, structure, and their dependence on the developmental conditions. *J Exp Mar Biol Ecol.* 2003;294:1-13.

[48] Bilan MI, Zakharova AN, Grachev AA, Shashkov AS, Nifantiev NE, Usov AI. Polysaccharides of Algae: 60. Fucoidan from the pacific brown alga *Analipus japonicus* (Harv.) Winne (Ectocarpales, Scytosiphonaceae). *Russ J Bioorg Chem.* 2007;33:38-46.

[49] Zhuang C. Itoh H, Mizuno T, Ito H. Antitumor active fucoidan from the brown seaweed, *Umitranoo* (*Sargassum thunbergii*). *Biosci Biotechnol Biochem.* 1995;59:563-567.

[50] Duarte ME, Cardoso MA, Noseda MD, Cerezo AS. Structural studies on fucoidans from the brown seaweed *Sargassum stenophyllum*. *Carbohydr Res.* 2001;333:281-293.

[51] Nishino T, Kiyohara H, Yamada H, Nagumo T. An anticoagulant fucoidan from the brown seaweed *Ecklonia kurome*. *Phytochemistry*. 1991;30:535-539.

[52] Nishino T, Nagumo T, Kiyohara H, Yamada H. Structural characterization of a new anticoagulant fucan sulfate from the brown seaweed *Ecklonia kurome*. *Carbohydr Res.* 1991;211:77-90.

[53] Ponce NM, Pujol CA, Damonte EB, Flores ML, Stortz CA. Fucoidans from the brown seaweed *Adenocystis utricularis*: extraction methods, antiviral activity and structural studies. *Carbohydr Res.* 2003;338:153-165.

[54] Lee JB, Hayashi K, Hashimoto M, Nakano T, Hayashi T. Novel antiviral fucoidan from sporophyll of *Undaria pinnatifida* (Mekabu). *Chem Pharm Bull.* 2004;52:1091-1094.

[55] Hemmingson JA, Falshaw R, Furneaux RH, Thompson K. Structure and antiviral activity of the galactofucan sulfates extracted from *Undaria pinnatifida* (Phaeophyta). *J Appl Phycol.* 2006;18:185-193.

[56] Rocha HA, Moraes FA, Trindade ES, Franco CR, Torquato RJ, Veiga SS, Valente AP, Mourão PA, Leite EL, Nader HB, Dietrich CP. Structural and hemostatic activities of a sulfated galactofucan from the brown alga *Spatoglossum schroederi. J Biol Chem.* 2005;280:41278-41288.

[57] Harada N, Maeda M. Chemical structure of antithrombin-active rhamnan sulfate from *Monostroma nitidum. Biosci Biotechnol Biochem.* 1998;62:1647-1652.

[58] Lee JB, Yamagaki T, Maeda M, Nakanishi H. Rhamnan sulfate from cell walls of *Monostroma latissimum. Phytochemistry.* 1998;48:921-925.

[59] Zhang HJ, Mao WJ, Fang F, Li HY, Sun HH, Chen Y, Qi XH. Chemical characteristics and anticoagulant activities of a sulfated polysaccharide and its fragments from *Monostroma latissimum. Carbohydr Polym.* 2008;71:428-434.

[60] Mao W, Li Y, Wu L, Wang H, Zhang Y, Zang X, Zhang H. Chemical characterization and radioprotective effect of polysaccharide from *Monostroma angicava* (Chlorophyta). *J Appl Phycol.* 2005;17:349-354.

[61] Carlberg GE, Percival E. The carbohydrates of the green seaweeds *Urospora wormskioldii* and *Codiolum pusillum. Carbohydr Res.* 1977;57:223-234.

[62] Lahaye M, Robic A. Structure and functional properties of ulvan, a polysaccharide from green seawees. *Biomacromolecules.* 2007;8:1765-1774.

[63] Chattopadhyay K, Mandal P, Lerouge P, Driouich A, Ghosal P, Ray B. Sulphated polysaccharides from Indian samples of *Enteromorpha compressa* (Ulvales, Chlorophyta): isolation and structural features. *Food Chem.* 2007;104:928-935.

[64] Rao EV, Ramana KS. Structural studies of a polysaccharide isolated from the green seaweed *Chaetomorpha antenina. Carbohydr Res.* 1991;217:163-170.

[65] Ramana KS, Rao EV. Structural features of the sulphated polysaccharide from a green seaweed, *Cladophora socialis. Phytochemistry.* 1991;30:259-262.

[66] Chattopadhyay K, Adhikari U, Lerouge P, Ray B. Polysaccharides from *Caulerpa racemosa*: purification and structural features. *Carbohydr Polym.* 2007;68:407-415.

[67] Love J, Percival E. The polysaccharides of the green seaweed *Codium fragile.* Part II. The water-soluble sulphated polysaccharides. *J Chem Soc.* 1964:3338-3345.

[68] Uehara T, Takeshita M, Maeda M. Studies on anticoagulant-active arabinan sulfates from the green alga, *Codium latum. Carbohydr Res.* 1992;235:309-311.

[69] Siddhanta AK, Shanmugam M, Mody KH, Goswami AM, Ramavat BK. Sulphated polysaccharides of *Codium dwarkense* Boergs. from the west coast of India: chemical composition and blood anticoagulant activity. *Int. J. Biol. Macromol.* 1999;26:151-154.

[70] Matsubara K, Matsuura Y, Bacic A, Liao ML, Hori K, Miyazawa K. Anticoagulant properties of a sulfated galactan preparation from a marine green alga, *Codium cylindricum. Int J Biol Macromol.* 2001;28:395-399.

[71] Bilan MI, Vinogradova EV, Shashkov AS, Usov AI. Structure of a highly pyruvylated galactan sulfate from the Pacific green alga *Codium yezoense* (Bryopsidales, Chlorophyta). *Carbohydr Res.* 2007;342:586-596.

[72] Farias EH, Pomin VH, Valente AP, Nader HB, Rocha HA, Mourão PA. A preponderantly 4-sulfated, 3-linked galactan from the green alga *Codium isthmocladum. Glycobiology.* 2008;18:250-259.

[73] Matsubara K, Matsuura Y, Hori K, Miyazawa K. An anticoagulant proteoglycan from the marine green alga, *Codium pugniformis*. *J Appl Phycol.* 2000;12:9-14.

[74] Bourin MC, Lindahl U. Glycosaminoglycans and the regulation of blood coagulation. *Biochem J.* 1993;289:313-330.

[75] Hayakawa Y, Hayashi T, Hayashi K, Hayashi T, Ozawa T, Niiya K, Sakuragawa N. Heparin cofactor II-dependent antithrombnin activity of calcium spirulan. *Blood Coagulation Fibrinolysis.* 1996;7:554-560.

[76] Hayakawa Y, Hayashi T, Lee JB, Ozawa T, Sakuragawa N. Activation of heparin cofactor II by calcium spirulan. *J Biol Chem.* 2000;275:11379-11382.

[77] Hayakawa Y, Hirashima Y, Yamamoto H, Kurimoto M, Hayashi T, Lee JB, Endo S. Mechanism of activation of heparin cofactor II by calcium spirulan. *Arch Biochem Biophys.* 2003;416:47-52.

[78] Pereira MG, Benevides NM, Melo MR, Valente AP, Melo FR, Mourão PA. Structure and anticoagulant activity of a sulfated galactan from the red alga, *Gelidium crinale*. Is there a specfic structural requirement for the anticoagulant action? *Carbohydr Res.* 2005;340:2015-2023.

[79] Mauray S, de Raucourt E. Talbot JC, Dachary-Prigent J. Jozefowicz M. Fischer AM. Mechanism of factor IXa inhibition by antithrombin in the presence of unfractionated and low molecular weight heparins and fucoidan. *Biochim Biophys Acta.* 1998; 1387:184-194.

[80] Pereira MS, Mulloy B, Mourão PA. Structure and anticoagulant activity of sulfated fucans. *J Biol Chem.* 1999;274:7656-7667.

[81] Kuznetsova TA, Besednova NN, Mamaev AN, Momot AP, Shevchenko NM, Zvyagintseva TN. Anticoagulant activity of fucoidan from brown algae *Fucus evanescens* of the Okhotsk sea. *Bull Exp Biol Med.* 2003;5:471-473.

[82] Church FC, Maeda JB, Treanor RE, Whinna HC. Antithrombin activity of fucoidan. *J Biol Chem.* 1989;264:3618-3623.

[83] Jung WK, Athukorala Y, Lee YJ, Cha SH, Lee CH, Vasanthan T, Choi KS, Yoo SH, Kim SK, Jeon YJ. Sulfated polysaccharide purified from *Ecklonia cava* accelerates antithrombin III-mediated plasma proteinase inhibition. *J Appl Phycol.* 2007; 19:425-430.

[84] Nishino T, Aizu Y, Nagumo T. Antithrombin activity of a fucan sulfate from the brown seaweed *Ecklonia kurome*. *Thromb Res.* 1991;62:765.

[85] Yoon SJ, Pyun YU, Hwang JK, Mourão PA. A sulfated fucan from the brown alga *Laminaria cichorioide* has mainly heparin cofactor II-dependent anticoagulant activity. *Carbohydr Res.* 2007;342:2326-2330.

[86] Hayakawa Y, Hayashi T, Lee JB, Srisomporn P, Maeda M, Ozawa T, Sakuragawa N. Inhibition of thrombin by sulfated polysaccharides isolated from green algae. *Biochim Biophys Acta.* 2000;1543:86-94.

[87] Mao W, Zang X, Li Y, Zhang H. Sulfated polysaccharides from marine green algae *Ulva conglobata* and their anticoagulant activity. *J Appl Phycol.* 2006;18:9-14.

[88] Jurd KM, Rogers DJ, Blunden G, McLellan DS. Anticoagulant properties of sulphated polysaccharides and proteoglycans from *Codium fragile* ssp. *atlanticum*. *J Appl Phycol.* 1995;7:339-345.

[89] Chandía NP, Matsuhiro B. Characterization of a fucoidan from *Lessonia vadosa* (Phaeophyta) and its anticoagulant and elicitor properties. *Int J Biol Macromol.* 2008;42:235-240.

[90] Haroun-Bouhedja F, Ellouali M, Sinquin C, Boisson-Vidal C. Relationship between sulfate groups and biological activities of fucans. *Thromb Res.* 2000;100:453-459.

[91] Carlucci MJ, Pujol CA, Ciancia M, Noseda MD, Matulewicz MC, Damonte EB, Cerezo AS. Antiherpetic and anticoagulant properties of carrageenans from the red seaweed *Gigartina skottsbergii* and their cyclized derivatives: correlation between structure and biological activity. *Int J Biol Macromol.* 1997;20:97-105.

[92] Dace R, McBride E, Brooks K, Gander J, Buszko M, Doctor VM. Comparison of the anticoagulant action of sulfated and phosphorylated polysaccharides. *Thromb Res.* 1997;87:113-121.

[93] Chevolot L, Fourcault A, Chaubet F, Kervarec N, Sinquin C, Fisher AM, Boisson-Vidal C. Further data on the structure of brown seaweed fucans: relationships with anticoagulant activity. *Carbohydr Res.* 1999;319:154-165.

[94] Doctor VM, Hill C, Jackson GJ. Effect of fucoidan during activation of human plasminogen. *Thromb Res.* 1995;79:237-247.

[95] Nishino T, Yamauchi T, Horie M, Nagumo T, Suzuki H. Effects of a fucoidan on the activation of plasminogen by u-PA and t-PA. *Thromb Res.* 2000;99:623-634.

[96] Hall G, Lang D, Qiu X, Doctor V. Effect of native fucoidan, sulfated fucoidan, heparin and 6-aminohexanoic acid on the activation of glutamic-plasminogen by urokinase: role of NaCl. *Blood Coagul Fibrinolysis* 2006;17:277-281.

[97] Hayakawa Y, Hayashi T, Hayashi K, Ozawa T, Niiya K, Sakuragawa N. Calcium spirulan as an inducer of tissue-type plasminogen activator in human fetal lung fibroblasts. *Biochim Biophys Acta.* 1997;1355:241-247.

[98] Yamamoto C, Nakamura A, Shimada S, Kaji T, Lee JB, Hayashi T. Differential effects of sodium spirulan on the secretion of fibrinolytic proteins from vascular endothelial cells: enhancement of plasminogen activator activity. *J Health Sci.* 2003;49:405-409.

[99] Leite EL, Medeiros MG, Rocha HA, Farias GG, da Silva LF, Chavante SF, de Abreu LD, Dietrich CP, Nader HB. Structure and pharmacological activities of sulfated xylofucoglucuronan from the alga *Spatoglossum schröederi. Plant Sci.* 1998;132:215-228.

[100] Kaji T, Shimada S, Yamamoto C, Fujiwara Y, Lee JB, Hayashi T. Inhibition of the asociation of proteoglycans with cultured vascular endothelial cell layers by calcium and sodium spirulan. *J Health Sci.* 2002;48:250-255.

[101] Baba M, Snoeck R, Pauwels R, De Clercq E. Sulfated polysaccharides are potent and selective inhibitors of various enveloped viruses, including herpes simplex virus, cytomegalovirus, vesicular stomatitis virus, and human immunodeficiency virus. *Antimicrob Agents Chemother.* 1988;32:1742-1745.

[102] Hayashi T, Hayashi K, Maeda M, Kojima I. Calcium spirulan, an inhibitor of enbeloped virus replication, from a blue-green alga *Spirulina platensis. J Nat Prod.* 1996;59:83-87.

[103] Rechter S, König T, Auerochs S, Thulke S, Walter H, Dörnenburg H, Walter C, Marschall M. Antiviral activity of *Arthrospira*-derived spirulan-like substances. *Antiviral Res.* 2006;72:197-206.

[104] Duarte ME, Nosede DG, Noseda MD, Tulio S, Pujol CA, Damonte EB. Inhibitory effect of sulfated galactans from the marine alga *Bostrychia montagnei* on herpes simplex virus replication in vitro. *Phytomedicine.* 2001;8: 53-58.

[105] Pujol CA, Estevez JM, Carlucci MJ, Ciancia M, Cerezo AS, Damonte EB. Novel DL-galactan hybrids from the red seaweed *Gymnogongrus torulosus* are potent inhibitors of herpes simplex virus and dengue virus. *Antiviral Chem Chemother.* 2002;13:83-89.

[106] Zhu W, Chiu LC, Ooi VE, Chan PK, Ang Jr PO. Antiviral property and mode of action of a sulphated polysaccharide from *Sargassum patens* against herpes simplex virus type 2. *Int J Antimicrob Agents.* 2004;24:81-85.

[107] Ghosh P, Adhikari U. Ghosal PK, Pujol CA, Carlucci MJ, Damonte EB, Ray B. In vitro anti-herpetic activity of sulfated polysaccharide fractions from *Caulerpa racemosa. Phytochemistry.* 2004;65:3151-3157.

[108] Talarico LB, Zibetti RG, Faria PC, Scolaro LA, Duarte ME, Noseda MD, Pujol CA, Damonte EB. Anti-herpes simplex virus activity of sulfated galactans from the red seaweed *Gymnogongrus griffithsiae* and *Cryptonemia crenulata. Int J Biol Macromol.* 2004;34:63-71.

[109] Lee JB, Hayashi K, Maeda M. Hayashi T. Antiherpetic activities of sulfated polysaccharides from green algae. *Planta Medica.* 2004;70:813-817.

[110] Talarico LB, Pujol CA, Zibetti RG, Faría PC, Noseda MD, Duarte ME, Damonte EB. The antiviral activity of sulfated polysaccahrides against dengue virus is dependent on virus serotype and host cell. *Antiviral Res.* 2005;66:103-110.

[111] de S.F-Tischer PC, Talarico LB, Noseda MD, Guimarães SM, Damonte EB, Duarte ME. Chemical structure and antiviral activity of carrageenans from *Meristiella gelidium* against herpes simplex and dengue virus. *Carbohydr Polym.* 2006;63:459-465.

[112] Matsuhiro B, Conte AF, Damonte EB, Kolender AA, Matulewicz MC, Mejías EG, Pujol CA, Zúñiga EA. Structural analysis and antiviral activity of a sulfated galactan from the red seaweed *Schizymenia binderi* (Gigartinales, Rhodophyta). *Carbohydr Res.* 2005;340:2392-2402.

[113] Zhu W, Chiu LC, Ooi VE, Chan PK, Ang Jr PO. Antiviral property and mechanisms of a sulphated polysaccharide from the brown alga *Sargassum patens* against herpes simplex type 1. *Phytomedicine.* 2006;13:695-701.

[114] Adhikari U, Mateu CG, Chattopadhyay K, Pujol CA, Damonte EB, Ray B. Structure and antiviral activity of sulfated fucans from *Stoechospermum marginatum. Phytochemistry.* 2006;67:2472-2482.

[115] Lee JB, Hayashi K, Hayashi T, Maeda M, Sankawa U. Antiviral activities against HSV-1, HCMV, and HIV-1 of rhamnan sulfate from *Monostroma latissimum. Planta Med.* 1999;65:439-441.

[116] Cáceres PJ, Carlucci MJ, Damonte EB, Matsuhiro B, Zúñiga EA. Carrageenans from chilean samples of *Stenogramme interrupta* (Phyllophoraceae): structural analysis and biological activity. *Phytochemistry.* 2000;53:81-86.

[117] Chattopadhyay K, Mateu CG, Mandal P, Pujol CA, Damonte EB, Ray B. Galactan suflate of *Grateloupia indica*: isolation, structural features and antiviral activity. *Phytochemistry.* 2007;68:1428-1435.

[118] Wang SC, Bligh SW, Shi SS, Wang ZT, Hu ZB, Crowder J, Brandord-White C. Vella C. Structural features and anti-HIV-1 activity of novel polysaccahrides from red algae

Grateloupia longifolia and *Grateloupia filicina*. *Int J Biol Macromol.* 2007;41:369-375.

[119] Baba M, Pauwels R, Balzarini J, Arnout J, Desmyter J, De Clercq E. Mechanism of inhibitory effect of dextran sulfate and heparin on replication of human immunodeficiency virus. *Proc Natl Acad Sci USA.* 1988;85:6132-6136.

[120] Baba M, Snoeck R, Pauwels R, De Clercq E. Sulfated polysaccharides are potent and selective inhibitors of various enveloped viruses, including herpes simplex virus, cytomegalovirus, vesicular stomatitis virus, and human immunodeficiency virus. *Antimicrob Agents Chemother.* 1988;32:1742-1745.

[121] Hayashi K, Hayashi T, Kojima I. A natural sulfated polysaccharide, calcium spirulan, isolated from *Spirulina platensis*: *In vitro* and *ex vivo* evaluation of anti-herpes simplex virus and anti-human immunodeficiency virus activities. *AIDS Res Hum Retroviruses.* 1996;12:1463-1471.

[122] Carlucci MJ, Ciancia M, Matulewicz MC, Cerezo AS, Damonte EB. Antiherpetic activity and mode of action of deverse structural types. *Antiviral Res.* 1999;43:93-102.

[123] Hayashi T. Studies on evaluation of natural products for antiviral effects and their application. *Yakugaku Zasshi*, 2008;128:61-79. (in Japanese)

[124] Noda H, Amano H, Arashima K, Nisizawa K. Antitumor activity of marine algae. *Hydrobiologia.* 1990;204/205:577-584.

[125] Koyanagi S, Tanigawa N, Nakagawa H, Soeda S, Shimeno H. Oversulfation of fucoidan enhances its anti-angiogenic and antitumor activities. *Biochem Pharmacol.* 2003;65:173-179.

[126] Dias PF, Siqueira Jr JM, Vendruscolo LF, Neiva T, Gagliardi AR, Maraschin M, Ribeiro-do-Valle RM. Antiangiogenic and antitumoral properties of a polysaccharide isolated from the seaweed *Sargassum stenophyllum*. *Cancer Chemother Pharmacol.* 2005;56:436-446.

[127] Ozawa T, Yamamoto J, Yamagishi T, Yamazaki N, Nishizawa M. Two fucoidans in the holdfast of cultivated *Laminaria japonica*. *J Nat Med.* 2006;60:236-239.

[128] Zhou G, Sun YP, Xin H, Zhang Y, Li Z, Xu Z. In vivo antitumor and immunomodulation activities of different molecular weight lambda-carrageenans from *Chondrus ocellatus*. *Pharmacol Res.* 2004;50:47-53.

[129] Yuan H, Song J, Li X, Li N, Dai J. Immunodulation and antitumor activity of κ-carrageenan oligosaccharides. *Cancer Lett.* 2006;243:228-234.

[130] Bobek V, Kovarík J. Antitumor and antimetastatic effect of warfarin and heparins. *Biomed Pharmacother.* 2004;58:213-219.

[131] Mishima T, Murata J, Toyoshima M, Fujii H, Nakajima M, Hayashi T, Kato T, Saiki I. Inhibition of tumor invasion and metastasis by calcium spirulan (Ca-SP), a novel sulfated polysaccharide derived from a blue-green alga, *Spirulina platensis*. *Clin Exp Metastasis.* 1998;16:541-550.

[132] Alekseyenko TV, Zhanayeva SY, Venediktova AA, Zvyagintseva TN, Kuznetsova TA, Besednova NN, Korolenko TA. Antitumor and antimetastatic activity of fucoidan, a sulfated polysaccahride isolated from the Okhotsk Sea *Fucus evanescens* brown alga. *Bull Exp Biol Med.* 2007;143:730-32.

[133] Tzianabos AO. Polysaccharide immunomodulators as therapeutic agents: structural aspects and biologic function. *Clin Microbiol Rev.* 2000;13:523-533.

[134] Schepetkin IA, Quinn MT. Botanical polysaccharides: macrophage immunomodulation and therapeutic potential. *Int Immunopharmacol.* 2006;6:317-333.

[135] Choi EM, Kim AJ, Kim YO, Hwang JK. Immunomodulating activity of arabinogalactan and fucoidan *in vitro*. *J Med Food.* 2005;8:446-453.

[136] Nakamura T, Suzuki H, Wada Y, Kodama T, Doi T. Fucoidan induces nitric oxide production via p38 mitogen-activated protein kinase and NF-κB-dependent signaling pathways through macrophage scavenger receptors. *Biochem Biophys Res Commun.* 2006;343:286-294.

[137] Yang JW, Yoon SY, Oh SJ, Kim SK, Kang KW. Bifunctional effects of fucoidan on the expresseion of inducible nitric oxide synthase. *Biochem Biophys Res Commun.* 2006;346:345-350.

[138] Hayashi K, Nakano T, Hashimoto M, Kanekiyo K, Hayashi T. Defensive effects of a fucoidan from brown alga *Undaria pinnatifida* against herpes simplex virus infection. *Int Immunopharmacol.* 2008;8:109-116.

[139] Kim MH, Joo HG. Immunostimulatory effects of fucoidan on bone marrow-derived dendritic cells. *Immunol Lett.* 2008;115:138-143.

[140] Leiro JM, Castro R, Arranz JA, Lamas J. Immunomodulating activities of acidic sulphated polysaccharides obtained from the seaweed *Ulva rigida* C. Agardh. *Int Immunopharmacol.* 2007;7:879-888.

[141] McCaffrey TA, Falcone DJ, Vicente D, Du B, Consigli S, Borth W. Protection of transforming growth factor-beta 1 activity by heparin and fucoidan. *J Cell Physiol.* 1994;159:51-59

[142] Vischer P, Buddecke E. Different action of heparin and fucoidan on arterial smooth muscle cell proliferation and thrombospondin and fibronectin metabolism. *Eur J Cell Biol.* 1991;56:407-414.

[143] Logeart D, Prigent-Richard S, Jozefonvicz J, Letourneur D. Fucans, sulfated polysaccahrides extracted from brown seaweed, inhibit vascular smooth muscle cell proliferation I: comparison with heparin for antiproliferative activity, binding and internalization. *Eur J Cell Biol.* 1997;74:376-384.

[144] Religa P, Kazi M, Thyberg J, Gaciong Z, Swedenborg J, Hedin U. Fucoidan inhibits smooth muscle cell proliferation and reduces mitogen-activated protein kinase activity. *Eur J Endovasc Surg.* 2000;20:419-426.

[145] Deux JF, Meddahi-Pellé A, Le Blanche AF, Feldman LJ, Colliec-Jouault S, Brée F, Boudghéne F, Michel JB, Letourneur D. Low molecular weight fucidan prevents neointimal hyperplasia in rabbit iliac artery in-stent restenosis model. *Arterioscler Thromb Vasc Biol.* 2002;22:1604-1609.

[146] Kaji T, Okabe M. Shimada S. Yamamoto C, Fujiwara Y, Lee JB, Hayashi T. Sodium spirulan as a potent inhibitor of arterial smooth muscle cell proliferation in vitro. *Life Sci.* 2004;74:2431-2439.

[147] Kaji T, Fujiwara Y, Hamada C, Yamamoto C, Shimada S, Lee JB, Hayashi T. Inhibition of cultured bovine aortic endothelial cell proliferation by sodium spirulan, a new sulfated polysaccharide isolated from *Spirulina platensis*. *Planta Med.* 2002;68:505-509.

[148] Wickens AP. Ageing and the free radical theory. *Respir Physiol.* 2001;128:379-391.

[149] Nagai T, Yukimoto T. Preparation and functional properties of beverages made from sea algae. *Food Chem.* 2003;81:327-332.

[150] Rupérez P, Ahrazem O, Leal JA. Potential antioxidant capacity of sulfated polysaccharides from the edible marine brown seaweed *Fucus vesiculosus*. *J Agric Food Chem.* 2002;50:840-845.

[151] Zhang Q, Li N, Zhou G, Lu X, Xu Z, Li Z. In vivo antioxidant activity of polysaccharide fraction from *Porphyra haitanensis* (Rhodophyta) in aging mice. *Pharmacol Res.* 2003;48:151-155.

[152] Qi H, Zhang Q, Zhao T, Chen R, Zhang H, Niu X, Li Z. Antioxidant activity of different sulfate content derivatives of polysaccharide extracted from *Ulva pertus* (Chlorophyta) in vitro. *Int J Biol Macromol.* 2005;37:195-199.

[153] Josephine A, Amudha G, Veena CK, Preetha SP, Rajeswari A, Varalakshmi P. Beneficial effects of sulfated polysaccharides from *Sargassum wightii* against mitochondrial alterations induced by cyclosporine A in rat kidney. *Mol Nutr Food Res.* 2007;51:1413-1422.

[154] de Souza MC, Marques CT, Dore CM, da Silva FR, Rocha HA, Leite EL. Antioxidant activities of sulfated polysaccharides from brown and red seaweeds. *J Appl Phycol.* 2007;19:153-160.

[155] Zhao T, Zhang Q, Qi H, Zhang H, Niu X, Xu Z, Li Z. Degradation of porphyran from *Porphyra haitanensis* and the antioxidant activities of the degraded porphyrans with different molecular weight. *Int J Biol Macromol.* 2006;38:45-50.

[156] Teas J, Hebert JR, Fitton JH, Zimba PV. Algae – a poor man's HAART? *Med Hypohteses.* 2004;62:507-510.

[157] Li B, Wei XJ, Sun JL, Xu SY. Structural investigation of a fucoidan containing a fucose-free core from the brown seaweed *Hizikia fusiforme*. *Carbohydr Res.* 2006;341:1135-1146.

[158] Cooper R, Dragar C, Elliot K, Fitton JH, Godwin J, Thompson K. GFS, a preparation of Tasmanian *Undaria pinnatifida* is associated with healing and inhibition of reactivation of herpes. *BMC Complement Altern Med.* 2002;2:11.

[159] Shibata H, Iimuro M, Uchiya N, Kawamori T, Nagaoka M, Ueyama S, Hashimoto S, Yokokura T, Sugimura T, Wakabayashi K. Preventive effects of *Cladisiphon* fucoidan against *Helicobacter pylori* infection in Mongolian gerbils. *Helicobacter.* 2003;8:59-65.

[160] Doh-ura K, Kuge T, Uomoto M, Nishizawa K, Kawasaki Y, Iha M. Prophylactic effect of dietary seaweed fucoidan against enteral prion infection. *Antimicrob Agents Chemother.* 2007;51:2274-2277.

In: Algae: Nutrition, Pollution Control and Energy Sources
Editor: Kristian N. Hagen, pp. 177-193
ISBN: 978-1-60692-008-4

Chapter 9

SEAWEEDS AND THYROID GLAND – POTENTIAL SEQUELAE OF SEAWEED-DERIVED IODINE

Karsten Müssig*

Division of Endocrinology, Diabetology, Nephrology, Angiology, and Clinical Chemistry, Department of Internal Medicine, University Hospital of Tübingen, Germany

Abstract

Marine algae traditionally serve as food and medicine in East-Asian countries as well as in Celtic areas, such as Scotland, Ireland, and Brittany. In recent years the use of seaweeds in cuisine and as health foods has been adopted in many European and North American countries. An increasing number of studies support anti-cancer properties of seaweed. Furthermore, it has been found to be a very nutritious food source containing large amounts of antioxidants, vitamins, and trace elements. For instance, seaweeds are known to comprise high quantities of iodine. Iodine concentrations vary widely among seaweed species, with geographic and seasonal variations as well as post-harvest storage conditions as further contributing factors.

In humans, iodine plays a crucial role in thyroid hormone synthesis. Thus, iodine excess following the consumption of seaweeds or seaweed-based products may affect thyroid gland function. In countries where marine algae are traditionally consumed as food, regular seaweed intake appears to be frequently associated with thyroid enlargement, hypofunction, and Hashimoto's thyroiditis, a chronic inflammatory autoimmune disease of the thyroid gland. In contrast, in patients with underlying thyroid disease excessive consumption of seaweed or seaweed-containing dietary supplements may cause thyroid hyperfunction.

Introduction

Harvesting and utilization of seaweeds are part of millennial cultures in many maritime countries, especially in East-Asian regions and some Celtic areas, such as Scotland, Ireland, and Brittany. Seaweeds have been traditionally used not only as food for humans and animals,

* E-mail address: Karsten.Muessig@med.uni-tuebingen.de. Phone: +49-(0)7071-29-83670, Fax: +49-(0)7071-29-2784, Correspondence address: Dr. Karsten Müssig, M.D., Medizinische Klinik IV, Universitätsklinikum Tübingen, Otfried-Müller-Str. 10, 72076 Tübingen, Germany.

but also as materials for industry, traditional medicine, and fertilizers and manures for terrestrial agriculture. Ancient documents show the consumption and economical importance of seaweeds in East-Asian communities. In 600 BC, Sze Teu wrote in China, "Some algae are a delicacy fit for the most honoured guests, even for the King himself" (Indergaard, 1983). In records from the first century AD the export of tangle (*Laminaria*) from Japan to China is mentioned (Matsuzaki and Iwamura, 1981). Tax records of the eighth century AD indicate the use of diverse species of seaweed as tax payments to the Japanese government (Arasaki and Arasaki, 1983).

Three classes of edible marine algae can be distinguished — the red (rhodophyta), brown (phaeophyta), and green seaweeds (chlorophyta) — based on the pigment molecules in their chloroblasts (Levring et al., 1969). The majority of green algae live in freshwater, with only 10 % being marine species. The chlorophylls A and B containing chloroblasts have a bright green colour. Brown algae are mostly marine multicellular seaweeds, reaching their greatest abundance in cold-temperate waters. The largest seaweeds found in the oceans are used in numerous ways and are commercially important for food, cosmetics, pharmaceuticals, and in sciences. For example, algin is extracted from the cell walls of brown algae and used as a thickener or emulsifier. Chloroblasts of brown algae contain, in addition to chlorophyll, the pigment fucoxanthin which gives them a brown or olive-green colour. Fucoxanthin was found to have antiobesity (Maeda et al., 2005) and anticancer effects in human prostate cancer cells (Kotake-Nara et al., 2005) and in colon cancer cells (Das et al., 2005). In line with the latter finding prior epidemiological studies showed preventive effects of seaweeds against colorectal carcinogenesis (Hoshiyama et al., 1993). Anti-cancerous effects of seaweed appear also to be a crucial contributor to the markedly decreased breast cancer rate in Japan as compared to the United States. Animal studies as well as human studies suggest that algae intake causes longer menstrual cycles and lower serum estradiol levels (Skibola, 2004; Skibola et al., 2005), both of which have been shown to be associated with a lower risk for breast cancer (Beiler et al., 2003). Some algae appear to have potent post-coital contraceptive activity which may be mediated by a reduction of the ovarian progesterone release (Premakumara et al., 1995).

Red algae are a large group of mostly multicellular marine algae. In addition to chlorophyll, red algae contain the pigments phycocyanin and phycoerythrin, which give this group their red colouration. Agar and carrageenan, two polysaccharides found in the cell walls of red algae, are widely used in food industry and scientific research as thickening and stabilising agents. Several sulfated polysaccharides, including dextran and carrageenans, proved to be potent inhibitors of human viruses, such herpes simplex virus, cytomegalovirus, human immunodeficiency virus, and human papilloma virus (Baba et al., 1988; Buck et al., 2006).

Seaweeds are still part of the daily diets in many East-Asian countries. Japan, China, and Korea have the highest yearly algae consumption with approximately 97,000, 71,000, and 10,000 tonnes, respectively, of dry products in the late 1980s. In these countries the majority of harvested edible seaweed is directly consumed, whereas in North-America and Europe algae are mainly used in industry as a source of material for agar, carrageenan, and alginate processing (Mabeau and Fleurence, 1993). The average daily marine algae consumption in Japan is approximately 14.3 g per adult person. In general, women eat more seaweed than men and and algae consumption increases with age. Approximately 20-40 percent of the adult Japanese population report daily seaweed intake (Section of Nutrition, Bureau of Public

Health, Ministry of Health and Welfare, 2003; Iso et al., 2005). Seaweeds typically eaten in Japan comprise Kombu (Laminaria japonica), Wakame (Undaria pinnatifida), Hijiki (Hizikia fusiforme), Nori (Porphyra tenera), and Mozuku (Nemacystus decipiens).

Nutritional characteristics of edible marine algae comprise a significant protein content (Fleurence, 1999), a low fat content (Jeong et al., 1993), and a high concentrations of polysaccharides, vitamins, minerals, and trace elements (Dang and Hoang, 2004). Dietary fibers in seaweed positively affect gastrointestinal bacterial flora, metabolism of carbohydrate, fat, cholesterol, and bile acids, leading to an improvement of insulin sensitivity (Tokudome et al., 2004). The cholesterol-diminishing effects of seaweeds, their water-soluble extracts, or isolated algal polysaccharides have been documented by several animal studies (Jiménez-Escrig and Sánchez-Muniz, 2000).

Though some fresh seaweeds can be eaten raw in salads, most are not very tasty when eaten fresh. Cooked marine algae are used in the preparation of meat, fish, soups and also as a vegetable with rice. Dried seaweed can serve as a snack or fried like chips. Alternatively, it may be grained and used as a seasoning for sauces and soups. Sushi is one of the best known seaweed dishes with nori ingredient (Matsuzaki and Iwamura, 1981).

The spread of Japanese and Chinese cuisine throughout the world caused a markedly increased consumption of seaweed in Western countries in the last decade (Rupérez and Saura-Calixto, 2001). Furthermore, due to the growing interest in health foods by the general public, health food stores have vastly expanded their offer of exotic products, including marine algae and seaweed-derived dietary supplements (Darcy-Vrillon, 1993).

A recent Times Online article, published April 02, 2005, praised the nutrional benefits of seaweed with "We all need a little kelp" (Wyke, 2005).

Iodine Content of Edible Seaweeds

Brown algae are the most efficient iodine accumulators among all living organisms. Due to an accumulation by a concentration factor of 1.5×10^5 of this element from seawater (Küpper et al., 1998), the iodine content of some species, i.e. *Laminaria spp.*, reaches up to 5% of dry weight (Ar Gall et al., 2004). Though the biochemical pathways of iodine accumulation in detail have still to be determined, a previous study suggested that iodide is oxidized to hypoiodous acid or molecular iodine by cell wall haloperoxidases, allowing the oxidized iodine species to pass the algal cell wall by facilitated diffusion (Küpper et al., 1998).

As Table 2 shows, iodine concentrations vary widely among species. Brown algae, such as *Laminaria japonica* with a mean iodine content of 2288 μg/g, have substantially higher iodine concentrations than red algae, including *Porphyra sp.* with an average of 33.8 μg/g iodine, and green algae, such as *Ulva pertusa* containing an average of 23.1 μg/g iodine.

In addition, iodine content of algae species depends on geographical variations which may be due to differences in salinity, water temperature, depth of the seaweed, and distance from the equator (Saenko et al., 1978). For example, iodine concentrations in the brown seaweed *Undaria pinnatifida* were found to be 24-fold higher in samples harvested in China in comparison to the mean value of samples collected in Australia, Japan, and New Zealand. Furthermore, seasonal variations contribute to differences in iodine concentrations in marine algae. Lower concentrations in the initial and later growth stages and the highest concentrations in the middle growth stage have been shown (Hou and Yan, 1998).

Accordingly, Teas et al. (2004) found an almost two-fold higher iodine content in growing juvenile compared to adult seaweed, with higher iodine concentrations in the meristematic tissue, the growing area at the base of the seaweed blade, than in the stipe. In line with these findings a recent study revealed a marked, decreasing gradient from the meristoderm, the outer layer of the stipe, to the medulla, the inner part (Verhaeghe et al., 2007).

Diminished iodine concentrations in sun-bleached blades collected from the beach or from floating drifts of seaweed indicate the rapid decline in iodine content after growth discontinuation (Teas et al., 2004). Post-harvest storage conditions are additional factors contributing to a decrease in iodine content in marine algae. Storage of seaweed in open containers or in paper bags significantly diminished iodine content, in particular under humid conditions, whereas the storage in watertight bags or boxes did not affect iodine concentrations (Marchal et al,. 2000). The decrease in iodine concentration is most probably due the high degree of water solubility of iodine in seaweed, ranging from 40 % in the red alga *Sargassum Kjellmanianum* to 99.2 % in the brown alga *Laminaria japonica* (Hou et al., 1997). According to the high water solubility of iodine in seaweed, 99 % of iodine in kombu (*Laminaria japonica*) has been found in water after 15 minutes of boiling (Ishizuki et al., 1989). Marine algae do not only differ in iodine content, but also in the chemical species of iodine. Hou et al. found that the percentage of iodide (I^-) ranged from 61.0 % in the green alga *Codium fragile* to 92.7 % in the brown alga *Dictyopteris divaricata*, whereas the organic iodine content was high in *Codium fragile* with 37.4 % and low in *Dictyopteris divaricata* with 5.5 %. The percentages of iodate ($IO3^-$) ranged from 1.4 % in *Laminaria japonica* to 4.5 % in *Sargassum Kjellmanianum* (Hou et al., 1997). The majority of organic iodine is bound with protein as residues of mono- or di-iodo-tyrosines, part of iodine with pigment and polyphenol, and just a small part with polysaccharides, such as algin, fucoidan and cellulose (Hou et al., 2000).

In accordance with the remarkably different bioavailibility of various chemical species of iodine, Aquaron et al. (2002) found differences in the iodine bioavailability of diverse marine algae in humans. Iodine bioavailability was better from the red seaweed *Gracilaria verrucosa* compared to the brown seaweed *Laminaria hyperborea* (101 % vs. 90 % in Marseille, 85 % vs. 61.5 % in Brussels). Chemical species of iodine were 80 % of iodide in *Laminaria hyperborea* and 80 % of organic iodine in *Gracilaria verrucosa*. This discrepancy between mineral and organic iodine bioavailability in red and brown seaweed may be explained by the different nature of carbohydrates in red and brown algae. While soluble fibres in the cell walls of brown seaweed are alginates, the cell walls in red seaweed contain as soluble fibres carrageenans and agar. In the stomach, alginate and agar increase food viscosity by absorption of water, that might lead to iodide being trapped like starch and, consequently, to a decrease in intestinal iodide absorption.

Thyroid Hormone Synthesis

The adult human body contains 15 to 20 mg of iodine, with the majority (80 %) located in the thyroid gland. Thyroid hormones represent the only known biological function of iodine. The two principal thyroid hormones thyroxine (tetraiodothyronine, T4) and triiodotyronine (T3) play an important role in the regulation of crucial physiological processes, such as development, growth, and metabolism. T4 and T3 contain four and three iodine atoms per molecule, respectively, equalling 65 % of T4's weight and 59 % of T3's weight (Dunn, 1998).

Iodine is absorbed in the form of iodide in the small intestine. While the greater part of food-derived iodine is iodide, iodate, which is frequently used in iodized salt or as a baking additive, is rapidly reduced to iodide in the gastrointestinal tract. Iodide is then actively transported from the plasma into the thyroid by the sodium/iodide symporter (NIS) (Eskandari et al., 1997). It follows a passive transfer from the cytosol of thyrocytes to the lumen of the follicles which contain the large glycoprotein thyroglobulin, the principal storage form of thyroid hormones. Through the enzymatic activity of thyroid peroxidase (TPO), located at the apical plasma membrane, iodide is rapidly oxidized in the presence of H_20_2 and covalently bound to tyrosine residues of thyroglobulin resulting in formation of monoionotyrosine (MIT) and diiodotyrosine (DIT). In the subsequent coupling reaction, which is also catalyzed by TPO, two DIT molecules are linked to form T4 or one MIT and one DIT moiety to form T3 (Nakamura et al., 1984). Release of thyroid hormones into the circulation requires endocytosis of thyroid hormone-laden thyroglobulin into the thyrocytes and thyroglobulin digestion by proteases (Bernier-Valentin et al., 1990). Iodotyrosine residues of MIT and DIT which comprise 70 % of thyroglobulin iodine are enzymatically deiodinated enabling intrathyroidal iodide recycling (Dunn et al., 1996). Approximately 80 % of the biologically active hormone T3 is peripherally synthesized by an enzymatic de-iodination of the precursor T4, whereas the remaining 20 % of the hormone stems from production in the thyroid. Thyroid hormone synthesis and secretion is regulated by thyroid-stimulating hormone (TSH), a glycoprotein hormone that is synthesized in and secreted from the anterior lobe of the pituitary gland. TSH has been shown to modulate gene expression of key modulators in thyroid hormone synthesis, such as NIS, TPO, and thyroglobulin (Kogai et al., 1997). Besides TSH regulation T4 and T3 production depends on iodine availability. Deficiency as well as excess of iodine may affect thyroid hormone synthesis.

Thyroid Disorders in Iodine Deficiency

According to The Food and Nutrition Board of the Institute of Medicine, the recommended daily intake of iodine is 150 μg (Institute of Medicine, Food and Nutrition Board, 2001). During pregnancy and lactation, iodine requirement increases to 220-250 μg per day (Zimmermann, 2007). Iodine deficiency still constitutes a global health problem with approximately one third of the world's population at risk of insufficient iodine intake. Iodine-deficient areas are regions, where iodine has been washed out of the soil due to glaciation and flooding.

Inadequate iodine intake initially results in low T4 serum levels, the main product of thyroid secretion. Following the thyroid-pituitary feed-back mechanisms, diminished T4 levels act as a signal to the pituitary gland to enhance the secretion of TSH. Under the influence of TSH, thyroid hormone synthesis and secretion increases and, furthermore, follicular cell hypertrophy and hyperplasia develop. An enlargement of the thyroid gland, also known as goitre, is one of the most visible signs of iodine deficiency. After an initial diffuse thyroid enlargement with time the goitre may become nodular. Some of these thyroid nodules have the tendency to produce thyroid hormones autonomously, i.e. independently of the thyroid-pituitary feed-back mechanisms.

Upon failure of the thyroid gland to compensate for iodine deficiency, thyroid hypofunction develops. In adults, clinical signs of hypothyroidism comprise weakness, hoarseness, dry, flaky, inelastic skin, puffy face, thick tongue, bradycardia, decreased pulse pressure and cardiac output, delayed relaxation phase in deep tendon reflexes, poor memory, hearing loss, chills, anorexia, as well as menorrhagia and amenorrhea in women. Hypothyroidism *in utero* or in early infancy results in permanent mental retardation, neurological defects, and growth abnormalities, known as cretinism, and is the leading cause of preventable mental retardation worldwide (Dunn, 1997). Recently, moderate iodine deficiency has been found to be associated with a diminished intelligence quotient in school children (Pineda-Lucatero et al., 2008).

Thyroid Disorders in Iodine Excess

In accordance with the recommendations of The Food and Nutrition Board of the Institute of Medicine the highest level of daily iodine consumption, without risk of adverse health effects, is 1100 µg/day (Institute of Medicine, Food and Nutrition Board, 2001). Iodine is mainly excreted in urine and to a lesser degree in feces and sweat.

In healthy individuals, the thyroid gland has intrinsic autoregulatory mechanisms to adapt to excess iodine (for review, see Braverman 1994). When plasma iodide exceeds a certain level, iodide oxidation in the thyroid, an essential first reaction in thyroid hormone synthesis, is inhibited. Animal studies suggest that the acute thyroid protection from the adverse effects of excessive iodide, also known as Wolff-Chaikoff effect (Wolff and Chaikoff, 1948), is mediated by a decrease in the active transport of plasma iodide into the thyroid (Galton and Pitt-Rivers, 1959; Braverman and Ingbar, 1963). However, this acute inhibitory effect is only transient and after approximately 48 hours thyroid hormone production normalizes due to an escape from the acute Wolff-Chaikoff effect (Wolff et al., 1949).

Individuals with underlying thyroid disease may fail to adapt to excessive iodine resulting in hypothyroidism or hyperthyroidism. The occurrence of these two disparate responses to iodine excess may be explained with differences in the sensitivity to iodide-induced turn-off in hormone biosynthesis (Fradkin and Wolff, 1983). Subjects with an elevated sensitivity, such as patients previously treated for Graves' disease, an autoimmune-triggered hyperfunction of the thyroid, with radioactive iodine, subtotal thyroidectomy, or antithyroid drugs, and those with chronic autoimmune thyroiditis, are at increased risk to develop hypothyroidism after administration of excessive iodide due to failure to escape from the acute Wolff-Chaikoff effect (Braverman, 1994). Furthermore, in individuals with a genetic predisposition, chronic ingestion of large amounts of iodine may significantly reduce organic binding of iodine by the thyroid gland leading to goiter and hypothyroidism that are reversible after iodine withdrawal (Wolff, 1969). Additional consumption of foods containing goitrogenic substances, such as cassava or cabbage, further increases the risk of goiter development (Abuye et al., 1998). In contrast, in subjects with lowered sensitivity, such as patients with endemic goitre and iodide deficiency, and in iodine-rich regions, patients with nodular goitre containing autonomous nodules, iodide excess may result in hyperthyroidism (Fradkin and Wolff, 1983).

According to the complex effects of iodine on the human thyroid, marine algae, such as *Fucus vesiculosus* (bladderwrack), has traditionally been used to treat subjects with hypothyroidism as well as with hyperthyroidism (Mills, 1991).

Thyroid Hypofunction Following Seaweed Consumption

Animal and human studies suggest that dietary seaweeds rich in iodine protect the thyroid against radioactive iodine by preventing its uptake (Maruyama and Yamamoto, 1992). Due to the substantial iodine content in some marine algae, reaching up to 5 % of dry weight, individuals who regularly consume certain seaweeds are at risk to exceed the provisional maximum tolerable daily intake of 1000 µg (Lightowler and Davies, 1998). Thus, chronic overconsumption of iodine-rich diets may also have adverse effects. Previous studies clearly indicate an association between consumption of large amounts of kelp and thyroid disorders, such as goitre, hypothyroidism alone or in combination with thyroid enlargement, Hashimoto's thyroiditis, and neonatal hypothyroidism. The causative role of seaweed in the pathophysiology of these thyroid disorders is underlined by the rapid decrease in goitre size and normalisation of thyroid function after discontinuation of seaweed consumption.

On the coast of Hokkaido, the northern island of Japan, excessive intake of seaweed, with an average of 16.1 g per day, resulting in a markedly increased mean urinary iodine excretion of 23,727 µg/day, was associated with a higher prevalence of goitre compared to the in-land city of Sapporo. The decrease of goitre size after restriction of kelp intake suggests a crucial role of excessive consumption of iodine-rich seaweed in the development of the "endemic coast goitre"(Higuchi, 1964). In line with these findings, in euthyroid school children from the coast of Hokkaido, the prevalence of goitre, ranging from 2.6 to 9.0 %, exceeded the one in a peer group from Sapporo with only 1.3 %. Females were more often affected than males, with diffuse goitre in the majority of cases. Again, consumption of a large quantity of iodine-rich seaweeds was identified as a specific environmental factor, as reflected by the very high urinary iodine excretion exceeding 20 mg per day in some subjects, and withdrawal of kelp from the usual diet induced a pronounced reduction in goitre size (Suzuki et al., 1965).

A further study on the "endemic coast goitre" in Hokkaido described a marked decrease in the prevalence of goitre in two districts of the Japanese islands in the past ten years. A marked, though statistically not significant, reduction in kelp consumption from 31.5 to 13.5 g/day/person was identified as the only change in the alimentary habits of the residents (Suzuki and Mashimoto, 1973). A recent study confirmed the association between thyroid volume and excess dietary iodine intake in children. In coastal Hokkaido, mean thyroid volume was two-fold higher than mean thyroid volume from other sites (Zimmermann et al., 2005). The only case of kelp-induced goitre and hypothyroidism outside Asia was reported in Finland. A 30-year-old woman developed goitrous hypothyroidism after a prolonged intake of seaweed tablets. Histological investigation of a thyroid fine needle biopsy showed chronic lymphocytic thyroiditis. Symptoms and signs of hypothyroidism resolved spontaneously after discontinuation of iodine intake (Liewendahl and Turula, 1972). A similar case of a 44-year-old male with seaweed-induced goitrous hypothyroidism, histologically proven Hashimoto's thyroiditis, and complete recovery after seaweed restriction has been described in Japan (Okamura et al., 1978). The pathological histological findings in both patients were most likely already present on initiation of seaweed intake and were causative for the failure to escape from the acute Wolff-Chaikoff effect. However, excessive intake of iodine-rich kelp may have caused, in addition to goitre and hypothyroidism, also chronic lymphocytic thyroiditis. This assumption is in agreement with a previous study showing a statistically significant higher prevalence of juvenile chronic lymphcytic thyroiditis in Japanese school children from a seaside area as compared to an urban area (5.3 per 1000 vs. 1.4 per 1000, respectively, $p<0.005$). There was a considerable sex difference with a female to male ratio of 6.5:1 and an increase with age (Inoue et al., 1975). The authors suggested that the higher intake of seaweed in the seaside area as a potential causative factor for the increased prevalence of juvenile chronic lymphcytic thyroiditis. However, seaweed-induced hypothyroidism is not always associated with Hashimoto's disease. A 47-year-old female developed clinical and laboratory signs of hypothyroidism after consumption of more than 50 g of kombu daily for more than five years and at least 1 g of wakame daily for six months. Histological examination of a thyroid biopsy did not reveal characteristic features of Hashimoto's thyroiditis (Yamaguchi et al., 1984). In two further cases of kelp-induced hypothyroidism, occurring after excessive (40-50 g) daily intake of confectionary made with kombu during bulimic episodes in a 20-year-old female patient (Matsubayashi et al., 1998) and after daily consumption of 100-200 g cooked wakame in a 71-year-old woman (Moriyama et al., 2006), thyroid autoantibodies were negative. In all three cases, thyroid function was restored after stopping the ingestion of seaweed. Normalisation of thyroid function after avoidance of foods with very high iodine content in patients with hypothyroidism has also been shown by several observational studies. Yoshinari et al. (1983) found that patients who were newly diagnosed with primary hypothyroidism and had a goitre with a preserved thyroid radioactive iodine uptake were likely to become spontaneously euthyroid after dietary iodine restriction. In contrast, in patients with a normal sized thyroid gland and a low thyroid iodine uptake, primary hypothyroidism appeared to be irreversible and necessitated thyroid hormone replacement therapy. The existence of a reversible type of primary hypothyroidism responding to kelp discontinuation was confirmed by two other studies. The reversible type was characterised by a higher radioactive iodine uptake after one week of restricted iodine consumption and minor changes of lymphocytic thyroiditis histologically (Tajiri et al., 1986) or negative thyroid autoantibodies (Konno et al., 1993). In the latter study, patients with unsuspected hypothyroidism showed, independently on their thyroid autoantibody status, a statistically significant higher urinary iodine excretion than healthy subjects.

An additional study from the coastal regions of Japan revealed a higher prevalence of hypothyroidism in thyroid autoantibody-negative subjects with high urinary iodine excretion than with normal urinary iodine levels, whereas high urinary iodine levels were not significantly associated with hypothyroidism in thyroid autoantibody-positive subjects (Konno et al., 1994). A clinical study on childhood overt primary hypothyroidism reported a reversible type of hypothyroidism which responded to restriction of dietary iodine intake despite positive thyroid autoantibodies (Okamura et al., 1994). In this study, which is in accordance with previous studies (Yoshinari et al., 1983; Tajiri et al., 1986), characteristic findings of the reversible group were goitrous hypothyroidism, high non-hormonal iodine levels along with a history of eating excessive seaweed, high thyroidal radioactive iodine uptake compared to the irreversible group (58±19 %/24 h vs. 5±4 %/24 h), and positive perchlorate discharge test. In agreement with these findings, Kasagi et al. (2003) showed in more than half of the patients with primary hypothyroidism recovery of thyroid function after iodine restriction. Reversibility of hypothyroidism was not significantly associated with signs of Hashimoto's thyroiditis, such as thyroid autoantibody levels or echogenicity on ultarsound, but with increased thyroid ^{99m}technetium-uptake and elevated non-hormonal iodine levels due to habitually ingestion of seaweed products. A Korean cross-sectional study revealed significantly higher iodine intake (673.8±794.9 μg/day vs. 468.9±481.9 μg/day) and urinary iodine excretion (4.33±5.70 mg/l vs. 2.11±0.69 mg/l) in patients with thyroid diseases, including hypothyroidism and Hashimoto's thyroiditis, as compared to normal subjects, with iodine intake determined by the amount of seaweed consumption (Kim and Kim, 2000). The inhibitory effects of kelp supplementation on thyroid function in euthyroid subjects have been also studied in two randomized, placebo-controlled studies. In a double-blind prospective clinical trial, short-term low- and high-dose dietary supplementation with kelp was found to increase significantly both basal as well as poststimulation TSH, whereas serum total T_3 significantly decreased only after high-dose kelp supplementation (Clark et al., 2003). In line with these findings, in 25 healthy postmenopausal women, seven weeks of seaweed supplementation, containing 475 μg of iodine/day, significantly increased urinary iodine excretion as well as serum TSH levels, though the latter increase was only slight and peripheral thyroid hormone concentrations were not affected (Teas et al., 2007).

High intake of kombu or other seaweeds rich in iodine during pregnancy may also cause thyroid dysfunction in the newborn in terms of transient hypothyroidism or persistent hyperthyrotropinemia. Excess ingestion of iodine during pregnancy caused twelve from fifteen infants to be diagnosed with hyperthyrotropinemia and requirement of levo-thyroxine supplementation. Postnatal consumption of iodine was identified as an additional risk factor for persistent hyperthyrotropinemia (Nishiyama et al., 2004).

In summary, the majority of studies reporting on the association of seaweed intake with hypothyroidism, goitre, and chronic lymphocytic thyroiditis come from Japan, a country known for its high seaweed consumption. One study is from Korea, which is also a country where seaweed is traditionally used as food (Kim and Kim, 2000). Elsewhere, goitre and hypothyroidism following seaweed or seaweed-containing diets appear to occur infrequently. Only one case is reported in the literature on the development of diffuse goitre and signs of hypothyroidism after taking seaweed tablets outside Asia (Liewendahl and Turula, 1972). However, the incidence of kelp-induced hypothyroidism in Europe and North-America may increase with rising consumption of kelp-containing products. Recently, in two randomized, placebo-controlled clinical trials an increase of serum TSH was found after short-term kelp supplementation in euthyroid subjects (Clark et al., 2003; Teas et al., 2007).

Thyroid Hyperfunction Due to Ingestion of Seaweed or Seaweed-Derived Dietary Supplements

While kelp-induced goitre, hypothyroidism, and Hashimoto's thyroiditis appear not to be uncommon, at least in countries with substantial seaweed intake, such as Japan, hyperthyroidism following consumption of seaweed or seaweed-derived dietary supplements is rare as reflected by the limited number of case reports found in literature.

In six cases, seaweed tablets were causative for thyrotoxicosis. A 38-year-old female developed hyperthyroidism and goitre after ingestion of seaweed tablets with an estimated daily iodine intake of 1-2 mg for 2.5 years. Symptoms spontaneously resolved three months after withdrawal of excess iodine (Liewendahl and Gordin, 1974). A 72-year-old woman presented with characteristic features of thyrotoxicosis, including weight loss, increased sweating, fatigue, and bowel habit changes, after having taken 4-6 sea-kelp tablets, corresponding to 2.8-4.2 mg iodine, per day for one year. Thyroid function normalised six months after discontinuation of sea-kelp tablets (Shilo and Hirsch, 1986). A 50-year-old female showed clinical and laboratory signs of hyperthyroidism due to intake of six kelp-containing tablets per day for two months. Due to the great variation in the iodine content of the kelp-containing tablets the estimated daily iodine content was 1-12 mg. Treatment comprised discontinuation of iodine intake and temporarily administration of metoprolol and diazepam. Two months after stopping intake of kelp tablets signs and symptoms of hyperthyroidism completely subsided (De Smet et al., 1990). In a 27-year-old woman with laboratory signs of hyperthyroidism after consumption of two dietary supplements containing kelp for four weeks, thyroid function was normalised seven weeks after withdrawal of the dietary supplements (Eliason, 1998). A 38-year-old male developed tachycardic atrial fibrillation following ingestion of kelp tablets for eight months with a total dosage of 1 g of iodine. Six weeks after discontinuation of kelp tablets intake and initiation of antithyroid treatment with thiamazole and atenolol the patient became euthyroid (Henzen et al., 1999). A 28-year-old man suffered from clinical and laboratory signs of hyperthyroidism after intake of six kelp-containing tablets daily for one year. Signs and symptoms of hyperthyroidism resolved six months after withdrawal of iodine intake and initiation of antithyroid therapy comprising thiamazole and propanolol (Salas Coronas et al., 2002).

In two other cases, iodine-induced thyrotoxicosis was due to ingestion of dried powdered seaweed. Two male patients aged 36 and 52 years developed thyrotoxicosis after consumption of dried powdered sea-kelp with an estimated daily iodine intake of 2-10 mg for 2-12 months. In both patients, signs and symptoms of hyperthyroidism resolved spontaneously 9 and 11 months, respectively, after cessation of kelp intake (Skare and Frey, 1980).

In two cases, hyperthyroidism occurred after consumption of kombu. Two Japanese women, 42 and 59 years old, developed signs of thyrotoxicosis one month and one year, respectively, after having eaten kombu with an estimated total iodine intake of 28 and 140 mg, respectively. Thyroid function normalised one and three months, respectively, after withdrawal of seaweed consumption (Ishizuki et al., 1989).

Finally, a 39-year-old female with pre-existing multinodular goitre developed thyrotoxicosis after consumption of seaweed-containing tea. The patient became euthyroid seven months after discontinuation of kelp intake and initiation of antithyroid treatment with thiamazole and propanolol (Müssig et al., 2006).

Seven of the eleven cases of kelp-induced thyrotoxicosis occurred in Europe, two patients came from Japan, one from Israel and one from the United States. Gender distribution was seven females and four males. Mean age on presentation was 44±13 years (mean±standard deviation). Mean iodine exposition was 25±37 months with a mean daily iodine dosage of 10±11 mg. In four cases, thyroid glands were normal on palpation or ultrasound. In five patients, diffuse thyroid enlargement was found and in one patient multinodular goitre. Signs and symptoms on admission were characteristic for hyperthyroidism: fine finger tremor, fatigue, weight loss (in six cases), increased sweating, nervousness (in four cases), bowel habit changes, palpitations (in three cases), heat intolerance, tachycardia, lid lag (in two cases), enhanced deep tendon reflexes, pretibial edema, tachycardic atrial fibrillation, secondary amenorrhoera, chest pain, and muscle atrophy (in one case). In the majority of cases, hyperthyroidism was mild to moderate, as reflected by the slight elevated T_4, fT_4 (in seven patients), and T_3 values (in four patients) with 1.3±0.3-, 1.5±0.2-, and 1.5±0.6-fold, respectively, over the upper limit of the reference range. In seven of the nine available values, TSH was suppressed. Thyroid autoantibodies were tested in all patients and proved to be negative, except for one patient with slightly positive microsomal antibodies. Urinary iodine excretion was markedly increased with 16301±20503 nmol/day and thyroid uptake of ^{131}iodine at 24 hours was diminished with 2.2±1.3 %; both values were present in four patients. In the majority of patients, the only treatment was discontinuation of iodine intake, without other measures (seven patients). In the remaining cases, treatment regimens comprised anti-thyroid therapy with thiamazole (three patients) and beta-blockers, such as propanolol (two cases), atenolol, or metoprolol (one case). One patient received temporarily diazepam. Thyroid function was restored 16±14 weeks after initiation of treatment. Only in one patient a previous history of thyroid disease was reported. The 39-year-old female had been diagnosed with multinodular goiter with a moderate autonomy in the right upper lobe before intake of a kelp-containing tea was started. Thyrotoxicosis in this patient was more pronounced compared to the others, as reflected by the highest fT_3 and fT_4 levels (1.9-fold higher than the upper value of the reference range), the largest amounts of anti-thyroid therapy, and the longest duration of treatment until normalisation of thyroid function, despite the lowest iodine exposure and the lowest daily iodine dosage (Müssig et al., 2006).

Conclusion

Seaweeds are a traditional part of daily diets in several East-Asian countries due to the beneficial nutritional characteristics. Marine algae are a known important source of iodine, though iodine concentrations and chemical species of iodine vary widely between algae species. Excessive iodine intake may affect thyroid function and in particular thyroid hormone synthesis. Regular consumption of large amounts of seaweeds is frequently associated with thyroid enlargement, hypothyroidism, and Hashimoto's thyroiditis. In contrast, thyrotoxicosis due to ingestion of seaweed or seaweed-derived dietary supplements appears to be rare. As consumption of seaweed markedly increased in Western countries in the last decade, clinicians have to be aware of the adverse effects of seaweed on thyroid function.

Acknowledgements

The author thanks Mary-Ann Schneider for proof reading the manuscript and Yuko Yokoyama for assistance with the articles written in Japanese.

References

Abuye, C., Kelbessa, U., Wolde-Gebriel, S. (1998). Health effects of cassava consumption in south Ethiopia. *East Afr. Med. J.*, **75**, 166-170.

Aquaron, R., Delange, F., Marchal, P., Lognone, V., Ninane, L. (2002). Bioavailability of seaweed iodine in human beings. *Cell. Mol. Biol.,* **48**, 563-569.

Arasaki, S. & Arasaki, T. *Vegetables from the Sea*. Tokyo; Japan Publications Inc.; 1983.

Ar Gall, E., Küpper, F.C., Kloareg, B. (2004). A survey of iodine content in Laminaria digitata. *Bot. Mar.*, **47**, 30-37.

Baba, M., Snoeck, R., Pauwels, R., de Clercq, E. (1988). Sulfated polysaccharides are potent and selective inhibitors of various enveloped viruses, including herpes simplex virus, cytomegalovirus, vesicular stomatitis virus, and human immunodeficiency virus. Antimicrob. *Agents. Chemother.*, **32**, 1742-1745.

Beiler, J.S., Zhu, K., Hunter, S., Payne-Wilks, K., Roland, C.L., Chinchilli, V.M. (2003). A case-control study of menstrual factors in relation to breast cancer risk in African-American women. *J. Natl. Med. Assoc.*, **95**, 930-938.

Bernier-Valentin, F., Kostrouch, Z., Rabilloud, R., Munari-Silem, Y., Rousset, B. (1990). Coated vesicles from thyroid cells carry iodinated thyroglobulin molecules. First indication for an internalization of the thyroid prohormone via a mechanism of receptor-mediated endocytosis. *J. Biol. Chem.*, **265**, 17373-17380.

Braverman, L.E. (1994). Iodine and thyroid: 33 years of study. *Thyroid*, **4**, 351-356.

Braverman, L.E., Ingbar, S.H. (1963). Changes in thyroidal function during adaptation to large doses of iodide. *J. Clin. Invest.*, **42**, 1216-1231.

Buck, C.B., Thompson, C.D., Roberts, J.N., Muller, M., Lowy, D.R., Schiller, J.T. (2006). Carrageenan is a potent inhibitor of papillomavirus infection. PloS. Pathog., 2, e69.

Clark, C.D., Bassett, B., Burge, M.R. (2003). Effects of kelp supplementation on thyroid function in euthyroid subjects. *Endocr. Pract.,* **9**, 363-369.

Dang, D.H., Hoang, T.M. (2004). Nutritional analysis of Vietnamese seaweeds for food and medicine. *Biofactors*, **22**, 323-325.

Darcy-Vrillon B. (1993). Nutritional aspects of the developing use of marine macroalgae for the human food industry, *Int. J. Food Sci. Nutr.* **44**, (Suppl. 1), S23-S35.

Das, S.K., Hashimoto, T., Shimizu, K., Yoshida, T., Sakai, T., Sowa, Y., Komoto, A. and Kanazawa, K. (2005). Fucoxanthin induces cell cycle arrest at G0/G1 phase in human colon carcinoma cells through up-regulation of p21WAF1/Cip1. *Biochim. Biophys. Acta.*, **1726**, 328-335.

De Smet, P.A., Stricker, B.H., Wilderink, F., Wiersinga, W.M. (1990). Hyperthyroidism during treatment with kelp tablets. *Ned. Tijdschr. Geneeskd.*, **134**, 1058-1059.

Dunn, A.D., Myers, H.E., Dunn, J.T. (1996). The combined action of two thyroidal proteases releases T4 from the dominant hormone-forming site of thyroglobulin. *Endocrinology*, **137**, 3279-3285.

Dunn, J.T. Adverse effects of iodine deficiency and its eradication by iodine supplementation. In: Bravermann, L.E., editor. *Contemporary endocrinology: diseases of the thyroid.* Totowa, NJ; Humana Press, Inc.; 1997; 349-360.

Dunn, J.T. (1998). What's happening to our iodine? *J. Clin. Endocrinol. Metab.,* **83**, 3398-3400.

Eliason, B.C. (1998). Transient hyperthyroidism in a patient taking dietary supplements containing kelp. *J. Am. Board Fam. Pract.,* **11**, 478-480.

Eskandari, S., Loo, D.D., Dai, G., Levy, O., Wright, E.M., Carrasco, N. (1997). Thyroid Na+/I- symporter. Mechanism, stoichiometry, and specificity. *J. Biol. Chem.*, **272**, 27230-27238.

Fleurence, J. (1999). Seaweed proteins: biochemical, nutritional aspects and. potential uses. *Trends. Food Sci. Tech.*, **10**, 25-28.

Fradkin, J.E., Wolff, J. (1983). Iodide-induced thyrotoxicosis. *Medicine (Baltimore),* **62**, 1-20.

Galton, V.A., Pitt-Rivers, R. (1959). The effect of excessive iodine on the thyroid gland of the rat. *Endocrinology*, **64**, 835-839.

Guiry, M.D. and Guiry, G.M. (2007). AlgaeBase version 4.2. World-wide electronic publication, National University of Ireland, Galway. http://www.algaebase.org; searched on 02 May 2007.

Henzen, C., Buess, M., Brander, L. (1999). [Iodine-induced hyperthyroidism (iodine-induced Basedow's disease): a current disease picture]. *Schweiz. Med. Wochenschr.*, **129**, 658-664.

Higuchi, T. (1964). The study of endemic seashore goiter in Hokkaido. *Folia Endocrinol. Jpn.*, **40**, 982-995.

Hoshiyama, Y., Sekine, T., Sasaba, T. (1993). A case-control study of colorectal cancer and its relation to diet, cigarettes, and alcohol consumption in Saitama Prefecture, Japan. *Tohoku J Exp Med.*, **171**, 153-165.

Hou, X., Chai, C., Qian, Q., Yan, X., Fan, X. (1997). Determination of chemical species of iodine in some seaweeds (I). *Sci. Total Environ.*, **204**, 215-221.

Hou, X. and Yan, X. (1998). Study on the concentration and seasonal variation of inorganic elements in 35 species of marine algae. *Sci. Total Environ.* **222**, 141-156.

Hou, X., Yan, X., Chai, C. (2000). Chemical species of iodine in some seaweeds. II. Iodine-bound biological macromolecules. *J. Radioanal. Nucl. Chem.*, **245**, 461-467.

Indergaard, M. The aquatic resource. I. The wild marine plants: a global bioresource. In: Cote, W.A., editor. *Biomass utilization.* New York; Plenum Publishing Corporation; 1983; 137-168.

Inoue, M., Taketani, N., Sato, T., Nakajima, H. (1975). High incidence of chronic lymphocytic thyroiditis in apparently healthy school children: epidemiological and clinical study. *Endocrinol. Japon.*, **22**, 483-488.

Institute of Medicine, Food and Nutrition Board. *Dietary Reference Intakes for Vitamin A, Vitamin K, Arsenic, Boron, Chromium, Copper,Iodine, Iron, Manganese, Molybdenum, Nickel, Silicon, Vanadium, and Zinc.* Washington, DC; National Academy Press; 2001; 258.

Ishizuki, Y., Yamauchi, K., Miura, Y. (1989). Transient thyrotoxicosis induced by Japanese Kombu. *Folia Endocrinol.*, **65**, 91-98.

Iso, H., Date, C., Noda, H., Yoshimura, T., Tamakoshi, A. (2005). Frequency of Food Intake and Estimated Nutrient Intake among Men and Women: The JACC Study. *J Epidemiol.,* **15** (Supplement I), S24-S42.

Jiménez-Escrig, A., Sánchez-Muniz, F.J. (2000). Dietary fibre from edible seaweeds: chemical structure, physicochemical properties and effects on cholesterol metabolism. *Nutr. Res.*, **20**, 585-598.

Jeong, B.Y., Cho, D.M., Moon, S.K. and Pyeum, J.H. (1993). Quality factors and functional components in the edible seaweeds. *J. Korean Soc. Food Nutr.*, **22**, 612-628.

Kasagi, K., Iwata, M., Misaki, T., Konishi, J. (2003). Effect of iodine restriction on thyroid function in patients with primary hypothyroidism. *Thyroid*, **13**, 561-567.

Kim, J.Y., Kim, K.R. (2000). Dietary iodine intake and urinary iodine excretion in patients with thyroid diseases. *Yonsei Med. J.,* **41**, 22-28.

Kogai, T., Endo, T., Saito, T., Miyazaki, A., Kawaguchi, A., Onaya, T. (1997). Regulation by thyroid-stimulating hormone of sodium/iodide symporter gene expression and protein levels in FRTL-5 cells. *Endocrinology*, **138**, 2227-2232.

Kotake-Nara, E., Asai, A., Nagao, A. (2005). Neoxanthin and fucoxanthin induce apoptosis in PC-3 human prostate cancer cells. *Cancer Lett.*, **220**, 75-84

Konno, N., Yuri, K., Taguchi, H., Miura, K., Taguchi, S., Hagiwara, K., Murakami, S. (1993). Screening for thyroid diseases in an iodine sufficient area with sensitive thyrotrophin assays, and serum thyroid autoantibody and urinary iodide determinations. *Clin. Endocrinol.(Oxf.),* **38**, 273-281.

Konno, N., Makita, H., Yuri, K., Iizuka, N., Kawasaki, K. (1994). Association between dietary iodine intake and prevalence of subclinical hypothyroidism in the coastal regions of Japan. *J. Clin. Endocrinol. Metab.*, **78**, 393-397.

Küpper, F., Schweigert, N., Ar Gall, E., Legendre, J.M., Vilter, H., Kloareg, B. (1998). Iodine uptake involves extracellular haloperoxidase - mediated oxidation of iodide. *Planta*, **207**, 163–171.

Lee, S.M., Lewis, J., Buss, D.H., Holcombe, G.D., Lawrance, P.R. (1994). Iodine in British foods and diets. *Br. J. Nutr.*, **72**, 435-446.

Levring, T, Hoppe H.A., Schmid, O.J. *Marine Algae – A Survey of Research and Utilization.* Hamburg; Cram, de Gruyter & Co.; 1969.

Liewendahl, K., Turula, M. (1972). Iodide-induced goitre and hypothyroidism in a patient with chronic lymphocytic thyroiditis. *Acta Endocrinol. (Copenh.),* **71**, 289-296.

Liewendahl, K., Gordin, A. (1974). Iodine-induced toxic diffuse goitre. *Acta Med. Scand.,* **196**, 237-239.

Lightowler, H.J., Davies, G.J. (1998). Iodine intake and iodine deficiency in vegans as assessed by the duplicate-portion technique and urinary iodine excretion. *Br. J. Nutr.*, **80**, 529-535.

Lu, Y., Suliman, S.,A, Hansen, H.R., Feldmann, J. (2006). Iodine excretion and accumulation in seaweed-eating sheep from Orkney, *Scotland. Environ. Chem*.**,** **3**, 338–344.

Mabeau, S., Fleurence, J. (1993). Seaweed in food products: biochemical and nutritional aspects. *Trends. Food. Sci. Technol.*, **4**, 103–107.

Maeda, H., Hosokawa, M., Sashima, T., Funayama, K., Miyashita, K. (2005). Fucoxanthin from edible seaweed, Undaria pinnatifida, shows antiobesity effect through UCP1 expression in white adipose tissues. *Biochem. Biophys. Res. Commun.*, **332**, 392-397.

Marchal, P., Lognoné, V., Fuselier, M., Bonabèze, E., Brault, D., Blondel, J.M., Franc, M., Ninane, L., Schwartz, D., Ménager, M., Delange, F., Aquaron, R. Iodized salt for sustaining IDD elimination. In: Geertman, R.M., editor. 8th *World Salt Symposium.* Amsterdam; Elsevier Inc.; 2000; 1015-1020.

Maruyama, H., Yamamoto, I. (1992). Suppression of 125I-uptake in mouse thyroid by seaweed feeding: possible preventative effect of dietary seaweed on internal radiation injury of the thyroid by radioactive iodine. *Kitasato Arch. Exp. Med.*, **65**: 209-216.

Matsubayashi, S., Mukuta, T., Watanabe, H., Fuchigami, H., Taniguchi, J., Chinen, M., Ninomiya, H., Sasaki, H. (1998). Iodine-induced hypothyroidism as a result of excessive intake of confectionery made with tangle weed, Kombu, used as a low calorie food during a bulimic period in a patient with anorexia nervosa. *Eat. Weight Disord.*, **3**, 50-52.

Matsuzaki, S. & Iwamura, K. Application of seaweeds to human nutrition and medicine. In: Noelle, H., editor. *Nahrung aus dem Meer – Food from the Sea*. New York; Springer-Verlag; 1981; 162-185.

Mills, S.Y. *Out of the Earth: The Essential Book of Herbal Medicine*. Middlesex, UK; Viking-Arkana; 1991.

Moriyama, K., Kanamoto, N., Hataya, Y., Nanbu, T., Hosoda, K., Arai, H., Nakao, K. (2006). A case of type 2 diabetes mellitus developing hypothyroidism discovered as a result of a discrepancy between glycated hemoglobin and glycated albumin values. *Diabetes Res. Clin. Pract.*, **71**, 227-232.

Müssig, K., Thamer, C., Bares, R., Lipp, H.P., Häring, H.U., Gallwitz, B. (2006). Iodine-induced thyrotoxicosis after ingestion of kelp-containing tea. *J. Gen. Intern. Med.*, **21**, C11-C14.

Nakamura, M., Yamazaki, I., Nakagawa, H., Ohtaki, S., Ui, N. (1984). Iodination and oxidation of thyroglobulin catalyzed by thyroid peroxidase. *J. Biol. Chem.*, **259**, 359-364.

Nishiyama, S., Mikeda, T., Okada, T., Nakamura, K., Kotani, T., Hishinuma, A. (2004). Transient hypothyroidism or persistent hyperthyrotropinemia in neonates born to mothers with excessive iodine intake. *Thyroid*, **14**, 1077-1083.

Okamura, K., Inoue, K., Omae, T. (1978). A case of Hashimoto's thyroiditis with thyroid immunological abnormality manifested after habitual ingestion of seaweed. *Acta Endocrinol. (Copenh.)*, **88**, 703-712.

Okamura, K., Sato, K., Ikenoue, H., Nakagawa, M., Kuroda, T., Yoshinari, M., Fujishima, M. (1994). Primary hypothyroidism manifested in childhood with special reference to various types of reversible hypothyroidism. *Eur. J. Endocrinol.*, **131**, 131-137.

Phaneuf, D., Côté, I., Dumas, P., Ferron, L.A., LeBlanc, A. (1999). Evaluation of the contamination of marine algae (Seaweed) from the St. Lawrence River and likely to be consumed by humans. *Environ Res.*, **80**, S175-S182

Pineda-Lucatero, A., Avila-Jiménez, L., Ramos-Hernández, R.I., Magos, C., Martínez, H. (2008). Iodine deficiency and its association with intelligence quotient in schoolchildren from Colima, Mexico. *Public Health Nutr.*, **11**, 1-9.

Premakumara, G.A., Ratnasooriya, W.D., Tillekeratne, L.M. (1995). Studies on the post-coital contraceptive mechanisms of crude extract of Sri Lankan marine red algae, Gelidiella acerosa. *Contraception.*, **52**, 203-207.

Rupérez, P., Saura-Calixto, F. (2001). Dietary fibre and physicochemical properties of edible Spanish seaweeds. *Eur. Food Res. Technol.*, **212**, 349-354.

Saenko, G.N.; Kravtsova, Y.Y.; Ivaneneko, V.V. and Sheludko, S.I. (1978). Concentration of iodine and bromine by plants in seas of Japan and Okhotsk. *Mar. Biol.*, **47**, 243-250.

Salas Coronas, J., Cruz Caparros, G., Laynez Bretones, F., Diez Garcia, F. (2002). [Hyperthyroidism secondary to kelp tablets ingestias]. *Med. Clin. (Barc.)*, **118**, 797-798.

Science and Technology Agency, Japan. *Standard Tables of Food Composition in Japan*. The Fifth Edition. Tokyo; Ministry of Finance; 2001 (in Japanese).

Section of Nutrition, Bureau of Public Health, Ministry of Health and Welfare, Japan. *Annual Report of the National Nutrition Survey in 2003*. Tokyo; Daiichi Shuppan Publishing Co. Ltd.; 2005 (in Japanese).

Shilo, S., Hirsch, H.J. (1986). Iodine-induced hyperthyroidism in a patient with a normal thyroid gland. *Postgrad. Med. J.,* **62**, 661-662.

Skare, S., Frey, H.M.M. (1980). Iodine induced thyrotoxicosis in apparently normal thyroid glands. *Acta Endocrinol. (Copenh.),* **94**, 332-336.

Skibola, C.F. (2004). The effect of Fucus vesiculosus, an edible brown seaweed, upon menstrual cycle length and hormonal status in three pre-menopausal women: a case report. BMC Complement. *Altern. Med.*, **4**, 10.

Skibola, C.F., Curry, J.D., VandeVoort, C., Conley, A., Smith, M.T. (2005). Brown kelp modulates endocrine hormones in female sprague-dawley rats and in human luteinized granulosa cells. *J. Nutr.*, **135**, 296-300.

Socolow, E.L., Suzuki, M. (1964). Possible goitrogenic effects of selected Japanese foods. *J. Nutr.*, **83**, 20-26.

Suzuki, H., Higuchi, T., Sawa, K., Ohtaki, S., Horiuchi, Y. (1965). "Endemic coast goitre" in Hokkaido, Japan. *Acta Endocrinol. (Copenh.),* **50**, 161-176.

Suzuki, H., Mashimo, K. (1973). Further studies of "endemic coast goiter" in Hokkaido, Japan. Iodine Metab. *Thyroid Funct*., **6**, 143-156.

Tajiri, J., Higashi, K., Morita, M., Umeda, T., Sato, T. (1986). Studies of hypothyroidism in patients with high iodine intake. *J. Clin. Endocrinol. Metab.*, **63**, 412-417.

Teas, J., Pino, S., Critchley, A. and Braverman, L.E. (2004). Variability of iodine content in common commercially available edible seaweeds. *Thyroid*, **14**, 836-841.

Teas, J., Braverman, L.E., Kurzer, M.S., Pino, S., Hurley, T.G., Hebert, J.R. (2007). Seaweed and soy: companion foods in Asian cuisine and their effects on thyroid function in American women. *J. Med. Food*, **10**, 90-100.

Tokudome, S., Tokudome, Y., Goto, C., Suzuki, S., Moore, M.A. (2004). Seaweed as a beneficial iodine food source. *Asian Pac. J. Cancer Prev.*, **5**, 89.

Van Netten, C., Hoption Cann, S.A., Morley, D.R., van Netten, J.P. (2000). Elemental and radioactive analysis of commercially available seaweed. *Sci. Total Environ.*, **255**, 169-175.

Verhaeghe, E.F., Fraysse, A., Guerquin-Kern, J.L., Wu, T,D,, Devès, G., Mioskowski, C., Leblanc, C., Ortega, R., Ambroise, Y., Potin, P. (2008). Microchemical imaging of iodine distribution in the brown alga Laminaria digitata suggests a new mechanism for its accumulation. *J. Biol. Inorg. Chem.*, **13**, 257-269.

Wolff, J., Chaikoff, I,L. (1948). Plasma inorganic iodide as a homeostatic regulator of thyroid function. *J. Biol. Chem.*, **174**, 555-564.

Wolff, J., Chaikoff, I.L., Goldberg, R.C., Meier, J.R. (1949). The temporary nature of the inhibitory action of excess iodide on organic iodide synthesis in the normal thyroid. *Endocrinology*, **45**, 504-513.

Wolff, J. Iodide goiter and the pharmacologic effects of excess iodide. *Am J Med* 1969; 47:101–124.

Wyke, N. We all need a little kelp [online]. TIMESONLINE April 02, 2005. Available from: http://www.timesonline.co.uk/article/0,,8126-1549366_2,00.html

Yamaguchi, K., Fukushima, H., Uzawa, H., Iseki, T., Shiroozu, A., Inoue, K. (1984). A case of iodide myxedema observed for three years under a low iodide diet—especially on the restoration of the escape mechanism from the Wolff-Chaikoff effect. *Folia Endocrinol.*, **60**, 79-88.

Yoshinari, M., Okamura, K., Tokuyama, T., Shiroozu, A., Nakashima, T., Inoue, K. and Omae, T. (1983). Clinical importance of reversibility in primary goitrous hypothyroidism. *Br. Med. J. (Clin. Res. Ed.),* **287**, 720-722.

Zimmermann, M.B., Ito, Y., Hess, S.Y., Fujieda, K., Molinari, L. (2005). High thyroid volume in children with excess dietary iodine intakes. *Am. J. Clin. Nutr.*, **81**, 840-844.

Zimmermann, M.B. (2007). The adverse effects of mild-to-moderate iodine deficiency during pregnancy and childhood: a review. *Thyroid*, **17**, 829-835.

Reviewed by Prof. Thomas Remer, Department of Nutrition and Health, Research Institute of Child Nutrition, Dortmund, Germany

In: Algae: Nutrition, Pollution Control and Energy Sources
Editor: Kristian N. Hagen, pp. 195-199
ISBN 978-1-60692-008-4

Chapter 10

ALGAE FROM THE VIEWPOINT OF MATHEMATICS

Hiroshi Yoshida*
Faculty of Mathematics, Kyushu University,
Hakozaki 6-10-1, Higashi-ku, FUKUOKA 812-8581 Japan

Abstract

The multicellular organisms we shall treat in this chapter range from unicellular *Gonium* to *Volvox* having thousands of cells. By introducing a seminal notion of complexity hierarchy, we have constructed a simple classification of extracellular matrices in the framework of formal language theory. Then, this classification has enabled us to understand the necessity of various forms of multicells.

We have also constructed another model for generalized Anabaena, which is a genus of filamentous cyanobacteria or blue-green algae, using a Lindenmayer system (an L-system). An L-system is a parallel rewriting system that was introduced originally to model the development of multicellular organisms. The aim of this model is the derivation of exact algebraic equations between the proliferation and transition rates for high cell-type diversity. For simplicity, we have studied 'generalized Anabaena' having three cell types: A, B and C on the assumption of $A \Rightarrow B \Rightarrow C$ cell lineage. In this model, the high cell-type diversity corresponds to the condition that A, B and C cells are well mingled. This model has revealed that the condition for high cell-type diversity can be described as an elliptic curve under a certain condition that is deeply related to 'Fermat's last theorem.'

1. Evolution of Green Algae from the Viewpoint of Mathematics

Here, we deal with *Gonium pectorale*, *Eudorina elegans* and *Volvox globator* on the evolutionary pathway of green algae.

The basic idea is that a cell is regarded as a rewriting system in formal languages. Rewriting rules of the system ordinarily mean transformations over strings. By contrast, in this chapter, the rules mean transformations over chemical substances. Thus, if rewriting rules can be treated over chemical substances, the complexity of a system with such rules can be evaluated.

*E-mail address: phiroshi@math.kyushu-u.ac.jp. Tel/fax: +81-92-642-7396.

An alga consists of cells that output some chemical substances in response to received chemical substances and the states of neighbouring cells. Here, the cell is regarded as a rewriting system with a grammar of the following matrix form:

$$(A_1 \to \omega_1, A_2 \to \omega_2, \ldots, A_k \to \omega_k), \tag{1}$$

where the received chemical substances are denoted by A_i and the chemical substances to be output are denoted by ω_i $(1 \leq i \leq k)$. This matrix means that if all the chemical substances $A_1, A_2, \ldots, A_k$ are received, the cell outputs the chemical substances $\omega_1, \omega_2, \ldots, \omega_k$.

The complexity a certain system can construct depends on the number k in Formula (1). Interestingly, the rewriting rule of *Eudorina elegans* corresponds to the case of $k = 1$, while that of *Volvox globator* corresponds to the case of $k \geq 2$. $k \geq 2$ designates that the constituent cells cooperate with one another. By contrast, *Gonium pectorale* corresponds to the finite application of the rule (1) because *Gonium pectorale* has no space closed to the environment due to gaps in its extracellular matrices. Because of finiteness of application, *Gonium pectorale* can construct only structures of low complexity.

From the viewpoint of formal language theory, the finiteness yields Ps$\mathcal{FIN}$, while $k = 1$ and $k \geq 1$ yield Ps$\mathcal{CF}$ and Ps$\lambda\mathcal{MAT}$, respectively. The complexity hierarchy of these three classes is as follows:

$$\text{Ps}\mathcal{FIN} \subset \text{Ps}\mathcal{CF} \subset \text{Ps}\lambda\mathcal{MAT}. \tag{2}$$

Ps$\mathcal{CF}$ can generate only compositions in which the relation between chemical substances (corresponding to extracellular matrices) is linear, whereas Ps$\lambda\mathcal{MAT}$ can generate some nonlinearly related compositions such as *exponent*, *square* or *cube*. The extracellular matrices of representative *Volvox globator* are divided into four zones: the flagellar zone, the boundary zone, the cellular zone and the deep zone [1, 2]. The volumes of the boundary zone and the deep zone are nearly proportional to the square and the cube, respectively, of the radius of the Volvox spheroid. This fact indicates that the compositions of the extracellular matrices are nonlinear.

Therefore, from the viewpoint of evolutionary processes, we can state the following scenario: *Eudorina elegans* could change from *Gonium pectorale*'s flat random colony formation to a somewhat ordered spheroidal formation by having a closed space to acquire higher complexity. Furthermore, *Volvox globator* could change to a more ordered formation than *Eudorina elegans*. That is, *Volvox*'s cells aggregate regularly on the surface of the spheroid with the cells mutually cooperating to acquire higher complexity [5]. This complexity hierarchy is shown in Fig. 1.

2. Elliptic Curves over the Anabaena Model

The development of a multicellular organism is a dynamic process. Starting with one or a few cells, the organism develops into different types of cells with distinct functions to survive. Such a process is called *differentiation* and can be described as cell lineage. Hereby, we have constructed a simple model with cell number increase and *cell-type order conservation*. This model represents the generalized system of Anabaena, and describes a probabilistic Lindenmayer system for three types of cells.

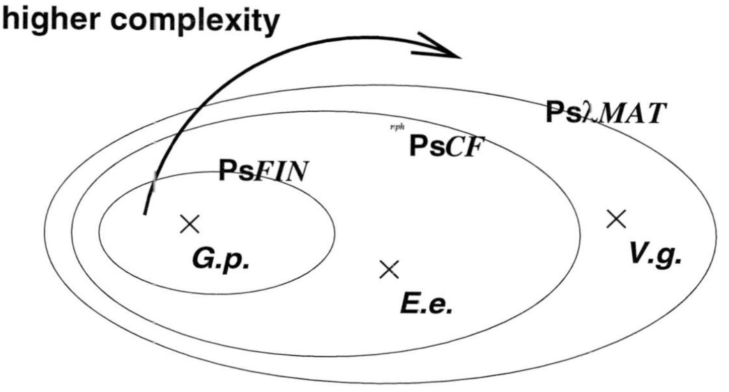

Figure 1. Complexity hierarchy; G.p., E.e. and V.g. denote *Gonium pectorale*, *Eudorina elegans* and *Volvox globator*, respectively.

In this chapter, we assume the following cell differentiation rule:

$$A \Rightarrow B \Rightarrow C.$$

Using this system, the proliferation and transition rates of each cell type A, B, C are defined as follows:

$$
\begin{aligned}
A &\rightarrow \begin{cases} AA & p_{1,1} \\ B & p_{1,2} \\ A & 1 - p_{1,1} - p_{1,2} \end{cases}, \\
B &\rightarrow \begin{cases} BB & p_{2,2} \\ C & p_{2,3} \\ B & 1 - p_{2,2} - p_{2,3} \end{cases}, \\
C &\rightarrow \begin{cases} CC & p_{3,3} \\ C & 1 - p_{3,3} \end{cases}
\end{aligned}
\tag{3}
$$

with $0 < p_{i,i} < 1$ $(1 \leq i \leq 3)$, $0 < p_{i,i+1} < 1$ $(1 \leq i \leq 2)$, $p_{i,i} + p_{i,i+1} < 1$ $(1 \leq i \leq 2)$. In addition to the rewriting rules above, we further adopt a rewriting rule, a cell-type order conservation rule:

$$AC \rightarrow ABC,\ CA \rightarrow CBA, \tag{4}$$

which guarantees the contiguity of cell types. The one-dimensional cell-chain becomes longer as these rules are applied.

Using this model, we estimate the cell-type diversity by calculating the following composition in the cell-chain:

$$AA, AB, BA, (AC, CA), BB, BC, CB, CC,$$

where (AC, CA) does not exist when the cell-type order conservation rule (4) is adopted. Under the high-diversity constraint:

$$gN(AA) = N(BB) = N(CC) \ \&\ gN(AB) = N(BC), \tag{5}$$

Table 1. The exact relation between f and g.

Without the cell-type order conservation rule (4)
$\begin{cases} 2(g+1)f^4 + g(g+1)f^3 - g(2g+5)f^2 - g^2 f + 2g^2 = 0^{\dagger}, \\ 2(3g-1)f^2 + (3g-1)f - 3g = 0 \end{cases}$
With the rule
$\begin{cases} 2(1-g)f^3 + g(3+g)f^2 - gf - g^2 = 0^{\ddagger}, \\ 2gf^2 + (g-1)f - g = 0 \end{cases}$

the cell-type diversity can be described as the maximum value of $N(AB)/N(AA)$.

Let f denote the maximum value of $N(AB)/N(AA)$. We have successfully derived the exact algebraic equations between f and g as shown in Table 1. The curve † in Table 1 is transformed into the following Weirstraß minimal model:

$$y^2 + xy + y = x^3 - x^2 - 9x + 9 \tag{6}$$

by the bi-rational transformation:

$$\begin{cases} f &= (3x+y-1)/(x+y-3), \\ g &= (3x+y-1)^2(x^2-4x+3-2y)/(2(x+y-3)(x-y+5)(2x+y-2)), \end{cases}$$

$$\begin{cases} x &= (f^2(3f+1)+2(f+1)(f-2)g)/(f^2(f-1)), \\ y &= -2(f+1)(-2f^2+(f-2)(f-3)g)/(f^2(f-1)^2). \end{cases}$$

Likewise, the curve ‡ is transformed into the minimal model:

$$y^2 + xy = x^3 - 7x + 9 \tag{7}$$

by:

$$\begin{cases} f &= (x-2)/(x+2), \\ g &= (2-x)(3+y)/(2x(2+x)), \end{cases}$$

$$\begin{cases} x &= 2(1+f)/(1-f), \\ y &= (-3f+3f^2-4g-4fg)/((1-f)f). \end{cases}$$

Interestingly, the curves (6) and (7) are *elliptic curves* that are deeply related to 'Fermat's last theorem.'

The cell-chain pattern is periodic when f is a rational number. Periodic patterns seem like a basic pattern in nature. Using the feature of an elliptic curve, we easily find the rational points of (g, f) on the elliptic curve ‡:

$$\begin{aligned}(g, f) = {} & (1/4, 1/3), (725/3588, 145/483), (8832/9019493, 552/25201), \\ & (3466764335501/11363197974660, 127769297/347639747)\cdots \end{aligned} \tag{8}$$

The alga's structure corresponding to the above rational points will be revealed in the near future [3,4].

3. Summary

In this chapter, we have investigated algae from the viewpoint of mathematics, especially formal language theory. First, using complexity hierarchy, we explain the evolutionary process of the algae *Gonium pectorale*, *Eudorina elegans* and *Volvox globator*. Second, from exact algebraic equations such as an elliptic curve, we have found rational points that may shed profound insights into the structure of algae.

Acknowledgment

We wish to express our gratitude to Prof. C. W. Brown for helpful calculations and useful suggestions on QEPCAD-B and to Prof. M. Kaneko and Ms. E. Yoshida for valuable discussions and suggestions. The author was partly supported by a Grant-in-Aid for Young Scientists (B) (grant 19790881) from the Ministry of Education, Culture, Sports, Science and Technology of Japan (MEXT). This study was also supported by the Program for Improvement of Research Environment for Young Researchers from Special Coordination Funds for Promoting Science and Technology (SCF) commissioned by the Japan Science and Technology Agency (JST).

References

[1] Kirk, D. L., Birchem, R. and King, N.: The extracellular matrix of *Volvox*: A comparative study and proposed system of nomenclature, *J. Cell Sci.*, Vol. 80 (1986), 207–231.

[2] Sumper, M. and Hallmann, A.: Biochemistry of the extracellular matrix of *Volvox*, *Int. Rev. Cyt.*, Vol. 180 (1998), 51–85.

[3] Yoshida, H., Anai, H. and Horimoto, K.: Derivation of rigorous conditions for high cell-type diversity by algebraic approach, *Biosystems*, Vol. 90 (2007), 486–495, (doi:10.1016/j.biosystems.2006.11.008).

[4] Yoshida, H., Horimoto, K. and Anai, H.: Inference of probabilities over a stochastic IL-system by quantifier elimination, special issue on *Modeling and Analysis of Complex systems in Mathematics in Computer Science*, Vol. 1(3) (2008), 473–485, (doi:10.1007/s11786-007-0037-z).

[5] Yoshida, H., Yokomori, T. and Suyama, A.: A simple classification of the volvocine algae by formal languages, *Bull. Math. Biol.*, Vol. 67 (2005), 1339–1354, (doi:10.1016/j.bulm.2005.03.001).

In: Algae: Nutrition, Pollution Control and Energy Sources ISBN: 978-1-60692-008-4
Editor: Kristian N. Hagen, pp. 201-246

Chapter 11

ION TRANSPORT IN MARINE ALGAE: FROM UNIFORM TO SELF-ORGANIZED PROCESSES

Fabrice Homblé[1] and Marc Léonetti[2]
[1]Structure et Fonction des Membranes Biologiques, Université Libre de Bruxelles, Campus Plaine (CP 206/2), B – 1050 Brussels, Belgium
[2]Institut de Recherche sur les Phénomènes Hors-Equilibre, Université Aix-Marseille, Technopôle de Château-Gombert, F-13384 Marseille Cedex 13, France

Abstract

Cells and organelles are surrounded by at least one membrane that controls the exchange of energy and matter between the cytoplasm and its environment. Ion transport through membranes is an essential process for life. For instance, algae nutrition, osmotic and hydrostatic pressure regulation and cell signaling are typical cellular functions that are directly controlled by the transport of ions through the plasma membrane. The quantitative description of membrane transport through algae (and plants in general) is usually restricted to the uniform steady state case. However, spatial and temporal dynamics arising from the nonlinear properties of ion transport have recently been revealed to be of prime importance for cell signaling and developmental axis emergence. We will review the mechanisms of ion transport (active and passive transport) in marine algae and their implication in cell physiology, morphology and homeostasis. The basic principles of ion transport will be explain and we will show how the nonlinear coupling between different ion transport systems such as ATPases, channels and co-transports can give rise to self-organized spatiotemporal events such as action potentials or stationary patterns of transcellular currents observed in marine algae.

Introduction

Algae are the main photosynthetic organisms of the ocean. Most of them are autotrophic organisms: they produce organic compounds from inorganic molecules using the energy from their environment. Some algae are auxotrophic since they depend on the absorption of vitamins (e.g., B_{12}, thiamine, biotine) for their growth and physiology [Pedersen 1969; Croft,

Warren and Smith 2006]. Thus, the absorption of mineral element from their environment is essential for their life.

Planktonic and benthic marine algae absorb mineral nutrients through their entire thallus surface in contact with sea water. The holdfast of benthic algae serves mainly for their anchorage. Some genera such as Caulerpa, Halimeda, Udotea and Penicillus have a complex rhizoidal holdfast that might suggest a root plant [Anderson, Close, DeWreede, et al. 2006]. However, the experimental evidences indicating a specific role for the holdfast in mineral nutrition are scarce. Williams have shown that the rhizoids of Caulerpa are involved in the uptake of NH_4^+ [Williams 1984].

The membrane that surrounds each cell is the site of exchange between its cytoplasm and its external environment. This surface must allow sufficient exchange to support the contents of the cell. As a cell increases in size its volume increases as the cube of its linear dimensions while surface area increases as the square. Any other conditions being equal, if the cell grows beyond a certain limit, not enough material will be able to cross the membrane fast enough to accommodate the increased cellular volume. Thus, the surface area-to-volume ratio (S/V) of a cell imposes an upper limit on size, limiting the rate at which the cell ionic status can be controlled efficiently. For a sphere, the S/V ratio is 3/r, where r is the radius. Small cells have a large surface-area-to-volume ratio, which quickly decreases as the radius increases (figure 1A). Moreover, the S/V ratio depends on the cell geometry (figure 1B). The size of marine algae encompasses a broad range of values, from small unicellular organisms to large multicellular seaweeds. For instance, planktonic diatom cells are most commonly in the size range of 10 - 200 μm whereas giant kelps range in size from 1 meter to more than 100 meters (Laminaria, Macrocystis, Nerocystis). A spherical cell of 10^{-4} m radius and a blade of 1m in length, 0.1m width and 10^{-3} m thick have an S/V ratio of $3 10^4$ m^{-1} and 10^3 m^{-1}, respectively. Thus, despite their large size difference (4-5 orders of magnitude) the S/V ratio of algae changes by only 1 to 2 orders of magnitude as a consequence of their shape. This illustrates how an organism can gain advantage of growing larger while avoiding the inefficiencies of increased size.

The major abiotic limiting factors for the growth of marine algae are light, temperature, hydrodynamic flow of water, salinity and nutrient availability. There are two parameters in salinity that affect the growth and development of marine algae: the osmotic potential and the ionic composition. The osmotic potential is one component of the hydrostatic potential which controls the water flux between algae and their environment. The nutrient composition of the seawater is significantly different depending on whether one considers, for instance, the deep sea, the surface or the estuary. Thus, the nutrition of algae will depend on their habitat. The hydrodynamic movement of the sea water is important to provide a continuous renewal of the nutrients around algae and the mixing of nutrient-rich bottom water with nutrient-depleted surface water.

The diversity of algae and of their environmental conditions is reflected in their nutrient requirements and in their uptake kinetics. The nutrition of marine algae can be different between algal groups and even within a species it can be affected by age, nutrient status or developmental stage. Thus, caution must be taken before drawing conclusion by analogy between unicellular algae, siphonous algae and seaweeds. All the naturally-occurring elements (~ 90) can be found in tissue. However, only about a quarter elements are truly essential for completion of algal growth, development and reproduction. Essential mineral elements display similar distribution in the water column of oceanic ecosystems.

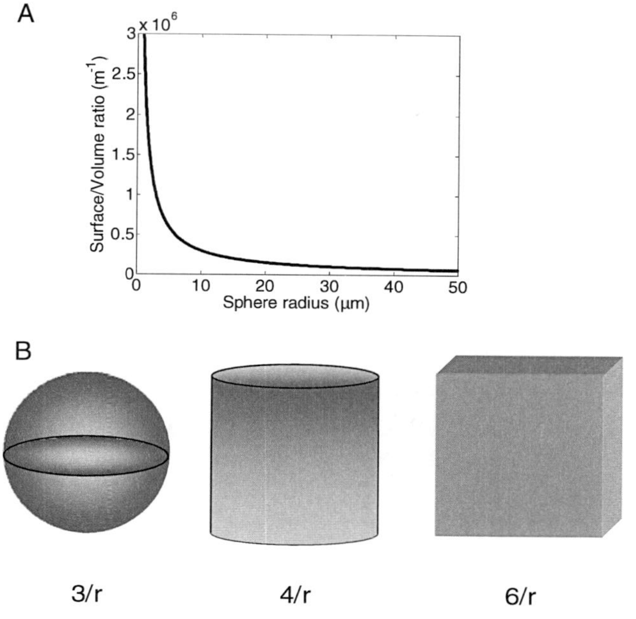

Figure 1. Effect of size and geometry on the Surface/Volume (S/V) ratio. A) The S/V ratio decreases nonlinearly as the size of a sphere increases. B) Comparison of the S/V ratio of a sphere (radius=r), a cylinder (radius=r, height=r) and a cube (length=r). The S/V ratio is indicated below each shape.

In the upper layer, photosynthetic activity and ion uptake deplete algal nutrients while at depth the nutrient concentration increases because of decomposition of organic matter and remineralization of sedimenting death organisms. This cycling of mineral nutrients is to a large part related to the life cycle of marine microorganisms. Essential elements common to all algae are C, H, O, N, S, P, K, Ca, Mg, Cu, Zn, Mn, some (V, Mo, Se, Si, B, Co, Cl, Cd, I) are required by some algae. A typical feature of some algae is the possibility to replace some of the essential elements by chemically related elements: K by Na, Ca by Sr, Fe by Cd...). The function of essential elements in the metabolism can be found in textbooks [Mengel and Kirkby 1987; Dring 1991; Marschner 1998; Lobban and Harisson 2000]. Here, we mention some elements which confer specificity to marine algae:

- Co is required for the synthesis of cobamides: cobalt-containing corinoids such as vitamin B_{12}. Algal vitamin B_{12} is sometimes called pseudo-B_{12} to means that it does not have biological activity in human. The structure of vitamin B_{12} is based on a corrin ring, consisting of four pyrrole cycles, and a central Co atom [Croft, Warren and Smith 2006].
- Si is a hardening component of the frustule, or silica cell wall, of diatoms [Armbrust, Berges, Bowler, et al. 2004]. Three genes coding for putative silicic acid transporters have been identified in Thalassiosira pseudonana. After absorption silica precipitates in silica-deposition vesicles in the presence of silaffins (phosphoproteins). These vesicles are targeted to the plasma membrane where they pour their content outside the cell.

- Cd is usually known for its toxicity. But it is an essential element in marine diatoms where it is used as a catalytic metal atom in a specific carbonic anhydrase [Lane and Morel 2000; Lane, Saito, George, et al. 2005; Xu, Feng, Jeffrey, et al. 2008]. The carbonic anhydrase catalyses the reversible hydration of CO_2.

Essential elements exist in sea water in a soluble form which can be absorbed by algae. Small neutral or polar molecules (e.g., CO_2, O_2, NH_3, H_2O) can cross biological membrane by free diffusion through the lipid bilayer. Though a selective transport of ions occurs through the lipid bilayer [Fuks and Homble 1994; Fuks and Homble 1996], its magnitude is negligible compare to that through biological membranes containing transmembrane proteins such as carriers or channels. For instance, the permeability coefficient of K^+ is 10^{-4}-10^{-3} nm s^{-1} for a lipid bilayer whereas it is larger than 10 nm s^{-1} for a biological membrane. Membrane transporters of ions can be classified into two families on the basis of their kinetic properties: carriers (e.g., pumps and cotransporters) and channels [Sanders 1990]. One difference between these transporters is the rate at which they can transport ion: 10^3 ion s^{-1} for carriers and 10^7 ion s^{-1} for channels [Hille 1992]. Experimental evidences for a specific absorption of an element have often been obtained through the study of its uptake kinetics. The rate of uptake measured as a function of the concentration, also called isotherm for ion absorption, has a hyperbolic shape: at low concentration the uptake rate increases linearly with concentration but saturated at high concentration (Figure 2).

Though the isotherm of absorption is a good way to quantify and to compare uptake rate in different conditions or between different organisms it provides little information about the mechanism of ion transport and its regulation. For instance, both ion channels and carriers can display a saturation of ion flux as the ion concentration is increased.

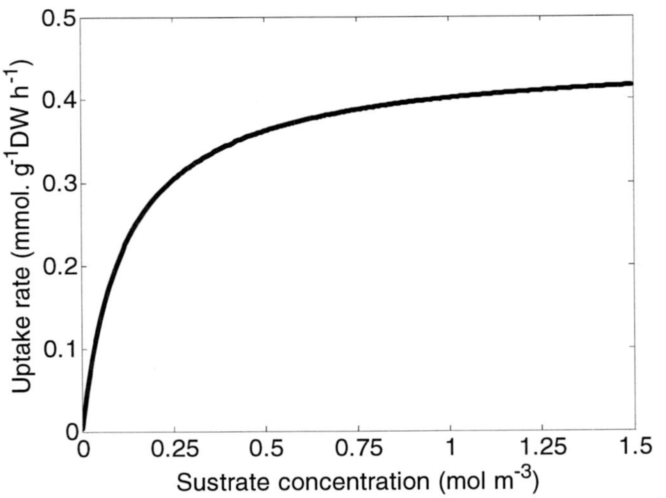

Figure 2. Isotherm for ion absorption. The curve has been drawn assuming a maximal uptake rate of 0.45 and 0.12 for the ion concentration corresponding to half the maximal uptake rate. These values are in the range of the values reported for the rate of NH_4 uptake in Ulva [Runcie, Ritchie and Larkum 2003].

We mention here above that the nutrition of marine algae is variable with time and space, it depends on the state of developmental and growth as well as environmental factors that are continuously changing. Nonetheless, the general principles of the physiology of algae nutrition are to some extend conserved among algae. Some differences exist between algae, notably the mechanism of osmoregulation, the presence of some ion transport processes or the causal effect of the coexistence of specific transport systems. In this chapter we will review the mechanisms of ion transport through algae membranes. First, we will describe the basic and classical principles that are used to describe ion transport processes through algae membranes. In this classical view, each transporter is seen as an independent unit. This approach allows the quantitative characterization of ion transporters and the quantitative description of the mineral nutrition in terms of ion fluxes. Then, we will show how various types of complexity emerge in algae due to the collective behavior of ion channels and carriers. In this integrated approach, the collective behavior of ion transporters is put forward to highlight their role in the dynamics of algae physiology and morphology.

Ion Transport in the Uniform Steady State

The conceptual frame generally used to describe and quantify solute transport through biological membrane derives in large part from theories that describe such phenomenon in homogeneous phase. As a matter of fact, this approach has been successfully applied for more than half a century indicating that despite its inherent simplicity, it is relevant as long as the size of the system investigated is large with respect to the atomic scale (thermodynamic limit).

Solute uptake by cells or tissues involves the movement of molecules in a direction perpendicular to the cell membrane. When the number of molecules moving in a given direction exceeds that of molecules flowing in the opposite direction, then there will be a net transport of molecules through the membrane. The magnitude of this net transport is described by the flux defined as the number of molecules crossing a unit area of membrane per unit time.

This part is intended to give an overview of the physicochemical principles of the ion transport through biological membranes. For the sake of simplicity, we only consider molecule motion along one spatial axis called x axis. It is implicitly assumed that the transport of ions is uniform over the whole cell or tissue surface. The purpose is to derive some basic equations which can be easily used to derive information about the process involved in nutrient uptake. Our approach is based on thermodynamics theories which make no assumption on the specific molecular properties of the transport process and provides boundary conditions for any models of ion transport through a biological membrane.

1. The Flux Equation

1.1. The Electrochemical Potential

Each ion species j of a solution is characterized by its electrochemical potential $\overline{\mu_j}$. The electrochemical potential is a molar free energy viz., it is a thermodynamic way to describe the Gibbs free energy (G) associated to one mole of ion (n_j) and available to perform a work [Dill and Bromberg 2003]. Its mathematical expression is:

$$\frac{\partial G}{\partial n_j} = \overline{\mu_j} = \mu_j^* + RT \ln a_j + z_j F \phi \,, \qquad [1]$$

where μ_j^* is the standard chemical potential, a_j is the ion activity, z_j is the charge of the ion, ϕ is the electric potential, R is the gas constant, T is the temperature and F is the faraday. The ion activity is the 'effective' concentration of ion in solution. It is related to the ion concentration C_j by the activity coefficient γ:

$$a_j = \gamma\, C_j \,, \qquad [2]$$

where $0 < \gamma \leq 1$ [Lide 2005]. In ideal (dilute) solution ($< 1mM$) γ is close to unity. As the solution is getting more concentrated the free energy of ions decreases due to interactions between ions. The extent to which γ deviates from unity is a measure of the non-ideality of the solution. For the sake of simplicity we will use the concentration instead of the activity throughout this chapter.

1.2. Active and Passive Transports

Biological membranes separate solutions of different composition. First, let us consider the case where ions are at equilibrium across the membrane (Figure 3A). By definition this means that the electrochemical potential of j is the same on both side of the membrane. Thus, the difference in electrochemical potential for j^{th} ion species between each compartment is:

$$\Delta\overline{\mu_j} = \overline{\mu}_j^i - \overline{\mu}_j^o = 0 \,, \qquad [3]$$

where the superscripts i and o stand for inside and outside, respectively. Using equation [1] on each side of the membrane, we obtain a simple relation that relates the electric membrane potential difference to the ion concentration on each side of the membrane:

$$\phi^i - \phi^o = \Delta E_j^{eq} = \frac{RT}{z_j F} \ln \frac{C_j^o}{C_j^i} \,. \qquad [4]$$

The electric membrane potential difference corresponding to the equilibrium state is usually known as the equilibrium potential, ΔE_j^{eq}, and equation [4] is named the Nernst's equation. In the derivation of equation [4] we made the assumption that μ_j^* is the same in both side of the membrane. This is true if the solutions differ only in ion concentration and electric potential.

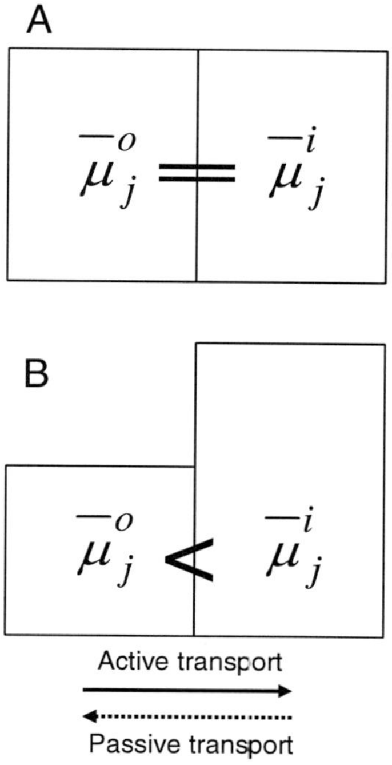

Figure 3. Schematic diagram of the evolution of the electrochemical potential of an ion species crossing a biological membrane. A) the electrochemical potential is identical on both side of the membrane. This corresponds to the equilibrium state. B) A change of the electrochemical potential occurs when crossing the membrane and both active and passive transport can occur through the membrane.

Second, let us consider the case were $\overline{\mu}_j^i > \overline{\mu}_j^o$ (Figure 3B). Then, $\Delta\overline{\mu_j} \neq 0$ and the thermodynamics teaches us that an ion transport will spontaneously occur from the membrane side of higher $\overline{\mu_j}$ to the membrane side of lower $\overline{\mu_j}$. The spontaneous transport of an ion along the gradient of electrochemical potential is called the passive transport. The active transport of an ion is the one that occurs against the gradient of electrochemical potential. The change of electrochemical potential does not only provide qualitative information about the direction of transport, it also permits to calculate the change of energy associated to the transport of one mole of ions of j

$$\Delta\overline{\mu_j} = \mu_j^i - \mu_j^o = RT\ln\frac{C_j^i}{C_j^o} + z_jF(\phi^i - \phi^o) = z_jF(\Delta E^{mb} - \Delta E_j^{eq}), \qquad [5]$$

where ΔE^{mb} is the electric membrane potential difference, usually abbreviated to V_m and named the membrane potential. In practice the outside is taken as a reference and set at

$\phi^o = 0$ so that V_m is a measure of the electric potential on the inner side of the membrane. $\Delta\overline{\mu_j}$ is the minimum energy that must be supply to one mole of j to be actively transported through the membrane; or, *mutatis mutandis*, this is the energy experienced by one mole of j when it is passively transported.

Case Study: Active and Passive Transports Across the Plasma Membrane

We report in Table 1 the values of cytoplasmic concentrations and the membrane potential recorded across the plasma membrane (ΔE^{c-o}) of different marine algae. In all cases, the $\Delta\overline{\mu_{Na}}$ for Na^+ ions is negative indicating an active (passive) transport outside (inside) the cell.

In Valonia and Acetabularia, there is a significant active Cl^- influx. In Acetabularia, the chloride active transport is electrogenic and contributes more than half of the plasma membrane potential (–170mV) when the cell is in the light. When put in the dark the membrane potential almost immediately switches to –80mV [Saddler 1970] indicating that the active influx of Cl^- is closely linked with photosynthesis.

Table 1. Examples of calculation of the equilibrium potential and of the electrochemical potential difference. Concentrations are written according to the International System of units (1 mol m^{-3}= 1mM)

		K^+ (mol m^{-3})	Na^+ (mol m^{-3})	Cl^- (mol m^{-3})	ΔE^{c-o} (mV)
Acetabularia[1, 2]	out	10	470	550	
	cyt	360	60	490	-170
ΔE_j^{eq} (mV)		-90	+52	-3	
$\Delta\overline{\mu_j}$ (kJ mol^{-1})		-7.7	-21.4	+16.1	
Valonia[3, 4]	out	10	432	549	
	cyt	431	38	134	-70
ΔE_j^{eq} (mV)		-95	+11	-35	
$\Delta\overline{\mu_j}$ (kJ mol^{-1})		+2.4	-7.8	+3.4	
Fucoid eggs[5, 6]	out	10	452	538	
	cyt	170	27	87	-50
ΔE_j^{eq} (mV)		-71	+71	-46	
$\Delta\overline{\mu_j}$ (kJ mol^{-1})		+1.9	-11.7	+0.4	

Source of data: [1][Saddler 1970],[2] [Amtmann and Gradmann 1994],[3] [Gutknecht 1966],[4] [Davis 1981],[5] [Taylor and Brownlee 1993],[6] [Allen, Jacobsen and Jaffe 1972].

The active transport of K^+ is directed toward the cytoplasm in Valonia and Fucus eggs but it occurs in the opposite direction in Acetabularia. Chloride is close to the equilibrium in Fucoids eggs. After fertilization, the zygote of Fucoids accumulated Cl^- in the cytoplasm (up to 320 mol m^{-3} in Silvetia (formerly Pelvetia) [Allen, Jacobsen and Jaffe 1972] and the membrane potential range between -70 to -80 mV. This indicates that a significant active Cl^- influx operates at the plasma membrane of zygotes. It is worth to mention that Ca^{2+} is also outside equilibrium: The calcium concentration of sea water is about 10 mol m^{-3} whereas its cytoplasmic concentration very low (<10^{-3} mol m^{-3}) [Amtmann, Klieber and Gradmann 1992]. Therefore, an active (passive) calcium efflux (influx) is expected at the plasma membrane.

1.3. Fluxes Are Driven by Forces

In solution, when a solute molecule of mass m experiences a force F, it will move and will collide with the solvent molecules and therefore it will be subject to a friction force F_f proportional to its velocity v:

$$F_f = -fv, \qquad [6]$$

where f is the coefficient of friction and the minus sign indicates that the friction force acts in the opposite direction of displacement. Thus, the total force acting on the solute particle is

$$F + F_f = F - fv, \qquad [7]$$

and the equation of movement of the particle is given by the second law of Newton:

$$m\frac{dv}{dt} = F - fv. \qquad [8]$$

In the stationary state, $\frac{dv}{dt} = 0$, the relation between the stationary velocity, v_s, and the force is

$$v_s = \frac{F}{f} = uF, \qquad [9]$$

where u $(= 1/f)$ is define as the mobility of the solute molecule.

Let us now consider an ensemble of solute molecules uniformly distributed in the solution and moving with a stationary velocity under the influence of a force F through surface of area A placed a right angle to the direction of movement Figure 4.

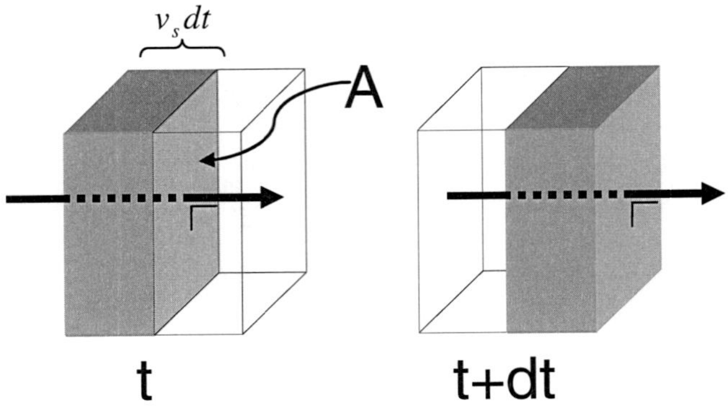

Figure 4. Diagram showing the flux of particles moving perpendicular to a surface of area A. All the particles enclosed at time in a volume will cross the surface A after a time. Figure 4. Diagram showing the flux of particles moving perpendicular to a surface of area A. All the particles enclosed at time in a volume will cross the surface A after a time.

After a short time dt, all the solute molecules will have moved a distance $v_s dt$. Thus, the number of solute molecules enclosed in the volume $Av_s dt$ at a time t will cross the surface after $t + dt$. This number of molecules, dN, is equal to

$$dN = CAv_s dt, \qquad [10]$$

where C is the concentration of the solute. The flux, J, is number of solute molecules crossing a unit area per unit time

$$J = \frac{1}{A}\frac{dN}{dt}. \qquad [11]$$

Combining equations [9], [10] and [11] we get

$$J = uCF. \qquad [12]$$

Equation [12] describes the flux of solute in response to an external force which is also called the migration flux.

1.4. The Driving Force Acting on an Ion and the Nernst-Planck Equation

The force acting on an ion j is the negative gradient of the electrochemical potential:

$$F_j = -\frac{d\overline{\mu_j}}{dx} \qquad [13]$$

and therefore combining equations [12] and [13] we get the flux equation for an ion j:

$$J_j = -u_j C_j \frac{d\overline{\mu_j}}{dx}. \qquad [14]$$

The minus sign that appears in equations [13] (and [14]) is introduced because the direction of the electrochemical potential gradient is opposite to that of the flux (or force): a molecule moves spontaneously to decrease its free energy. Thus, taking the negative gradient permits to define a positive flux (force) in the direction of increasing spatial coordinate (x) (Figure 5).

Introducing equations [1], [2] and Einstein's relation, $D_j = u_j RT$ [Einstein 1956] in equation [14] we get

$$J_j = -D_j C_j \frac{d \ln C_j}{dx} - z_j F C_j \frac{D_j}{RT} \frac{d\phi}{dx}, \qquad [15]$$

where D_j is the diffusion coefficient of the jth ion species. The derivative $d \ln C_j / dx$ is equal to $(1/C_j)(dC_j/dx)$ and equation [15] can be written

$$J_j = -D_j \left[\frac{dC_j}{dx} + \frac{z_j F C_j}{RT} \frac{d\phi}{dx} \right]. \qquad [16]$$

The first term on the right hand side of equation [16] is a diffusion term and is expressed with the concentration, which is valid for ideal solutions. For more concentrate solution the concentration term can be corrected empirically using the activities that are tabulated [Robinson and Stokes 1959; Bockris and Reddy 1970; Lide 2005].

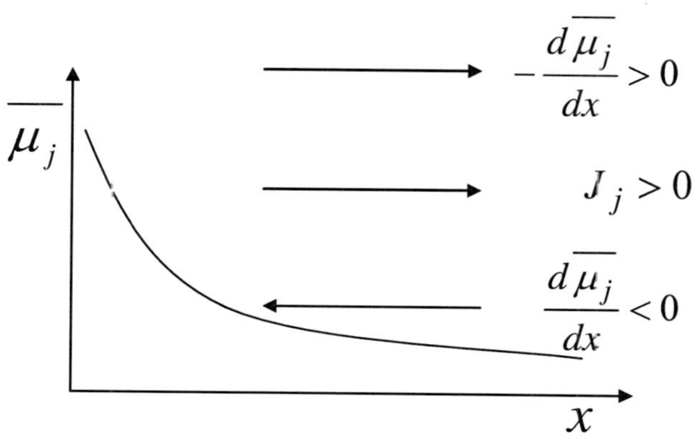

Figure 5. Schematic representation of the spatial variation of the electrochemical potential of an ion species and the respective direction of the flux, the gradient of electrochemical potential and its negative counter part which is the driving force for the flux. The flux is defined as positive when it points to the positive spatial direction. The negative sign in the driving force reflects the physical convention: both flux and force must have the same direction.

Note that this diffusion term is not a force *sensu stricto* (no external force). In fact, it is a pseudo force that describes the random walk process that arises spontaneously in an uneven distribution of solute molecules [Einstein 1956], and therefore, it reflects the entropy of mixing that occurs between the two sides of the membrane. The second term on the right side accounts for the electric force acting on ions. Contrary to the diffusion term, the electric driving force will impart a net velocity to ions, which will drift their movement in the direction of the electric field. As the equation of ionic flux involves both diffusion and electric terms it is often named equation of electrodiffusion or the Nernst-Planck equation. Neutral solutes will not sense the electric force so that their flux will depend only on the diffusion term and equation [16] will reduce to the classical Fick's equation.

1.5. Solving the Flux Equation

Integration of equation [16] will yield an expression of the steady state passive ion flux as function of both concentration and membrane potential. This equation cannot be integrated directly since it contains two unknown dependent variables, C_j and ϕ. A useful approximation of the solution was achieved by assuming that the membrane potential is a linear function of distance through the membrane, viz. that the electrical field is constant in the membrane [Goldman 1943; Hodgkin and Katz 1949]. Using the following identity

$$\frac{d}{dx}\left[C_j e^{z_j F\phi/RT}\right] e^{-z_j F\phi/RT} = \left[\frac{dC_j}{dx} + \frac{z_j F C_j}{RT}\frac{d\phi}{dx}\right] \quad [17]$$

equation [16] can be rewritten as follows:

$$J_j = -D_j \frac{d}{dx}\left[C_j e^{z_j F\phi/RT}\right] e^{-z_j F\phi/RT} \quad [18]$$

Integrating equation [18] from outside to inside over the membrane thickness δ gives:

$$J_j \int_o^{\delta} e^{z_j F\phi/RT} dx = -D_j \int_o^{\delta} \frac{d}{dx}\left[C_j e^{z_j F\phi/RT}\right] dx \quad [19]$$

Now using the arbitrary origin of potential at x=0 and the constant field approximation $\phi = \Delta E^{mb}\, x/\delta$ equation [19] can be integrated directly:

$$J_j \left[\frac{RT\delta}{z_j F V_m}(e^{z_j F V_m/RT} - 1)\right] dx = -D_j \left[C_j^{\delta} e^{z_j F V_m/RT} - C_j^{o}\right], \quad [20]$$

which after rearrangement gives the constant field flux equation:

$$J_j = \frac{P_j z_j F V_m}{RT} \frac{\left[C_j^o - C_j^\delta e^{z_j F V_m / RT} \right]}{e^{z_j F V_m / RT} - 1}, \qquad [21]$$

where $P_j = D_j / \delta$ is the permeability coefficient of species j. The concentration terms in equation [21] are the ones that prevail at the membrane solution interface, just inside the membrane. In their original work Hodgkin and Katz [Hodgkin and Katz 1949] introduced a partition coefficient ζ_j to account for any difference with the concentration in the bulk of solution ($C_j^o = C_j^{out}; C_j^\delta = C_j^{in}$). Taking this parameter into account the definition of the permeability coefficient is $P_j = D_j \zeta_j / \delta$ and the concentrations appearing in equation [21] are then the bulk concentrations. Equation [21] describes the net flux J_j of a species j that is transported passively through the membrane under the effect of an electrochemical potential gradient. If one assumes that the influx and the efflux are independent, the net flux is the difference between the unidirectional influx J_j^{in} of species j and the unidirectional efflux J_j^{out} : $J_j = J_j^{in} - J_j^{out}$ where

$$J_j^{in} = \frac{P_j z_j F V_m}{RT} \frac{C_j^{out}}{e^{z_j F V_m / RT} - 1}, \qquad [22]$$

and

$$J_j^{out} = \frac{P_j z_j F V_m}{RT} \frac{C_j^{in} e^{z_j F V_m / RT}}{e^{z_j F V_m / RT} - 1}. \qquad [23]$$

Equations [21] to [23] are important for the understanding of the passive transport of ions. Equation [22] is particularly useful to describe the initial passive unidirectional influx of a radioisotope of a species j. Initially the specific activity of the species j (its isotopic fraction with respect to the concentration), is equal to zero inside the cell or tissue. Thus, the efflux of radioisotope, J_j^{out}, is zero, and therefore, the initial net flux $J_j = J_j^{in}$. Only the initial flux will provide an accurate quantitatively determination of the radioisotope influx of species j since some of it will leak out after the isotope as entered inside the cell or tissue. Similarly, equation [23] can be used to describe the initial passive unidirectional efflux of a radioisotope of a species j out of a cell or tissue preloaded with its radioisotope. Initially, there is no radioisotope outside and therefore $J_j^{in} = 0$ so that the initial net flux $J_j = J_j^{out}$.

Case Study: Determination of P_j and Isotherm of Absorption

Gutknecht has measured the unidirectional flux for K^+, Na^+ and Cl^- in Valonia [Gutknecht 1966]. In the resting state the passive influx of K^+ is 890 nmol m^{-2} s^{-1}. Thus, using the data of Table 1 and equation [22] we can calculated that P_K = 30 nm s^{-1}. This value can be used to calculate (with equation [22]) the absorption isotherm viz., the passive influx as a function of the sea water K^+ concentration (Figure 6).

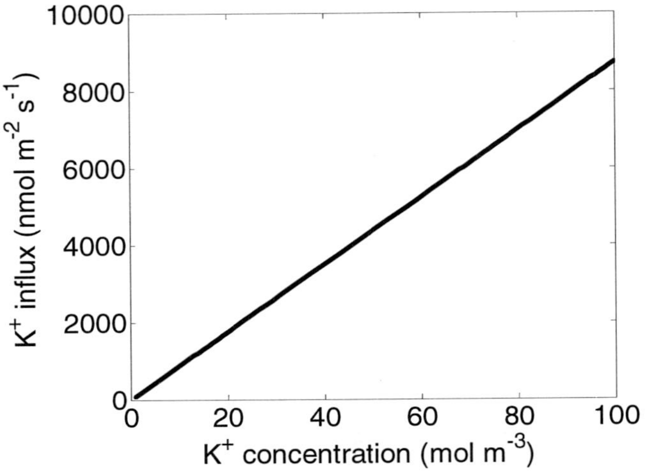

Figure 6. Isotherm for K^+ absorption in Valonia calculated using equation [22].

This isotherm is a straight line; no saturation exists as in Figure 2. This is a consequence of the Nernst-Planck model, which implicitly assumes that ions diffuse without interaction with the others or with a binding site and that there is no regulation of the influx as the external K^+ concentration increases.

2. The Electric Membrane Potential Difference

The electric membrane potential difference V_m is a consequence of a charge separation across the membrane due to the presence of active transport systems that sets up concentration differences between each side of the membrane. At rest, the membrane potential difference across the plasma membrane is negative (Table 1).

2.1. Passive Transport

The flux equation [21] can be used to derive an equation of the membrane potential as a function of the ion concentrations. This will be valid when the active transport is electroneutral, or when it has been inhibited, or else when its contribution to the net flux of ions through the membrane is negligible.

2.1.1. Membrane Permeable to Only Species

Let us consider the case where the ion flux through the membrane is dominated by one ion species only. To fulfill the steady state condition the flux $J_j = 0$, which led to the equilibrium condition and the membrane potential is given by equation [4].

2.2.2. Membrane Permeable to Several Species

When more than one ion species contribute to the net flux through the membrane then the steady state condition imposes that the membrane potential is constant in time. Thus, no net flux occurs through the membrane. This condition is fulfilled by considering the condition of electroneutrality $\sum_j zJ_j = 0$. When K^+, Na^+ and Cl^- are the main species flowing through the membrane the electroneutrality condition led to

$$J_K + J_{Na} - J_{Cl} = 0. \tag{24}$$

Substituting equation [21] for each ion species in equation [24] gives

$$P_K\left(C_K^{out} - C_K^{in}e^{z_jFV_m/RT}\right) + P_{Na}\left(C_{Na}^{out} - C_{Na}^{in}e^{z_jFV_m/RT}\right) + P_{Cl}\left(C_{Cl}^{out} - C_{Cl}^{in}e^{z_jFV_m/RT}\right) = 0$$

which after solving for $e^{z_jFV_m/RT}$ and taking the logarithm we get the following expression relating the membrane potential to the ion concentrations [Goldman 1943; Hodgkin and Katz 1949]:

$$\Delta E^{dif} = \frac{RT}{F}\ln\frac{P_KC_K^{out} + P_{Na}C_{Na}^{out} + P_{Cl}C_{Cl}^{in}}{P_KC_K^{in} + P_{Na}C_{Na}^{in} + P_{Cl}C_{Cl}^{out}}. \tag{25}$$

Equation [25] is a diffusion potential which is often named the Goldamn-Hodgkin-Katz (or GHK) equation. Note that when the membrane is much more permeable to K^+ than to the other ions, then $P_K >>> P_{Na}, P_{Cl}$ and equation [25] is reduces to equation [4].

2.2. Active Transport

In the presence of an electrogenic active transport, the electroneutrality condition is written

$$J_K + J_{Na} - J_{Cl} + z_jJ_j^{at} = 0. \tag{26}$$

where J_j^{at} is the flux of the actively transported ion species j and following the same procedure as that use here above, the membrane potential is given by [Moreton 1969]:

$$V_m = \frac{RT}{F} \ln \frac{P_K C_K^{out} + P_{Na} C_{Na}^{out} + P_{Cl} C_{Cl}^{in} - \frac{z_j RT}{F V_m} J_j^{at}}{P_K C_K^{in} + P_{Na} C_{Na}^{in} + P_{Cl} C_{Cl}^{out} - \frac{z_j RT}{F V_m} J_j^{at}}. \quad [27]$$

This is an implicit equation and its practical interest is limited. A powerful alternative to investigate the properties of the active transport is to use the electrophysiology for dissecting the various ionic currents that flow through the membrane (*vide infra*).

Basic Electrical Properties of Cell Membrane

The use of electricity to describe ion transport through biological membranes is natural from the biophysical point of view, since ions are charged particles, and that ion flux through the membrane generates a current and a change of membrane potential. The basic physical parameters that are measured to characterize a biological membrane are: the membrane potential, the membrane current, the membrane conductance and the membrane capacitance. The advance of the electrophysiological techniques and in particular of the patch clamp technique permits the electric characterization of the of ion transport through a the cell as well as at the level of a single molecule [Hille 1992; Sakmann and Neher 1995; Ashley 1995].

1. The Membrane Potential

The membrane potential (V_m) is the electric potential difference measured between the inside of a cell and seawater by means of micropipettes. It is measured in units of mV. The membrane potential is defined with respect to a reference point which is taken in the seawater.

To this end, the sea water is connected to the ground usually defined as the zero potential. Thus, the sign of the membrane potential refers to the electric potential inside the cell. In electrophysiology, the membrane potential is an important control parameter because carriers and channels may have voltage-dependent properties, which regulates the flux of ion through the membrane. The membrane potential of marine algae formed from giant cells can be measured either between the cytoplasm and the seawater ΔE^{c-o} or between the vacuole and the seawater ΔE^{v-o} (Table 2). In small cells where the vacuole occupies a large proportion of the intracellular volume, only ΔE^{v-o} can be measured.

The membrane potential of the plasma membrane is generally negative and larger (in absolute value) than that of the tonoplast. This is similar to what is known in other plant cells. It is so because the plasma membrane is the main barrier to the absorption of ions.

Ventricaria and Valonia are exceptions: the large membrane potential across their tonoplast is an unusual feature. The membrane potential of the tonoplast is as large as that of the plasma membrane and the two membranes have similar conductance [Davis 1981; Beilby and Bisson 1999]. The classical view according to which the plasma membrane and the tonoplast are two independent membranes in series has been questioned for Ventricaria and Valonia [Shepherd, Beilby and Bisson 2004]. These biophysical and cytological features constitute a limit to the quantitative interpretation of electrophysiological results in these algal genera [Bisson, Beilby and Shepherd 2006].

Table 2. Membrane potential across the plasma membrane ΔE^{c-o} and the tonoplast ΔE^{v-c}. The vacuole potential ΔE^{v-o} is the algebraic sum of the membrane potential across the plasma membrane and the tonoplast

	ΔE^{c-o} (mV)	ΔE^{v-c} (mV)	ΔE^{v-o} (mV)
Seawater Chlorophyta			
Acetabularia[1,2]	-170	0	-170
Bryopsis[3]	-70	0	-70
Ventricaria (Valonia)[4,5]	-70	+86	+16
Brackich water Chlorophyta			
Nitellopsis [3,6]	-130	+10	-120
Seawater Rhodophyta			
Griffithsia[7]	-60	-22	-82
Seawater Heterokontophyta			
Fucus egg[8]	-50	-	-
Diatom			
Coscinodiscus [9]	-70	-	-

Data source: [1][Saddler 1970],[2] [Gradmann 1978], [3] [Gutknecht and Dainty 1968],[4] [Gutknecht 1966],[5] [Davis 1981],[6] [Katsuhara and Tazawa 1986],[7] [Bauman and Jones 1986],[8] [Taylor and Brownlee 1993],[9] [Gradmann and Boyd 1999].

2. The Membrane Current

The membrane current (I_m) is the sum of all the ionic currents (I_j) flowing through the membrane:

$$I_m = \sum_j I_j \,. \qquad [28]$$

The ionic current is a direct measure of the ion fluxes through the membrane and the relation between the two values is given by:

$$I_j = z_j F J_j \,. \qquad [29]$$

By convention the current is positive when a cation is flowing from the cytoplasm to the seawater or an anion is flowing in the opposite direction. Thus, the current is positive when a positive voltage is applied on the cytoplasmic side, which agrees with the convention used in physics.

3. The Membrane Capacitance

A lipid bilayer separating two aqueous solutions behaves as a parallel plate capacitor. This analogy comes from the fact that the lipid bilayer is relatively impermeable to ions and acts as a thin insulator separating two conducting solutions. The total capacitance of a membrane (C) is proportional to the surface area of the membrane (A) and inversely proportional to the membrane thickness (d) and is given by:

$$C = \frac{\varepsilon_0 \varepsilon_m A}{d}, \qquad [30]$$

where ε_0 is the permittivity of the vacuum and ε_m is the dielectric constant (relative permittivity) of the membrane. It is measured in units of farad (F). Typically, ε_m=2.1 for the hydrocarbon chains forming the hydrophobic core of the lipid bilayer. The magnitude of the specific membrane capacitance (C_m) in unit of farad per unit area of biological membrane is of the order of 10^{-2} Fm^{-2}. The total capacitance ($C = C_m \times A$) is a measure of the amount of charges (Q_m) that are distributed on one side of the membrane to build up a membrane potential (V_m):

$$C = \frac{Q_m}{V_m}. \qquad [31]$$

For instance, for a fucoid egg of 60 μm diameter with a cytoplasmic Cl^- concentration of 538 mol m^{-3} (Table 1), an excess of about 40 10^{-18} mole of Cl^- would be required on the cis side of the lipid bilayer to build up a potential difference of -50 mV. This is a very tiny amount compared to the total mole number of anions (~60 10^{-12}) presents in the cytoplasm.

Differentiating equation [31] with respect to time we get an expression for the capacitive current (I_c):

$$I_c = \frac{dQ_m}{dt} = C_m \frac{dV_m}{dt}. \qquad [32]$$

Thus, the value of the membrane capacitance can be calculated from the measurement of the capacitive current in response to a change of applied voltage. Note that the capacitive current does not correspond to a charge movement through the membrane but rather to a change of charge distribution on each side of the membrane. The membrane capacitance of Valonia and Fucus eggs are in a good agreement with the value reported for a planar lipid bilayer (Table 3).

However, the value published for Acetabularia is rather high and far beyond the value usually reported for a biological membrane. It is inconceivable that such a high specific capacity in Acetabularia might be attributed to an inaccurate estimate of the membrane surface, as it should be assumed that the actual surface of the plasma membrane is about five times smaller than the cell surface. However, an overestimation of capacity might result from the presence of transport systems that have not been taken into account when analyzing the results [Wang, Zimmermann and Benz 1994].

Table 3. Electrical parameters of the plasma membrane

	Capacitance (10^{-2} F m^{-2})	Conductance (S m^{-2})
lipid bilayer[1]	0.8	10^{-5}
Acetabularia[2]	5	58
Valonia[3]	0.6	3 to 11
Fucus eggs[4]	0.7	1

Source of data: [1] [Fuks and Homble 1994], [2] [Gradmann and Klemke 1974; Gradmann 1975], [3] [Davis 1981; Benz and Zimmermann 1983; Wang, Wehner, Benz and Zimmermann 1993], [4][Taylor and Brownlee 1993].

4. The Membrane Conductance

The flux of ions through membrane is analogous to the flux of charged species through an electrical conductance. The specific membrane conductance is the inverse of the membrane resistance: $G_m=1/R_m$ and is measured in units of siemens (Sm^{-2}). According to Ohm's law, the relationship between the membrane current (I_m) and the membrane potential is given by

$$I_m = G_m(V_m - \Delta E^{rev}), \qquad [33]$$

where ΔE^{rev} is the reversal potential, viz., the membrane potential when $I_m = 0$. It is equal either to the resting potential when both electrogenic active transport and passive transport exist in the membrane, or to the diffusion potential given by Equation [25] when active transports are negligible and there is a passive transport of several ions through the membrane, or else to the equilibrium potential given by equation [4] when active transports are negligible and there is a passive transport of only one ion species through the membrane. In the membrane, each pump, carrier and channel is a conducting pathway, and all these conductances are in parallel. Thus, the ionic current I_j through each type of transporter is characterized by its specific conductance:

$$I_j = g_j(V_m - \Delta E^{rev}). \qquad [34]$$

At the steady state, when the capacitive current is negligible, the total current flowing through the membrane is the sum of the current flowing through each individual conductance. This means that it is possible to study the current flowing through one type of transporter when its contribution to the membrane conductance can be isolated. For instance, Gradmann and co-workers have used an original experimental protocol that allows the investigation of the electrogenic Cl^- pump in Acetabularia [Gradmann and Klemke 1974; Gradmann 1975].

The specific conductance of the plasma membrane is at least five orders of magnitude larger than that of a lipid bilayer (Table 3). This is in agreement with the negligible ionic permeability of the lipid bilayer. The high conductance found in Acetabularia arises from the strong electrogenic Cl- active transport in the plasma membrane. This active transport is activated in the light and, in the dark, the conductance drops by about 50% [Gradmann 1975]. The conductance of biological membranes is a direct consequence of the ion transport through carriers and channels. The current flowing through a single channel can be measured by several electrophysiological techniques such as the patch clamp or the planar lipid bilayer [Sakmann and Neher 1995; Smeyers, Leonetti, Goormaghtigh and Homble 2003]. The conductance of a single pore ranges between a few pS up to nS corresponding to the flux of 10^5-10^8 ions/s. Nowadays, the transport through a single pump or cotransporter cannot be measured because the ions flux through these proteins is too weak to produce a measurable current.

5. The Electric Equivalent Circuit

For the sake of simplicity, let us assume that the membrane contains one type of channels. Then, the membrane current (I_m) measured when a voltage is applied across a membrane is the sum of the capacitive current and conductive current

$$I_m = I_c + I_j = C_m \frac{dV_m}{dt} + g_j(V_m - \Delta E^{rev}). \qquad [35]$$

Accordingly, the membrane can be represented by the following electrical equivalent circuit shown in Figure 7.

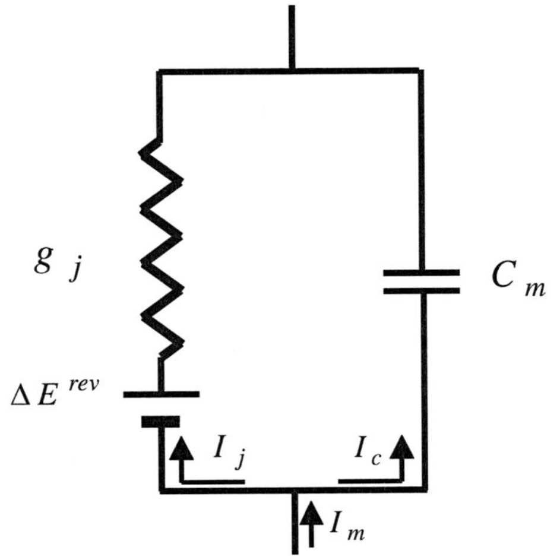

Figure 7. The electrical equivalent circuit for a biological membrane.

The conductance that appears in equations [33] to [35] and in Figure 7 is not necessarily constant, they can vary, for example, with time and voltage. This analysis in terms of electric equivalent has proved to be very fruitful for the study of ion transport through biological membranes from nearly a century. However, we must be aware that this approach implicitly assumes that the transport of ions through the membrane set up concentration gradients in the vicinity of the membrane, which do not contribute to the local electric field. This is true when the main ions diffusing through the membrane have similar diffusion coefficient in solution (this point will be discussed in the § 'transcellular currents').

6. The Current-Voltage Relationship

The recording of the steady state current flowing through the membrane at different values of membrane potential allows plotting the current-to-voltage (I/V) curve of the membrane. The I/V curve corresponding to the constant field flux equation [21] is shown in Figure 8 for the passive K^+ current of unfertilized Fucus eggs. Positive currents, also named outward currents, are defined with respect to cations: cations (anions) are flowing outside (inside) the cell.

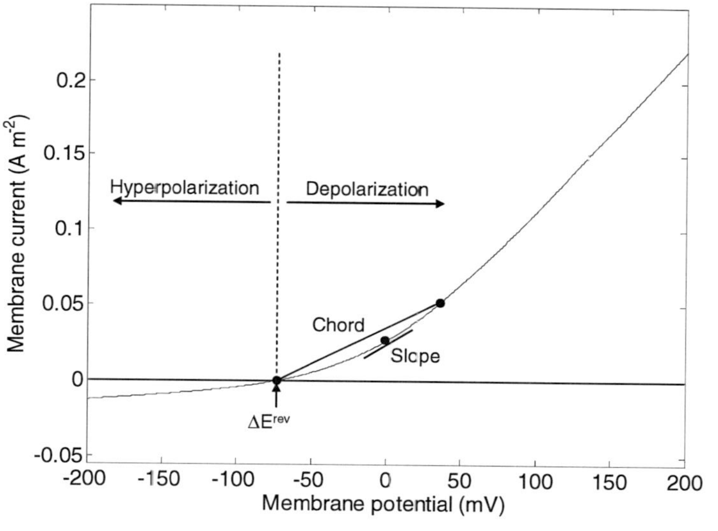

Figure 8. I/V plot for the Nernst-Planck constant field equation obtained by combining equations [21] and [29]. The curve was calculated using P_K=1.7 10^{-9} ms^{-1} using the data for unfertilized Fucus eggs [Robinson and Jaffe 1973; Taylor and Brownlee 1993].

Similarly, inward currents are negative current corresponding to cations (anions) moving inside (outside) the cell. The I/V curve shown in Figure 8 displays an outward rectification: the outward current is much larger than the inward current. This is a consequence of the asymmetric K^+ concentration across the plasma membrane.

The membrane conductance can be defined in two ways: the chord conductance and the differential conductance [Finkelstein and Mauro 1963].

6.1. The Chord Conductance

The chord conductance or integral conductance is the one defined by equation [33] or [34]. It is obtained after recording the current flowing through the membrane when the membrane potential is shifted from the resting state defines by ΔE^{rev} :

$$g_j^{chord} = \frac{I_j}{(V_m - \Delta E^{rev})}. \qquad [36]$$

The chord conductance is always positive since the current is positive when the membrane is depolarized, viz. $(V_m - \Delta E^{rev}) > 0$ and it is negative when the membrane is depolarized $(\Delta E^{mb} - \Delta E^{rev}) < 0$.

6.2. The Differential Conductance

The slope of the I/V curve can be defined at any point and it provides an operational definition of the differential conductance or slope conductance. In practice, it can be obtained by polarizing the membrane at a defined voltage V_m and by superimposing an infinitesimal voltage perturbation ∂V. By measuring the resulting infinitesimal current ∂I_j we can calculate the slope conductance at the voltage $V = V_m$ using

$$g_j^{slope} = \left(\frac{\partial I_j}{\partial V}\right)_{V=V_m}. \qquad [37]$$

The differential conductance can be positive or negative. A negative differential conductance may be source of instability which may lead to a complex dynamic behavior of the membrane (*vide infra*). The N-shaped I/V curve is a typical example that has a differential conductance with negative and positive values (Figure 9).

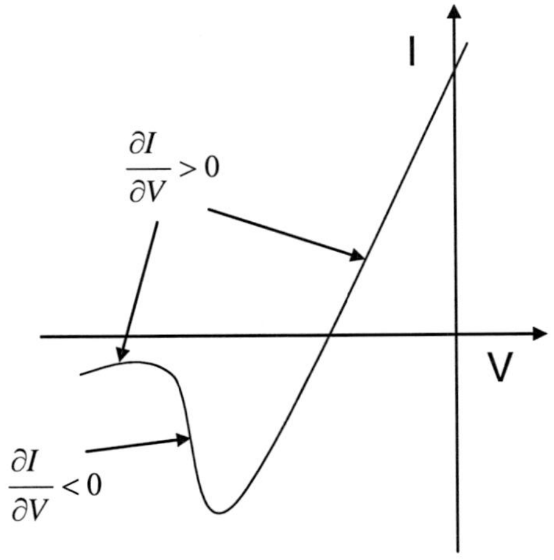

Figure 9. Schematic representation of an I/V curve displaying both negative and positive slope conductance.

This shape is observed notably when the voltage-dependence of ion channels occurs at voltage more negative than the diffusion potential of the ions crossing the channel [Hille 1992]. I/V curves with a negative slope were recorded for the calcium current in unfertilized Fucus eggs [Taylor and Brownlee 1993] and in Acetabularia after inhibition of the Cl^- pump[Gradmann 1975].

Kinetic Models in the Uniform Steady State

The Nernst-Planck equation describes the flux of ion in a homogeneous phase and we show in Figure 6 that it cannot account for saturating flux observed experimentally (Figure 2). An alternative approach is to consider the transport of ions as an enzymatic process. Then, the flux of ions will be described in terms of rate constants that characterize a sequence of states that occur during the ion translocation process. For instance, assuming that the transport is limited by binding of the ion X to a single site S in the protein according to the sequence

$$X_{out} + S \underset{k_{-1}}{\overset{k_1}{\Leftrightarrow}} XS \underset{k_{-2}}{\overset{k_2}{\Leftrightarrow}} S + X_{in},$$

then, the overall reaction follows the Michaelis-Menten kinetics and the flux is given by the saturating function:

$$J_X = \frac{J_X^{\max} X_{out}}{X_{out} + K}, \qquad [38]$$

where $K = (k_{-1} + k_2)/k_1$ is the concentration for half-maximal J_X.

The voltage-dependent properties can be taken into account using the Eyring's rate theory [Eyring 1935; Zwolinski, Eyring and Reese 1949; Eyring, Ma and Ueda 1981; Lauger 1987]. The transport of ions through carriers or channels can then be described using a kinetic model that account for the minimum states (or energy barriers) that control the ion flux. This minimal kinetic model, fitted to the I/V curve, permits the evaluation of the kinetic constants that control the ion flux. This approach has been used to describe ion transport through pumps and channel of marine algae [Mummert, Hansen and Gradmann 1981; Lauger 1987; Bertl and Gradmann 1987; Bertl, Klieber and Gradmann 1988; Beilby and Shepherd 2001]

Case Study: The Electrogenic Cl^- Pump

The electrogenic Cl^- pump of Acetabularia can be described by a two-states kinetic model connected by two reversible reaction steps (Figure 10) [Hansen, Gradmann, Sanders and Slayman 1981; Mummert, Hansen and Gradmann 1981; Tittor, Hansen and Gradmann 1983]. One branch accounts for the voltage-dependent translocation of the charged species across the membrane whereas the other branch lumps together all the electrically neutral reactions such as ATP hydrolysis and conformational changes. The membrane density of each form of the carrier is X_o^z and X_i^z with $X= X_o^z + X_i^z$. In Acetabularia z=-2 since the stoechiometric ratio $Cl^-/ATP=2/1$ [Gradmann 1975]. The rate constants are empirical rate constants that represent the product of the true rate constants and the related concentration.

The charge transfer through the membrane is assumed to involve only one pair of reactions between the two states X_o^z and X_i^z. The voltage-dependence is obtained by considering that the translocation of ions occurs through a single symmetric energy barrier [Lauger and Stark 1970]. The two rate constants of the voltage-dependent branch are given by

$$k_{io} = k_{io}^0 e^{-\frac{z_j F(V_m - \Delta E_{pump}^{rev})}{2RT}} \quad \text{and} \quad k_{oi} = k_{oi}^0 e^{\frac{z_j F(V_m - \Delta E_{pump}^{rev})}{2RT}} . \qquad [39]$$

At steady state, the differential equation for the two-state model is

$$\frac{dX_i^{2-}}{dt} = (k_{oi} + K_{oi})X_o^{2-} - (k_{io} + K_{io})X_i^{2-} = 0 . \qquad [40]$$

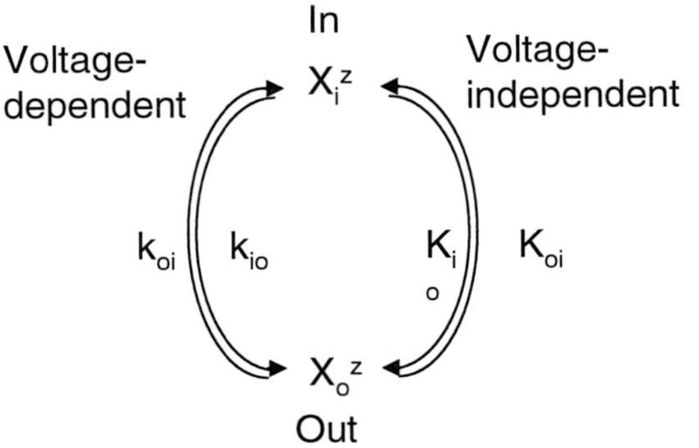

Figure 10. Two-states model for the Cl^- electrogenic pump of Acetabularia.

The current flowing through the pump is the difference between the outward flux of X_i^{2-} and the inward flux of X_o^{2-}

$$I_{pump} = zF(X_i^{2-}k_{io} - X_o^{2-}k_{oi}) . \qquad [41]$$

Combining Equations [39] to [41] we get

$$I_{pump} = zFX \frac{K_{oi}k_{io}^0 e^u - K_{io}k_{oi}^0 e^{-u}}{k_{io}^0 e^u + k_{oi}^0 e^{-u} + K_{oi} + K_{io}} , \qquad [42]$$

with $u = \dfrac{z_j F(V_m - \Delta E^{rev}_{pump})}{2RT}$. The total current through the membrane I_m is the sum of the pump current and of the passive transport current I_{leak} and is written as:

$$I_m = I_{pump} + I_{leak} = I_{pump} + g_{leak}(V_m - \Delta E^{rev}), \qquad [43]$$

where ΔE^{rev} is the reversal potential for the passive transport. When the pump is active the contribution of the leak current is negligible (Figure 11) which reflects the strong electrogenicity of the Cl^- pump in Acetabularia as expected from the data of Table 1.

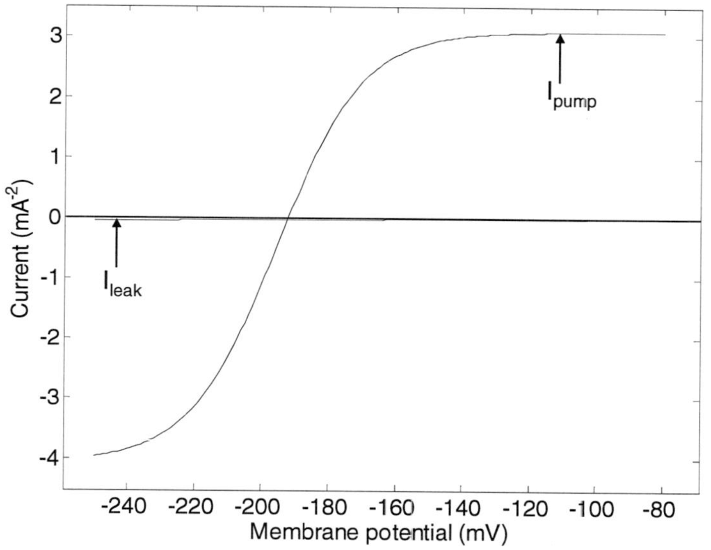

Figure 11. I/V curve of the leak current and the Cl^- pump current calculated using the data for Acetabularia [Gradmann 1975; Mummert, Hansen and Gradmann 1981].

Nonlinear Dynamics of Ion Transport

We learned from the steady state description of ion transport that the translocation of a charged species through the membrane is a nonlinear process, no matter the molecular mechanism involved. The parameters that control the nonlinearity are the membrane potential and the ion concentrations. The dynamics of ion transport takes also into accounts the time-dependence of the transport phenomena. It is now well established in physics, chemistry and biology that nonlinear dynamical systems are subjected to feedback regulations and may undergo self-generated changes leading to stationary or transient spatiotemporal patterns [Strogatz 1994; Goldbeter 1997; Keener and Sneyd 1998; Nicolis 1999; Murray 2002; Murray 2004]. In this part, we describe two nonlinear dynamic processes that have been observed in marine algae: the action potential and the pattern of transcellular currents.

1. The Action Potential

The action potential is a voltage wave that propagates over the cell membrane when it is depolarized beyond a threshold value. This is a self-generating process which is initiated by a positive feedback loop that suddenly and rapidly increases and amplifies an initial membrane depolarization. Once triggered, it follows its course regardless of the stimulus that provokes it. Notably, the shape and the rate do not depend on the excitation. In algae the lifetime of the action potential last from a few tens of millisecond in Fucus eggs to several hundred seconds in Acetabularia.

There is not enough data available to describe quantitatively the action potential in excitable cells of marine algae. However, the empirical description of the action potential of freshwater plants of the Characeae using the Hodgkin-Huxley formalism has shown that valuable information can be obtained from the model initially developed for the axon [Beilby and Coster 1979a; Beilby and Coster 1979b; Beilby 1984; Beilby 2007]. Thus, indicating that a qualitative analysis can be carried out provided that a relevant model can be proposed.

In Acetabularia the action potential plays a key role for the osmotic regulation [Mummert and Gradmann 1991b]. Contrary to other algae [Bisson, Beilby and Shepherd 2006], the turgor pressure of Acetabularia continuously increased due to the action of the Cl^- pump [Wendler, Zimmermann and Bentrup 1983]. Spontaneous action potential is accompanied by a large efflux of Cl^- [Gradmann, Wagner and Glasel 1973] that depolarizes the membrane, followed by a delayed K^+ efflux [Gradmann 1976; Bertl and Gradmann 1987; Bertl, Klieber and Gradmann 1988] that repolarizes the membrane to its resting level. Thus, the action potential leads to a net loss of salt with a concomitant cell turgor decrease. A fairly good simulation of the action was achieved considering the three major transport systems of the plasma membrane: Cl^- pumps, K^+ channels and Cl^- channels [Mummert and Gradmann 1991a].

Three different types of action potential were found in the diatom Coscinodiscus [Gradmann and Boyd 2000]. Their lifetime is comparable to that of Acetabularia. One type is activated by a membrane depolarization and lead to an efflux of Cl^- and K^+. The second is activated by hyperpolarization and also involves Cl^- and K^+ channels. The third type is also activated by hyperpolarization but requires a voltage-gated pump.

Action potentials of about 50 ms duration and triggered by a membrane depolarization have been recorded in the calcifying algae Coccolithus [Taylor and Brownlee 2003]. During the action potential the membrane depolarization phase is triggered by activation of the Cl^- inward-rectifying conductance.

The excitability of Fucus eggs is different from those described here above. The action potential of unfertilized eggs from Fucus lasts a few tens of milliseconds and it shares several properties in common with the excitable muscle fibers: 1) the fast initial depolarizing current is a calcium time- and voltage-dependent inward current, 2) the recovery phase is ensured by a potassium time- and voltage-dependent outward current, which is activated in the course of the membrane depolarization [Taylor and Brownlee 1993]. This prompts us to use the Morris-Lecar Model in order to analyze the dynamics of the action potential in Fucus eggs. The Morris-Lecar model is a two-dimensional simplified model applicable to systems having two non-inactivating voltage-sensitive conductances [Morris and Lecar 1981]. This is the case of Fucus eggs [Taylor and Brownlee 1993]. It would be presumptuous to think that a quantitative analysis of the plant action potential can be achieved by *in silico* modeling because our current knowledge of the dynamics of ion channels involved in the action potential is incomplete. Our purpose is to illustrate how a qualitative analysis of a nonlinear dynamical system can be used to understand the salient features of the action potential in Fucus. This qualitative approach is particularly useful to grasp the dynamics of a nonlinear physiological process for which no explicit solution exists. The model assumes that there is an inward Ca^{2+} current, an outward K^+ current and a leak current which is characterized by a constant conductance g_{leak} (Figure12). We further assume that the Ca^{2+} current is always in equilibrium so that its dynamics must not be described explicitly [Rinzel and Ermentrout 1998]. This is justified when the Ca^{2+} current is faster than the K^+ current. The calcium conductance is given by $g_{Ca} = \overline{g}_{Ca} m_\infty$ where m_∞ is the calcium activation parameter. The potassium conductance is given by $g_K = \overline{g}_K w$ where w is the potassium activation parameter. $\overline{g}_j$ is written for the maximal conductance of j. Therefore, the dynamics of ion transport is described by a set of two ordinary differential equations:

$$C_m \frac{dV_m}{dt} = -\overline{g}_{Ca} m_\infty (V_m - \Delta E_{Ca}^{eq}) - \overline{g}_K w (V_m - \Delta E_K^{eq}) - g_{leak}(V_m - \Delta E^{rev}) + I_{inj} \quad [44]$$

$$\frac{dw}{dt} = \phi \frac{w_\infty - w}{\tau} \quad [45]$$

where I_{inj} is the current injected in the cell to shift the membrane potential and

$$m_\infty = 0.5[1 + \tanh((V_m - V_1)/V_2)],$$

$$w_\infty = 0.5[1 + \tanh((V_m - V_3)/V_4)],$$

$$\tau = \frac{1}{\cosh((V_m - V_3)/V_4)}.$$

The steady state I/V curve is calculated by setting $w = w_\infty$ in equation [44]:

$$I_m = \overline{g}_{Ca} m_\infty (V_m - \Delta E_{Ca}^{eq}) + \overline{g}_K w_\infty (V_m - \Delta E_K^{eq}) + g_{leak}(V_m - \Delta E^{rev}). \quad [46]$$

The parameters were either obtained from the literature [Taylor and Brownlee 1993] or were determined empirically (Figure 12).

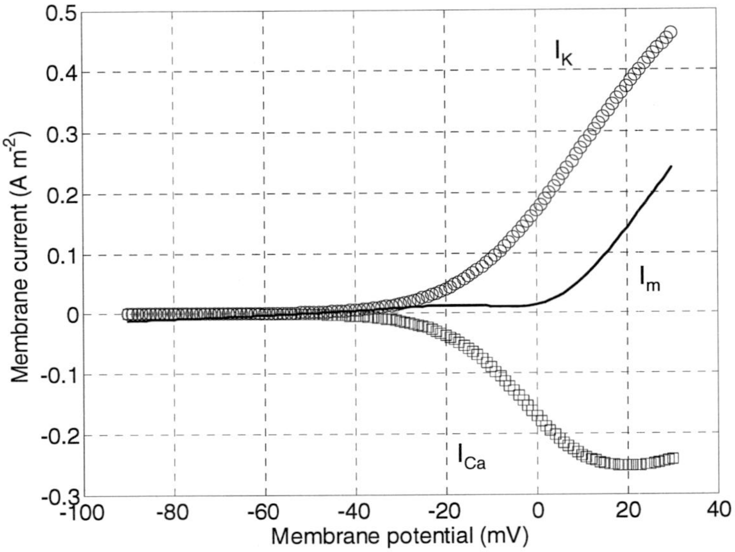

Figure 12. Steady state I/V curves for the total membrane current (I_m), the calcium current (I_{Ca}) and the potassium current (I_K) in Fucus.

Upon membrane depolarization there is an inward Ca^{2+} current that displays a negative differential conductance and an outward K^+ current that have a positive differential conductance. The I/V curve of the total membrane current has an N-shape which is typical of excitable cells (Figure 13). The system of equations [44] and [45] is nonlinear. There is no analytical solution to this system and its solution must be found numerically. However, valuable information about the dynamics of the system can be obtained with the graphical phase plane analysis where one dependent variable (w) is plotted against the other (V_m) (Figure 14). As the system evolved it will describe a trajectory in the phase plane. A family of trajectories will be obtained by changing the parameters of the model. The slope of a trajectory at a given point is dw/dV_m. This slope is nothing else than the ratio of dw/dt to dV_m/dt viz., a velocity vector given by the right hand side of equations [45] and [44]. Thus, the phase plane provides a global view of the dynamic evolution of the trajectory: the trajectory must be horizontal or vertical when $dw/dt = 0$ or $dV_m/dt = 0$, respectively. The functions $dw/dt = 0$ and $dV_m/dt = 0$ are called the nullclines. The V_m-nullcline and the w-nullcline are calculated using equations [44] and [45], respectively:

V_m-nullcline:
$$w = \frac{\bar{g}_{Ca} m_\infty (V_m - \Delta E_{Ca}^{eq}) + g_{leak}(V_m - \Delta E^{rev}) - I_{inj}}{\bar{g}_K (V_m - \Delta E_K^{eq})}, \qquad [47]$$

w-nullcline:
$$w = w_\infty. \qquad [48]$$

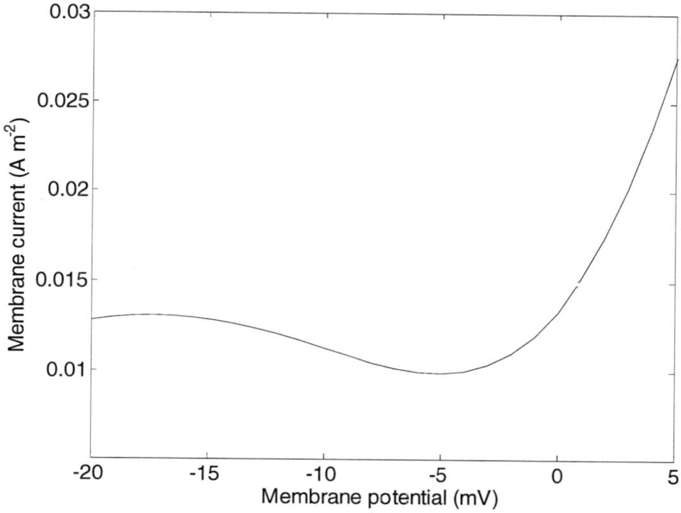

Figure 13. Enlargement of the steady state I/V curves for the total membrane current (I_m) showing the negative differential conductance region.

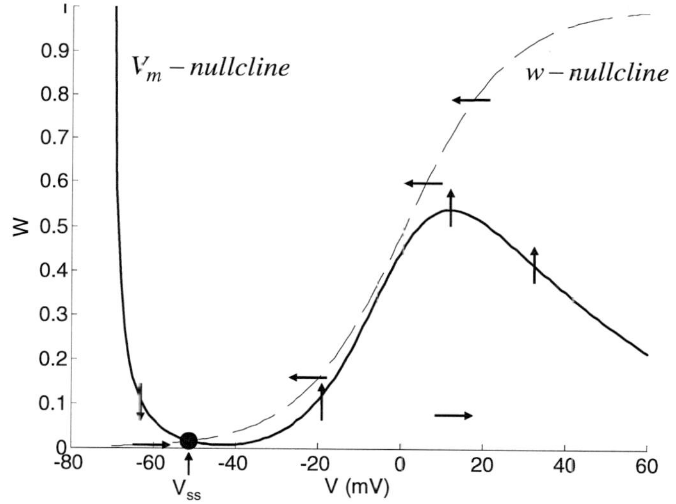

Figure 14. Phase plane showing the nullclines of the two ordinary differential equations used to describe the action potential in Fucus. The arrows indicate the direction of changes of the system (see text). V_{ss} is a stable stationary state.

The nullclines are shown in figure 14. The point of intersection between w-nullcline and V_m-nullcline is a stable stationary state: neither w nor V_m change with time. We can now analyze the direction of motion of w and V_m across the nullclines. When w is above the w-nullcline ($w > w_\infty$) equation [45] is negative and the direction of evolution of the trajectory will be downward whereas it will be upward when w is below the w-nullcline. Introducing equation [47] in equation [44] indicates that the direction of the trajectory will point to the left or right depending on whether we are above or below the V_m-nullcline , respectively (Figure 13).

The numerical integration of equations [44] and [45] permits to simulate the dynamics of the membrane potential of Fucus eggs (Figure 15). The action potential is triggered when the membrane potential is depolarized below a threshold value of -25mV which is slightly more positive than that reported in the literature [Taylor and Brownlee 1993]. When the membrane depolarized at voltages more negative than the threshold value then the membrane perturbation is quickly dissipated and the membrane potential returns to its resting state with a characteristic time given by the ratio of the membrane capacitance to the membrane conductance.

The corresponding trajectories in the phase plane are shown in Figure 16. After a brief substhreshold perturbation the membrane potential almost directly returns to the resting state following trajectory around the stable stationary state. A completely different trajectory occurs when the brief perturbation shifts the membrane potential above the threshold. It corresponds to the time course of the action potential shown in Figure 15. The system is unstable and the membrane potential suddenly diverges towards the rightmost branch of the V_m-nullcline, leading to the membrane depolarization. Then, system moves upward along the V_m-nullcline until it reaches the shoulder where it moves backward to the resting state. These trajectories occur as expected from the qualitative analysis of the phase plane (Figure 14).

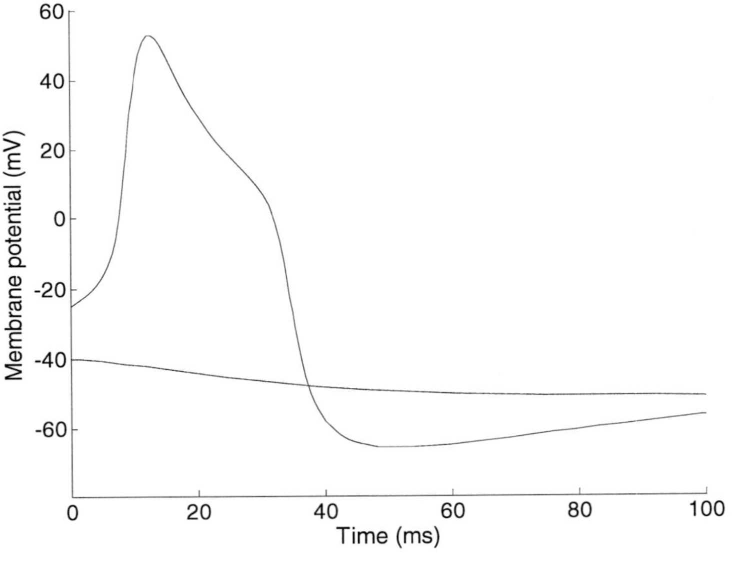

Figure 15. Simulation of the membrane voltage dynamics in response to an instantaneous membrane depolarization to either -40mV (solid line) or -25mV (dash-dot line).

The dependence of the dynamics on the initial conditions is one of the salient features of nonlinear systems. The initial direction of the evolution of the membrane potential depends on the relative magnitude of dw/dt and dV_m/dt the starting point. In the phase plane shown in Figure 16 the threshold arises because the V_m dynamics is much faster than that of the w-dynamics. Above the threshold, the voltage growth is exponential (Figure 15).

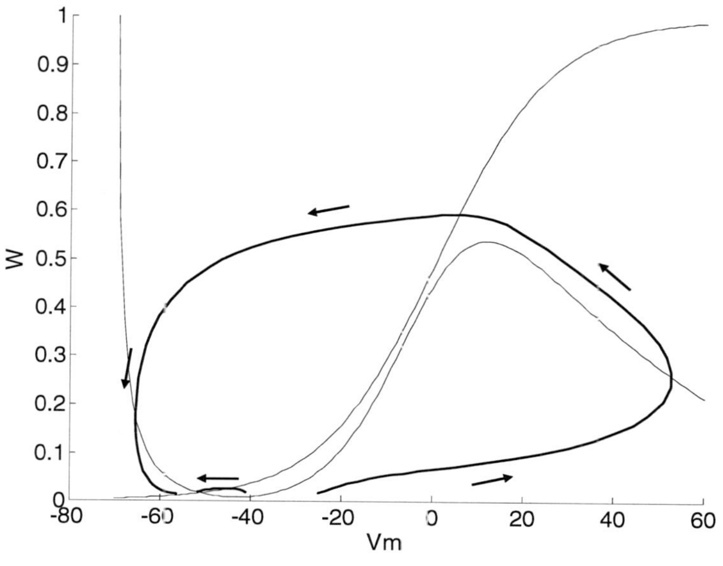

Figure 16. The phase plane. The heavy lines represent two solution trajectories starting from the initial condition of -40 and -25 mV.

This is a consequence of a positive feedback: the membrane depolarization activates the inward Ca^{2+} current which in turn further depolarizes the membrane and so on until the calcium conductance is maximal. Thus, the action potential is a typical self-organized non-stationary spatiotemporal structure that emerges from the interaction between different ion conductances which have nonlinear physical properties.

2. Transcellular Currents

Transcellular currents are ubiquitous in biology [Gow 1989; Hepler, Vidali and Cheung 2001] [Jaffe 1977; Nuccitelli 1988; Harisson 2005]and in plants they are notably correlated with the apical growth [Kropf, Lupa, Caldwell and Harold 1983; Cooke and Racusen 1986; Nawata, Hishinuma and Wada 1989; Feijo, Costa, Prado, et al. 2004]. Ionic currents leave the cell (or tissue) at one place; they flow around the cell and then enter the cell at another place, giving rise to current loops (Figure 17).

In algae transcellular currents have been extensively studied in brown algae [Jaffe 1966; Jaffe and Nuccitelli 1974; Nuccitelli and Jaffe 1974; Nuccitelli and Jaffe 1975; Brawley and Robinson 1985] and Acetabularia [Dazy, Borghi and Garcia 1986; Bowles and Allen 1986; O'Shea, Goodwin and Ridge 1990; Baumann 1991]. In Acetabularia, a pH pattern has been measured along the stalk. Contrary to what has been found in freshwater algae [Bulychev, Polezhaev, Zykov, et al. 2001] there is not a perfect correlation between the cell wall calcification and the alkaline region [Serikawa, Porterfield, Smith and Mandoli 2000]. Calcium pattern exists at the whorl and stalk growing tip region [Reiss and Herth 1979].

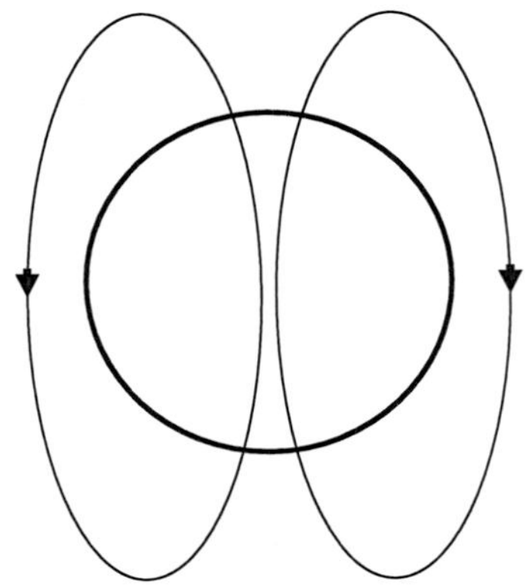

Figure 17. Schematic representation of transcellular currents.

An electrical pattern is established during regeneration after ablation of Acetabularia rhizoids [Novak and Bentrup 1972; Novak 1975] and extracellular gradient of both Ca^{2+} and K^{+} ions seem to play a role in the tip regeneration [Goodwin and Pateromichelakis 1979]. Transcellular currents are correlated to the early stage of the axial growth in Fucus and Silvetia zygote [Roberts, Berger and Brownlee 1993; Quatrano and Shaw 1997; Brownlee and Bouget 1998; Jaffe 1999; Kropf, Bisgrove and Hable 1999; Robinson, Wozniak, Pu and Messerli 1999; Homble and Leonetti 2007]. There are evidences indicating that in Fucoids the emergence of transcellular currents is self-generated[Leonetti, Dubois-Violette and Homble 2004; Homble and Leonetti 2007].

The mathematical description of transcellular current pattern formation must take into account both ion transport through the plasma membrane and electrodiffusion of ions in the solution adjacent to the membrane. For the sake of simplicity we assume that the cell has a cylindrical shape and we restrict ourselves to the one-dimensional case. This can be done without loss of the salient features of the model. Furthermore, we will write the equations for the cytoplasmic side only to keep equations to a minimum. This simplification is justified since the cytoplasmic volume is generally much lower than that of the extracellular medium and thus the driving force of electrochemical gradient is the strongest inside the cell. In other words, it is the cytoplamic compartment that drives the dynamics of pattern formation of both ion concentration and electric potential in the vicinity of the membrane.

The measurements of stationary patterns of extracellular currents around a cell or a tissue indicate that gradient of both ion concentration and electric potential exist around the plasma membrane. To describe the dynamics of ion concentrations and of the membrane potential difference, we need to refer to a relevant electrodiffusive model based on the Nernst-Planck equation with an additional capacitive relation [Quian and Sejnowski 1989; Qian and Sejnowski 1989; Leonetti and Dubois-Violette 1998]. Let us consider a small section of a cell of length, dx, and radius, R. The membrane can be viewed as an equivalent leaky cable consisting of a capacitance in parallel with conductances [Koch 1999]. The net flux of ion through the membrane is noted J_j^{mb}. In the solution, the net ion flux J_j is given by the Nernst-Planck equation (see equation [16]). Along the x axis, the continuity equation is:

$$\frac{\partial C_j}{\partial t} = -\frac{\partial J_j}{\partial x} - \frac{2J_j^{mb}}{r}. \qquad [49]$$

Introducing the Nernst-Planck equation in equation [49] gives:

$$\frac{\partial C_j}{\partial t} = D_j\left[\frac{\partial^2 C_j}{\partial x^2} + \frac{z_j F C_j}{RT}\frac{\partial}{\partial x}\left(\frac{d\phi}{dx}\right)\right] - \frac{2J_j^{mb}}{r}. \qquad [50]$$

The minus sign before the second term of the right side member arises from the convention according to which a positive current corresponds to an efflux of cations. In the cell, the electroneutrality principle must be conserved; viz., the excess of superficial charge must be equal to that of the volume:

$$\sigma 2\pi r dx = \pi r^2 F \sum_j z_j (C_j - C_{j0}) dx, \qquad [51]$$

where σ is the surface charge density and C_{j0} is the bulk concentration of the ion species j. The equivalent surface charge density is equal to the product of the specific membrane capacitance, C_m, and the membrane potential difference ($\sigma = C_m V_m$) and therefore the relationship between C_j and V is given by:

$$V_m = \frac{rF}{2C_m}\sum_j z_j (C_j - C_{j0}). \qquad [52]$$

Note that $V_m = (\phi^i - \phi^o)$ and as ϕ^o is by convention set to 0mV, V_m gives the value of the intracellular electric potential. Equations [50] and [52] allow a full description of the spatio-temporal evolution of ion concentrations and electric potential along the membrane. For the sake of simplicity, we only consider two cations 1 and 2. Therefore, our model will be described by a set of three partial differential equations:

$$\left.\begin{aligned}
\frac{\partial C_1}{\partial t} &= D_1\left[\frac{\partial^2 C_1}{\partial x^2} + \frac{z_1 F C_1}{RT}\frac{\partial}{\partial x}\left(\frac{dV_m}{dx}\right)\right] - \frac{2I_1^{mb}}{r z_1 F} \\
\frac{\partial C_2}{\partial t} &= D_2\left[\frac{\partial^2 C_2}{\partial x^2} + \frac{z_2 F C_2}{RT}\frac{\partial}{\partial x}\left(\frac{dV_m}{dx}\right)\right] - \frac{2I_2^{mb}}{r z_2 F} \\
V_m &= \frac{rF}{2C_m}\left[z_1(C_1 - C_{10}) + z_2(C_2 - C_{20})\right]
\end{aligned}\right\}, \qquad [53]$$

where $I_j^{mb} = z_j F J_j^{mb}$. It can be checked that in the absence of any concentration spatial gradient $\frac{\partial C_j}{\partial x} \approx 0$, the set of equations [53] reduces to the classical cable equation which describes the spatio-temporal dynamics of the voltage most animal cells [Keener and Sneyd 1998]:

$$C_m \frac{\partial V_m}{\partial t} = -\left[I_1^{mb} + I_2^{mb}\right] + \frac{r}{2\rho_s} \frac{\partial^2 V_m}{\partial x^2}, \qquad [54]$$

where ρ_s is the cytoplasmic resistivity. The comparison of equations [53] and [54] indicates the complexity that arises in systems where ion concentration gradients outside the membrane cannot be neglected. Contrary to animal cells, ion concentration gradients nearly always exist around plant cells and tissues.

The partial differential equations [53] are usually nonlinear because the current flowing through the membrane is generally a nonlinear function of the membrane potential. Thus, a solution to the set of equations [53] can only be obtained by numerical integration. Nonetheless, the instantaneous response of the model to a small perturbation $(\delta C_j, \delta V_m)$ may be grasped from the linear analysis of equations [53]. The linearized equations are:

$$\left.\begin{aligned} \frac{\partial \delta C_1}{\partial t} &= D_1 \left[\frac{\partial^2 \delta C_1}{\partial x^2} + \frac{z_1 F C_{10}}{RT} \frac{\partial^2 \delta V_m}{\partial x^2} \right] - \frac{2}{r z_1 F} \frac{\partial I_1^{mb}}{\partial V_m} \delta V_m \\ \frac{\partial \delta C_2}{\partial t} &= D_2 \left[\frac{\partial^2 \delta C_2}{\partial x^2} + \frac{z_2 F C_{20}}{RT} \frac{\partial^2 \delta V_m}{\partial x^2} \right] - \frac{2}{r z_2 F} \frac{\partial I_2^{mb}}{\partial V_m} \delta V_m \\ \delta V_m &= \frac{rF}{2C_m} \left[z_1 \delta C_1 + z_2 \delta C_2 \right] \end{aligned}\right\}. \qquad [55]$$

This linear set of partial differential equation can be completely solved assuming a solution of the form:

$$\delta C_j(x,t) = K_j e^{ikx + \omega t}, \qquad [56]$$

where K_j is the amplitude, k is the mode (wavenumber) of the perturbation and ω its growth rate. The choice of the function [56] is dictated by the basic feature of the spatio-temporal pattern of transcellular current: a spatial periodic wave that growth with time. Introducing [56] in [55] and solving the system of equations [55] for ω gives the dispersion (characteristic) equation:

$$\omega^2 + \frac{r}{2C_m\alpha}(k^2 - A^2)\omega + \frac{r}{2C_m\alpha}\frac{D_1D_2}{[D_1\beta_1 + D_2\beta_2]}k^2(k^2 - B^2) = 0, \qquad [57]$$

with

$$\alpha = \frac{1}{\varepsilon\chi^2[D_1\beta_1 + D_2\beta_2]},$$

where ε is the dielectric constant and χ^{-1} is the Debye length, $\beta = \frac{z_j^2 C_{j0}}{\sum_j z_j^2 C_{j0}}$,

$$A^2 = -\frac{2F}{r\varepsilon\chi^2}\frac{\left[\frac{\partial I_1^{mb}}{\partial V_m} + \frac{\partial I_2^{mb}}{\partial V_m}\right]}{[D_1\beta_1 + D_2\beta_2]} \text{ and } B^2 = -\frac{2F}{r\varepsilon\chi^2}\left[\frac{1}{D_1}\frac{\partial I_1^{mb}}{\partial V_m} + \frac{1}{D_2}\frac{\partial I_2^{mb}}{\partial V_m}\right].$$

The partial differentials that appear in A^2 and B^2 are the differential (slope) conductance of each ion species: g_1^{slope}, g_2^{slope} (see equation [37]). The roots of the dispersion equation [57] are the solutions of equations [55] and predict the dynamics of the system . When ω<0 the perturbation will be dissipated and the system decreases exponentially to its initial homogeneous resting state. This is the feature of a stable state. However, when ω>0 the system becomes unstable. Any perturbation will be amplified and initially its growth will be exponential. There are two conditions in which the system may be unstable (ω>0). First, A^2 is positive which implies that the total differential conductance is negative ($g_1^{slope} + g_2^{slope} < 0$), then the system is unstable and homogeneous (k=0). This unstable state is typical of excitable cells and an action potential is expected. Second, the system is unstable and generates a pattern of concentrations when A^2 is negative and B^2 positive [Leonetti, Dubois-Violette and Homble 2004], i.e., when:

$$\begin{cases} g_1^{slope} + g_2^{slope} > 0 \\ \frac{g_1^{slope}}{D_1} + \frac{g_2^{slope}}{D_2} < 0 \end{cases}. \qquad [58]$$

In this case, the system must fulfilled two simultaneous conditions to be unstable: the total differential conductance must be positive and if the differential conductance of species 1 is positive then the ion species 2 must have a negative differential conductance and its diffusion coefficient must be smaller than that of species 1. The stability conditions [58] do not depend on the type of transporters existing in the membrane. For instance, it can be achieved by coupling an electrogenic active transport with a voltage dependent ion channel [Leonetti and Dubois-Violette 1998] or an electrogenic active transport with a cotransport [Leonetti, Marcq, Nuebler and Homble 2005], or else with two ion channels [Leonetti, Dubois-Violette and Homble 2004].

The characteristic wavelength, λ_c, of the spatial pattern is given by:

$$\lambda_c \approx \left| \frac{2\pi}{\frac{2F}{r\varepsilon\chi^2}\left[\frac{g_1^{slope}}{D_1} + \frac{g_2^{slope}}{D_2}\right]} \right|, \qquad [59]$$

and the characteristic time, τ_c, is about: $\tau_c \approx r^2/D_2$ is a diffusive time which is much slower than the electrical time constant of the membrane ($= C_m/G_m$).

Case Study: Origin of Transcellular Current in Fucus

All organisms display some kind of symmetry[Harisson 2005]. How do symmetries emerge and change during development is still an open question. The cell (or tissue) growth is often correlated to tanscellular currents. For examples, in plants, this is the case in roots[Weisenseel, Becker and Ehlgotz 1992], pollen grains[Weisenseel, Nuccitelli and Jaffe 1975], wound healing tissues[Meyer and Weisenseel 1997], fungi hyphae[Gow, Kropf and Harold 1984; Kropf, Caldwell, Gow and Harold 1984; Kropf 1986a; Kropf 1986b]. Moreover, the apical growth may be reoriented in an electric field [Rathore and Goldsworthy 1985]. The pattern of transcellular current may be periodic. Periodic pattern have been well documented for the case of the freshwater plant Chara where alternating band of pH and electrical potential occurs when the cells are in the light[Lucas and SMITH 1973; Lucas 1975a; Lucas 1975b; Fisahn, Hansen and Lucas 1992; Fisahn and Lucas 1992; Bulychev, Polezhaev, Zykov, et al. 2001].

The case of fucoid eggs is interesting in several regards. It is a single wall-less cell with a simple spherical geometry. Before fertilization their cytological components are evenly distributed. Thus, there are characterized by an infinite number of symmetry axes. After fertilization, the zygote is polarized and has a single polar axis. Moreover, the relative large size (~80µm diameter) of fucoid egg cells makes them a convenient biological material for electrophysiology. After fertilization, a dipolar pattern of both electric field and transcellular ion currents are quickly formed around the zygote [Nuccitelli 1978], Ca^{2+} enters the cell at the nascent rhizoid pole and it flows out the cell at the opposite pole (the future thallus pole)[Jaffe, Robinson and Nuccitelli 1974; Robinson and Cone 1980; Brownlee and Wood 1986; Speksnijder, Miller, Weisenseel, et al. 1989; Taylor, Manison, Fernandez, et al. 1996; Jaffe 1999; Coelho, Taylor, Ryan, et al. 2002]. This happens before that any cytological reorganization may be detected. Therefore, fucoid eggs are simple systems to study how symmetry breaking might occurs at the early developmental stage and to investigate the origin and the role of transcellular current. This problem cannot be investigated on most organisms because their egg cells already have primary axis of symmetry before fertilization has occurred.

In fucoid zygotes, symmetry breaking occurs in the absence of any environmental cues [Robinson and Cone 1980; Robinson, Wozniak, Pu and Messerli 1999; Hable and Kropf 2000]. This is a reminiscent feature of symmetry breaking arising from a self-organizing process. It has been shown that a pattern of transcellular current may quickly self-organized around the zygote as a consequence of the non-linear dynamics of ion transport through the plasma membrane, even though membrane proteins themselves are homogeneously distributed in the membrane. A pattern of both ion concentrations and electric field can be established within a minute following fertilization, which polarizes the zygote with a single and labile axis of symmetry[Leonetti, Dubois-Violette and Homble 2004].

There are two ionic conductances (a Ca^{2+} and a K^{+}) that are activated during the membrane depolarization. In unfertilized eggs, the differential conductance of calcium is negative while that of potassium is positive but has small amplitude so that $g_K^{slope} + g_{Ca}^{slope} < 0$. As expected the unfertilized Fucus egg is excitable when the membrane is depolarized [Taylor and Brownlee 1993](*vide supra*). At fertilization, the membrane is quickly depolarized and there is a fast increase of the membrane permeability to potassium which indicates that the plasma membrane of the zygote is dominated by the potassium conductance. Then, $g_K^{slope} + g_{Ca}^{slope}$ becomes positive and as $D_K >> D_{Ca}$ the conditions required for the emergence of a pattern of transcellular current (see equations [56]) are fulfilled [Leonetti, Dubois-Violette and Homble 2004; Homble and Leonetti 2007]. The numerical solution of the electrodiffusive model for a circular geometry is shown in figure 18.

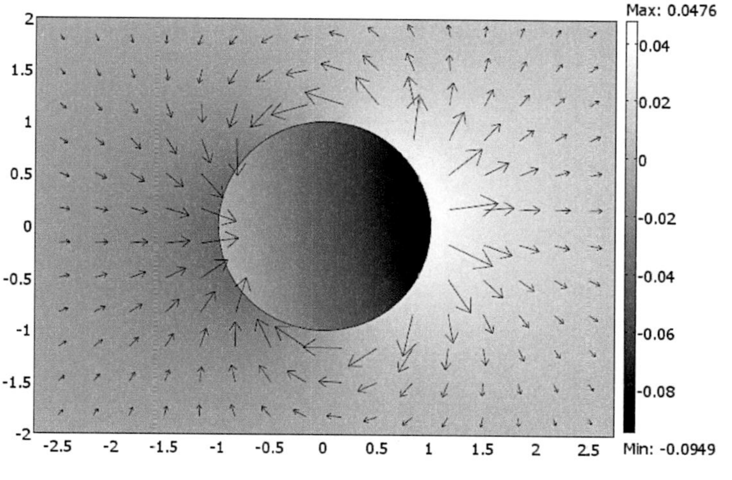

Figure 18. Transcellular ionic currents around a Fucus egg. Arrows indicate the magnitude and the direction of the external electric field. Calcium ions leave the cell on the right side and enter in the cell on the left side. The grey scale indicates the magnitude of a dimensionless electric potential outside the cell (zero at infinity) and, inside the cell, a quantity equal to the intracellular electric potential minus 1.06 to make the figure understandable. This numerical simulation was done using the data for Fucus [Leonetti, Dubois-Violette and Homble 2004].

In the absence of external cues, the position of the symmetry axis will be set by the sperm entry in fucoids and the symmetry axis, albeit in random directions [Knapp 1931; Robinson and Cone 1980; Robinson, Wozniak, Pu and Messerli 1999; Hable and Kropf 2000], determines the rhizoid–thallus axis growth. This indicates that symmetry breaking is an intrinsic property of the zygote. The initial axis is labile and reorientation is an intrinsic property of self-organization. Spontaneous polarization has been widely observed in biological cells and it has been suggested that the principle of self-organization might be central to the establishment of cell polarity [Kirschner, Gerhart and Mitchison 2000; Wedlich-Soldner and Li 2003]. In the presence of external clues, the orientation of the current loop axis will be affected by any macroscopic perturbation (applied gradient) that defines a preferential axis. For instance, in natural conditions the light reorients the symmetry axis so that the thallus develops toward the light. The emergence of trancellular current happens while all cellular components (including ion channels) are uniformly distributed in the cell. Such a self-organized structure is called 'dissipative structure' because energy must be supplied to the system through the imposition of a gradient to maintain the system even under conditions far from equilibrium [Nicolis and Prigogine 1977; Prigogine and Stengers 1986]. The emergence of the pattern requires a coupling between local autocatalytic (self-amplification) activation in response to a positive feedback and a long-range lateral inhibition [Gierer and Meinhardt 1972; Meinhardt and Gierer 2000; Homble and Leonetti 2007]. In this model, the amplification of the initial perturbation is due to the electrical field generated by the lateral electrodiffusion of ions along the membrane surface; and the long-range inhibition arises from the dissipation of the electric potential outside the self-organizing region (the so-called 'cable effect' [Keener and Sneyd 1998]) [Leonetti, Dubois-Violette and Homble 2004; Homble and Leonetti 2007]. It has been hypothesized that calcium channels aggregate at the rhizoid pole after fertilization [Jaffe 1977]. This lateral migration of ion channels in the membrane will be much slower than the formation of transcellular currents [Homble and Leonetti 2007]. The electric field set up by the transcellular currents might drive the lateral electromigration of membrane proteins proposed by Jaffe [Jaffe 1977]. Alternatively, self-aggregation of membrane channels might emerge after fertilization [Leonetti, Renversez and Dubois-Violette 1999; Leonetti, Nuebler and Homble 2006].

Conclusion

The steady state description of ion transport permits a quantification of ion transport across membrane and a detail kinetic analysis of the limiting steps that control the translocation of ions through the membrane. The nonlinear dynamics description of ion transport integrates the collective behavior of different transporters. As a consequence of the feedback regulation existing between the different transporters, new properties emerge such as stationary and non-stationary spatio-temporal patterns which might be involved in osmoregulation, transcellular currents and symmetry breaking observed in marine algae. Patterns emerging in a uniform distribution of channels and carriers are a consequence of their nonlinear behavior.

Acknowledgments

F.H. is a Research Director from the National Fund for Scientific Research (Belgium). This work was supported by a joint research program 'Tournesol' of the 'Communauté Française de Belgique' and the "Ministère français des affaires étrangères".

References

Allen,R.D., Jacobsen,J.J., and Jaffe,L.F. (1972). Ionic concentrations in developing Pelvetia eggs. *Dev.Biol.,* **27**, 538-545.

Amtmann,A., and Gradmann,D. (1994). Na+ Transport in Acetabularia Bypasses Conductance of Plasmalemma. *J. Membr. Biol.,* **139**, 117-125.

Amtmann,A., Klieber,H.G., and Gradmann,D. (1992). Cytoplasmic Free Ca2+ in the Marine Alga Acetabularia - Measurement with Ca2+-Selective Microelectrodes and Kinetic-Analysis. *J. Exp. Bot.,* **43**, 875-885.

Anderson,K., Close,L., DeWreede,R.E., Lynch,B.J., Ormond,C., and Walker,M. (2006). Biomechanical properties and holdfast morphology of coenocytic algae (Halimedales, Chlorophyta) in Bocas del Toro, Panama. *J. Exp. Marine Biol. Ecol.,* **328**, 155-167.

Armbrust,E.V., Berges,J.A., Bowler,C., Green,B.R., Martinez,D., Putnam,N.H., Zhou,S., Allen,A.E., Apt,K.E., Bechner,M., Brzezinski,M.A., Chaal,B.K., Chiovitti,A., Davis,A.K., Demarest,M.S., Detter,J.C., Glavina,T., Goodstein,D., Hadi,M.Z., Hellsten,U., Hildebrand,M., Jenkins,B.D., Jurka,J., Kapitonov,V.V., Kroger,N., Lau,W.W., Lane,T.W., Larimer,F.W., Lippmeier,J.C., Lucas,S., Medina,M., Montsant,A., Obornik,M., Parker,M.S., Palenik,B., Pazour,G.J., Richardson,P.M., Rynearson,T.A., Saito,M.A., Schwartz,D.C., Thamatrakoln,K., Valentin,K., Vardi,A., Wilkerson,F.P., and Rokhsar,D.S. (2004). The genome of the diatom Thalassiosira pseudonana: ecology, evolution, and metabolism. *Science,* **306**, 79-86.

Ashley,R.H. (1995). *Ion Channels.* Oxford: IRL Press.

Bauman,R.W., and Jones,B.R. (1986). Electrophysiological invetigations of the red alga Griffithsia pacifica Kyl. *J. Phycol.,* **22**, 49-56.

Baumann,S.B. (1991). The vibrating probe as a sensitive toxicological essay for the effects of tributylin on ionic currentsin Acetabularia mediterranea. *Toxicol.in Vitro,* **5**, 103-107.

Beilby,M.J. (1984). Calcium and Plant Action-Potentials. *Plant Cell Environ.,* **7**, 415-421.

Beilby,M.J., and Bisson,M.A. (1999). Transport systems of Ventricaria ventricosa: I/V analysis of both membranes in series as a function of [K(+)](o). *J. Membr. Biol.,* **171**, 63-73.

Beilby,M.J., and Coster,H.G.L. (1979a). Action Potential in Chara-Corallina .3. Hodgkin-Huxley Parameters for the Plasmalemma. *Aust. J. Plant Physiol.,* **6**, 337-353.

Beilby,M.J., and Coster,H.G.L. (1979b). Action Potential in Chara-Corallina .4. Activation Enthalpies of the Hodgkin-Huxley Gates. *Aust. J. Plant Physiol.,* **6**, 355-365.

Beilby,M.J., and Shepherd,V.A. (2001). Modeling the current-voltage characteristics of charophyte membranes. II. The effect of salinity on membranes of Lamprothamnium papulosum. *J. Membr. Biol.,* **181**, 77-89.

Beilby,M.J. (2007). Action Potential in Charophytes. In W.J.Kwang (Ed.), *International Review of Cytology: A Survey of Cell Biology.* (pp. 43-82). New York: Academic Press.

Benz,R., and Zimmermann,U. (1983). Evidence for the presence of mobile charges in the cell membrane of Valonia utricularis. *Biophys. J.*, **43**, 13-26.

Bertl,A., and Gradmann,D. (1987). Current-Voltage Relationships of Potassium Channels in the Plasmalemma of Acetabularia. *J. Membr. Biol.*, **99**, 41-49.

Bertl,A., Klieber,H.G., and Gradmann,D. (1988). Slow Kinetics of A Potassium Channel in Acetabularia. *J. Membr. Biol.*, **102**, 141-152.

Bisson,M.A., Beilby,M.J., and Shepherd,V.A. (2006). Electrophysiology of turgor regulation in marine siphonous green algae. *J. Membr. Biol.*, **211**, 1-14.

Bockris,J.O., and Reddy,A.K.N. (1970). *Modern Electrochemistry. Ionics.* New York: Plenum Press.

Bowles,E.A., and Allen,N.S. (1986). A vibrating probe analysis of light-dependent transcellular currents in Acetabularia. In R.Nuccitelli (Ed.), *Ionic Currents in Development.* (pp. 113-121). New York: Alan Liss.

Brawley,S.H., and Robinson,K.R. (1985). Cytochalasin treatment disrupts the endogenous currents associated with cell polarization in fucoid zygotes: studies of the role of F-actin in embryogenesis. *J. Cell Biol.*, **100**, 1173-1184.

Brownlee,C., and Bouget,F.Y. (1998). Polarity determination in Fucus: from zygote to multicellular embryo. *Semin.Cell Dev.Biol.*, **9**, 179-185.

Brownlee,C., and Wood,J.W. (1986). A gradient of cytoplasmic free calcium in growing rhizoid cells of Fucus serratus. *Nature*, **320**, 624-626.

Bulychev,A.A., Polezhaev,A.A., Zykov,S.V., Pljusnina,T.Y., Riznichenko,G.Y., Rubin,A.B., Jantoss,W., Zykov,V.S., and Muller,S.C. (2001). Light-triggered pH banding profile in Chara cells revealed with a scanning pH microprobe and its relation to self-organization phenomena. *J. Theor. Biol.*, **212**, 275-294.

Coelho,S.M., Taylor,A.R., Ryan,K.P., Sousa-Pinto,I., Brown,M.T., and Brownlee,C. (2002). Spatiotemporal patterning of reactive oxygen production and Ca(2+) wave propagation in fucus rhizoid cells. *Plant Cell*, **14**, 2369-2381.

Cooke,T.J., and Racusen,R.H. (1986). The role of electrical phenomena in tip growth, with special reference to the developmental plasticity of filamentous fern gametophytes. *Symp. Soc. Exp. Biol.*, **40** , 307-328.

Croft,M.T., Warren,M.J., and Smith,A.G. (2006). Algae Need Their Vitamins. *Eukaryotic Cell*, **5**, 1175-1183.

Davis,R.F. (1981). Electrical Properties of the Plasmalemma and Tonoplast in Valonia ventricosa. *Plant Physiol.*, **67**, 825-831.

Dazy,A.C., Borghi,H., and Garcia,E.P.-D.S. (1986). Transcellular current and morphogenesis in Acetabularia mediterranea grown in white, blue and red light. *Prog. Clin. Biol. Res.*, **210**, 123-130.

Dill,K.A., and Bromberg,S. (2003). *Molecular Driving Forces.* New York: Garland Science.

Dring,M.J. (1991). *The Biology of Marine Plants.* Cambridge: Cambridge University Press.

Einstein,A. (1956). *Investigations of the Theory of the Brownian Movement.* New York: Dover Publications.

Eyring,H. (1935). The activated complex in chemical reaction. *J. Chem. Phys.*, **3**, 107-115.

Eyring,H., Ma,S.M., and Ueda,I. (1981). Reaction kinetics in living systems. *Proc. Natl. Acad. Sci. USA.*, **78**, 5549-5553.

Feijo,J.A., Costa,S.S., Prado,A.M., Becker,J.D., and Certal,A.C. (2004). Signalling by tips. *Curr. Opin. Plant Biol.*, **7**, 589-598.

Finkelstein,A., and Mauro,A. (1963). Equivalent Circuits as Related to Ionic Systems. *Biophys. J.*, **3**, 215-237.

Fisahn,J., Hansen,U.P., and Lucas,W.J. (1992). Reaction Kinetic-Model of A Proposed Plasma-Membrane 2-Cycle H+-Transport System of Chara-Corallina. *Proc. Natl. Acad. Sci. USA.*, **89**, 3261-3265.

Fisahn,J., and Lucas,W.J. (1992). Direct Measurement of the Reversal Potential and Current-Voltage Characteristics in the Acid and Alkaline Regions of Chara-Corallina. *Planta*, **186**, 241-248.

Fuks,B., and Homble,F. (1994). Permeability and electrical properties of planar lipid membranes from thylakoid lipids. *Biophys. J.*, **66**, 1404-1414.

Fuks,B., and Homble,F. (1996). Mechanism of proton permeation through chloroplast lipid membranes. *Plant Physiol.*, **112**, 759-766.

Gierer,A., and Meinhardt,H. (1972). A theory of biological pattern formation. *Kybernetik*, **12**, 30-39.

Goldbeter,A. (1997). *Biochemical Oscillations and Cellular Rhythms.* Cambridge: Cambridge University Press.

Goldman,D.E. (1943). Potential, Impedance, and Rectification in membranes. *The Journal of General Physiology*, **27**, 37-60.

Goodwin,B.C., and Pateromichelakis,S. (1979). The role of electrical fields, ions, and the cortex in the morphogenesis of Acetabularia. *Planta*, **145**, 427-435.

Gow,N.A., Kropf,D.L., and Harold,F.M. (1984). Growing hyphae of Achlya bisexualis generate a longitudinal pH gradient in the surrounding medium. *J. Gen. Microbiol.*, **130**, 2967-2974.

Gow,N.A.R. (1989). Circulating ionic currents i micro-organisms. *Adv. Microb. Physiol.*, **30**, 89-123.

Gradmann,D. (1975). Analog Circuit of Acetabularia Membrane. *J. Membr. Biol.*, **25**, 183-208.

Gradmann,D. (1976). Metabolic Action Potentials in Acetabularia. *J. Membr. Biol.*, **29**, 23-45.

Gradmann,D. (1978). Green light (550 nm) inhibits electrogenic Cl-pump in the Acetabularia membrane by permeability increase for the carrier ion. *J. Membr. Biol.*, **44**, 1-24.

Gradmann,D., and Boyd,C. (1999). Electrophysiology of the marine diatom Coscinodiscus wailesii. II. Potassium currents. *J. Exp. Bot.*, **50**, 453-459.

Gradmann,D., and Boyd,C.M. (2000). Three types of membrane excitations in the marine diatom Coscinodiscus wailesii. *J. Membr. Biol.*, **175**, 149-160.

Gradmann,D., and Klemke,K. (1974). Current-voltage relationship of the electrogenic pump in Acetabularia mediterranea. In U.Zimmermann and J.Dainty (Eds.), *Membrane Transport in Plants.* (pp. 131-138). Berlin: Springer-Verlag.

Gradmann,D., Wagner,G., and Glasel,R.M. (1973). Chloride Efflux During Light-Triggered Action Potentials in Acetabularia-Mediterranea. *Biochim. Biophys. Acta*, **323**, 151-155.

Gutknecht,J. (1966). Sodium, potassium and chloride transport and membrane potentials in Valonia ventricosa. *Biol. Bull.*, **130**, 331-344.

Gutknecht,J., and Dainty,J. (1968). Ionic relations of marine algae. *Oceanogr. Marine Biol. Ann. Rev.*, **6**, 379-402.

Hable,W.E., and Kropf,D.L. (2000). Sperm entry induces polarity in fucoid zygotes. *Development*, **127**, 493-501.

Hansen,U.P., Gradmann,D., Sanders,D., and Slayman,C.L. (1981). Interpretation of Current-Voltage Relationships for Active Ion-Transport Systems .1. Steady-State Reaction-Kinetic Analysis of Class-I Mechanisms. *J. Membr. Biol.,* **63**, 165-190.

Harisson,L.G. (2005). *Kinetic Theory of Living Pattern.* Cambridge: Cambridge University Press.

Hepler,P.K., Vidali,L., and Cheung,A.Y. (2001). Polarized cell groth in higher plants. *Annu. Rev. Cell Dev. Biol.,* **17**, 159-187.

Hille,B. (1992). *Ionic Channels of Excitable membranes.* Sunderland: Sinauer Assocaited Inc.

Hodgkin,A.L., and Katz,B. (1949). The effect of sodium ions on the electrical activity of the giant axon of the squid. *The Journal of Physiology Online,* **108**, 37-77.

Homble,F., and Leonetti,M. (2007). Emergence of symmetry breaking in fucoid zygotes. *Trends Plant Sci.,* **12**, 253-259.

Jaffe,L.F. (1966). Electrical currents through the developing fucus egg. *Proc. Natl. Acad. Sci.USA.,* **56**, 1102-1109.

Jaffe,L.F. (1977). Electrophoresis along cell membranes. *Nature,* **265**, 600-602.

Jaffe,L.F. (1999). Organization of early development by calcium patterns. *Bioessays,* **21**, 657-667.

Jaffe,L.F., and Nuccitelli,R. (1974). An ultrasensitive vibrating probe for measuring steady extracellular currents. *J. Cell Biol.,* **63**, 614-628.

Jaffe,L.F., Robinson,K.R., and Nuccitelli,R. (1974). Local cation entry and self-electrophoresis as an intracellular localization mechanism. *Ann. NY. Acad. Sci.,* **238**, 372-389.

Katsuhara,M., and Tazawa,M. (1986). Salt tolerance inNitellopsis obtusa. *Protoplasma,* **135**, 155-161.

Keener,J., and Sneyd,J. (1998). *Mathematical Physiology.* New York: Springer Verlag.

Kirschner,M., Gerhart,J., and Mitchison,T. (2000). Molecular "vitalism". *Cell,* **100**, 79-88.

Knapp,E. (1931). Entwicklungsphysiologische Untersuchungen an Fucaceen-Eiern. *Planta,* **14**, 731-751.

Koch,C. (1999). *Biophysics of Computation.* Oxford: Oxford University Press.

Kropf,D.L. (1986a). Electrophysiological properties of Achlya hyphae: ionic currents studied by intracellular potential recording. *J. Cell Biol.,* **102**, 1209-1216.

Kropf,D.L. (1986b). Intracellular potential recording as a means to investigate the transhyphal current in Achlya. *Prog. Clin. Biol. Res.,* **210**, 97-104.

Kropf,D.L., Bisgrove,S.R., and Hable,W.E. (1999). Establishing a growth axis in fucoid algae. *Trends Plant Sci.,* **4**, 490-494.

Kropf,D.L., Caldwell,J.H., Gow,N.A., and Harold,F.M. (1984). Transcellular ion currents in the water mold Achlya. Amino acid proton symport as a mechanism of current entry. *J. Cell Biol.,* **99**, 486-496.

Kropf,D.L., Lupa,M.D.A., Caldwell,J.H., and Harold,F.M. (1983). Cell polarity: endogenous ion currents precede and predict branching in the water mold Achlya. *Science,* **220**, 1385-1387.

Lane,T.W., and Morel,F.M.M. (2000). A biological function for cadmium in marine diatoms. *Proc. Natl. Acad. Sci. USA.,* **97**, 4627-4631.

Lane,T.W., Saito,M.A., George,G.N., Pickering,I.J., Prince,R.C., and Morel,F.M.M. (2005). A cadmium enzyme from a marine diatom. *Nature,* **435**, 42.

Lauger,P. (1987). Dynamics of ion transport systems in membranes. *Physiol. Rev.*, **67**, 1296-1331.

Lauger,P., and Stark,G. (1970). Kinetics of carrier-mediated ion transport across lipid bilayer membranes. *Biochim. Biophys. Acta*, **211**, 458-466.

Leonetti,M., and Dubois-Violette,E. (1998). Theory of electrodynamic instabilities in biological cells. *Phys. Rev. Lett.*, **81**, 1977-1980.

Leonetti,M., Dubois-Violette,E., and Homble,F. (2004). Pattern formation of stationary transcellular ionic currents in Fucus. *Proc. Natl. Acad. Sci. USA.*, **101**, 10243-10248.

Leonetti,M., Marcq,P., Nuebler,J., and Homble,F. (2005). Cotransport-induced instability of membrane voltage in tip-growing cells. *Phys. Rev. Lett.*, **95**, 208105.

Leonetti,M., Nuebler,J., and Homble,F. (2006). Parity-breaking bifurcation and global oscillation in patterns of ion channels. *Phys. Rev. Lett.*, **9621**, 8101.

Leonetti,M., Renversez,G., and Dubois-Violette,E. (1999). Influence of specific ionic diffusion on the protein self-aggregation instability. *Europhys. Lett.*, **46**, 107-113.

Lide,D.R. (2005). *CRC Handbook of Chemistry and Physics.* CRC Press Inc.

Lobban,C.S., and Harisson,P.J. (2000). *Seaweed Ecology and Physiology.* Cambridge: Cambridge University Press.

Lucas,W.J. (1975a). Analysis of the Diffusion Symmetry Developed by the Alkaline and Acid Bands which form at the Surface of Chara corallina Cells. *J. Exp. Bot.*, **26**, 271-286.

Lucas,W.J. (1975b). The Influence of Light Intensity on the Activation and Operation of the Hydroxyl Efflux System of Chara corallina. *J. Exp. Bot.*, **26**, 347-360.

Lucas,W.J., and SMITH,F.A. (1973). The Formation of Alkaline and Acid Regions at the Surface of Chara corallina Cells. *J. Exp. Bot.*, **24**, 1-14.

Marschner,H. (1998). *Mineral Nutrition of Higher Plants.* London: Academic Press.

Meinhardt,H., and Gierer,A. (2000). Pattern formation by local self-activation and lateral inhibition. *Bioessays*, **22**, 753-760.

Mengel,K., and Kirkby,E.A. (1987). *Principles of Plant Nutrition.* Bern: International Potash Institute.

Meyer,A.J., and Weisenseel,M.H. (1997). Wound-Induced Changes of Membrane Voltage, Endogenous Currents, and Ion Fluxes in Primary Roots of Maize. *Plant Physiol.*, **114**, 989-998.

Moreton,R.B. (1969). An Investigation of the Electrogenic Sodium Pump in Snail Neurones, Using the Constant-Field Theory. *J. Exp. Biol.*, **51**, 181-201.

Morris,C., and Lecar,H. (1981). Voltage oscillations in the barnacle giant muscle fiber. *Biophys.J.*, **35**, 193-213.

Mummert,H., and Gradmann,D. (1991a). Action-Potentials in Acetabularia - Measurement and Simulation of Voltage-Gated Fluxes. *J. Membr. Biol.*, **124**, 265-273.

Mummert,H., and Gradmann,D. (1991b). Ion Fluxes in Acetabularia - Vesicular Shuttle. *J.Membr.Biol.*, **124**, 255-263.

Mummert,H., Hansen,U.P., and Gradmann,D. (1981). Current-Voltage Curve of Electrogenic Cl- Pump Predicts Voltage-Dependent Cl- Efflux in Acetabularia. *J. Membr. Biol.*, **62**, 139-148.

Murray,J.D. (2002). *Mathematical Biology. I: An Introduction.* New York: Springer-Verlag.

Murray,J.D. (2004). *Mathematical Bioloy. II: Spatial Models and Biomedical Applications.* New York: Springer-Verlag.

Nawata,T., Hishinuma,T., and Wada,S. (1989). Ionic currents during regeneration of thallus and rhizoid from cell segments isolated from the marine alga Bryopsis. *Biol. Bull.,* **176**, 41-45.

Nicolis,G. (1999). *Introduction to Nonlinear Science.* Cambridge: Cambridge University Press.

Nicolis,G., and Prigogine,I. (1977). *Self-Organization in Non-Equilibrium Systems.* New York: John Wiley.

Novak,B. (1975). Periodical signals in the spatial differentiation of plant cells. In G.Nicolis and R.Lefever (Eds.), *Advances in Chemical Physics: Membrane, Dissipative Structure and Evolution.* (pp. 281-299). San Francisco: John Wiley.

Novak,B., and Bentrup,F.W. (1972). An electrophysiological study of regeneration in Acetabularia mediterranea. *Planta,* **108**, 227-244.

Nuccitelli,R. (1978). Ooplasmic segregation and secretion in the Pelvetia egg is accompanied by a membrane-generated electrical current. *Dev. Biol.,* **62**, 13-33.

Nuccitelli,R. (1988). Ionic currents in morphogenesis. *Cellular and Molecular Life Sciences (CMLS),* **44**, 657-666.

Nuccitelli,R., and Jaffe,L.F. (1974). Spontaneous current pulses through developing fucoid eggs. *Proc. Natl. Acad. Sci.USA.,* **71**, 4855-4859.

Nuccitelli,R., and Jaffe,L.F. (1975). The pulse current pattern generated by developing fucoid eggs. *J. Cell Biol.,* **64**, 636-643.

O'Shea,P., Goodwin,B., and Ridge,I. (1990). A vibrating electrode analysis of extracellular ion current in Acetabularia acetabulum. *J. Cell. Sci,* **97**, 503-508.

Pedersen,M. (1969). Marine Brown Algae Requiring Vitamin B12. *Physiol. Plantarum,* **22**, 977-983.

Prigogine,I., and Stengers,I. (1986). *Order out of Chaos.* London: Flamingo.

Qian,N., and Sejnowski,T. (1989). An electro-diffusion model for computing membrane potentials and ionic concentrations in branching dendrites, spines and axons. *Biological Cybernetics,* **62**, 1-15.

Quatrano,R.S., and Shaw,S.L. (1997). Role of the cell wall in the determination of cell polarity and the plane of cell division in Fucus embryos. *Trends Plant Sci.,* **2**, 15-21.

Quian,N., and Sejnowski,J. (1989). An Electro- Diffusion Model for Computing Membrane Potentials and Ionic Concentrations in Branching Dendrites, Spines, and Axons. *Biol. Cyber.,* **62**, 1-15.

Rathore,K.S., and Goldsworthy,A. (1985). Electrical control of shoot regeneration in plant tissue cultures. *Biotechnology,* **3**, 1107-1109.

Reiss,H.D., and Herth,W. (1979). Calcium gradients in tip growing plant cells visualized by chlorotetracycline fluorescence. *Planta,* **146**, 615-621.

Rinzel,J., and Ermentrout,B.B. (1998). Analysis of neural excitability and oscillations. In C.Koch and I.Segev (Eds.), *Methods in Neuronal Modeling.* (pp. 251-292). Cambridge: MIT Press.

Roberts,S., Berger,F., and Brownlee,C. (1993). The role of Ca^{2+} in signal transduction following fertilization in Fucus serratus. *J. Exp. Biol.,* **184**, 197-212.

Robinson,K.R., and Cone,R. (1980). Polarization of fucoid eggs by a calcium ionophore gradient. *Science,* **207**, 77-78.

Robinson,K.R., and Jaffe,L.F. (1973). Ion movements in a developing fucoid egg. *Dev. Biol.,* **35**, 349-361.

Robinson,K.R., Wozniak,M., Pu,R., and Messerli,M. (1999). Symmetry breaking in the zygotes of the fucoid algae: controversies and recent progress. *Curr. Top. Dev. Biol.*, **44**, 101-125.

Robinson,R.A., and Stokes,R.H. (1959). *Electrolyte Solutions*. London: Butterworths.

Runcie,J.W., Ritchie,R.J., and Larkum,A.W.D. (2003). Uptake kinetics and assimilation of inorganic nitrogen by Catenella nipae and Ulva lactuca. *Aquatic Bot.*, **76**, 155-174.

Saddler,H.D.W. (1970). The Ionic Relations of Acetabularia mediterranea. *J. Exp. Bot.*, **21**, 345-359.

Sakmann,B., and Neher,E. (1995). *Single-Channel Recording*. New York: Plenum Press.

Sanders,D. (1990). Kinetic Modeling of Plant and Fungal Membrane Transport Systems. *Ann. Rev. Plant Physiol. Plant Molec. Biol.*, **41**, 77-107.

Serikawa,K.A., Porterfield,D.M., Smith,P.J., and Mandoli,D.F. (2000). Calcification and measurements of net proton and oxygen flux reveal subcellular domains in Acetabularia acetabulum. *Planta*, **211**, 474-483.

Shepherd,V.A., Beilby,M.J., and Bisson,M.A. (2004). When is a cell not a cell? A theory relating coenocytic structure to the unusual electrophysiology of Ventricaria ventricosa (Valonia ventricosa). *Protoplasma*, **223**, 79-91.

Smeyers,M., Leonetti,M., Goormaghtigh,E., and Homble,F. (2003). Structure and function of plant membrane ion channels reconstituted in planar lipid bilayers. In H.T.TIEN and A.Ottova-Leitmannova (Eds.), *Planar Lipid Bilayers (BLMs) and Their Applications*. (pp. 449-478). Amsterdam: Elsevier.

Speksnijder,J.E., Miller,A.L., Weisenseel,M.H., Chen,T.H., and Jaffe,L.F. (1989). Calcium buffer injections block fucoid egg development by facilitating calcium diffusion. *Proc.Natl.Acad.Sci.U.S.A.*, **86**, 6607-6611.

Strogatz,S.H. (1994). *Nonlinear dynamics and chaos: With applications to physics, biology, chemistry, and engineering*. Cambridge MA: Perseus Books.

Taylor,A.R., and Brownlee,C. (2003). A novel Cl- inward-rectifying current in the plasma membrane of the calcifying marine phytoplankton Coccolithus pelagicus. *Plant Physiol.*, **131**, 1391-1400.

Taylor,A.R., Manison,N., Fernandez,C., Wood,J., and Brownlee,C. (1996). Spatial Organization of Calcium Signaling Involved in Cell Volume Control in the Fucus Rhizoid. *Plant Cell*, **8**, 2015-2031.

Taylor,A., and Brownlee,C. (1993). Calcium and potassium currents in the Fucus egg. *Planta*, **189**, 109-119.

Tittor,J., Hansen,U.P., and Gradmann,D. (1983). Impedance of the Electrogenic Cl- Pump in Acetabularia - Electrical Frequency Entrainements, Voltage-Sensitivity, and Reaction Kinetic Interpretation. *J. Membr. Biol.*, **75**, 129-139.

Wang,J., Zimmermann,U., and Benz,R. (1994). Contribution of electrogenic ion transport to impedance of the algae Valonia utricularis and artificial membranes. *Biophys. J.*, **67**, 1582-1593.

Wang,J., Wehner,G., Benz,R., and Zimmermann,U. (1993). Harmonic system analysis of the algae Valonia utricularis: contribution of an electrogenic transport system to gain and phase-shift of the transfer function. *Biophys. J.*, **64**, 1657-1667.

Wedlich-Soldner,R., and Li,R. (2003). Spontaneous cell polarization: undermining determinism. *Nature Cell Biol.*, **5**, 267-270.

Weisenseel,M.H., Nuccitelli,R., and Jaffe,L.F. (1975). Large electrical currents traverse growing pollen tubes. *J. Cell Biol.,* **66**, 556-567.

Weisenseel,M.H., Becker,H.F., and Ehlgotz,J.G. (1992). Growth, Gravitropism, and Endogenous Ion Currents of Cress Roots (Lepidium sativum L.) : Measurements Using a Novel Three-Dimensional Recording Probe. *Plant Physiol.,* **100**, 16-25.

Wendler,S., Zimmermann,U., and Bentrup,F.W. (1983). Relationship between cell turgor pressure, electrical membrane potential, and chloride efflux inAcetabularia mediterranea. *J. Membr. Biol.,* **72**, 75-84.

Williams,S.L. (1984). Uptake of sediment ammonium and translocation in a marine green macroalga *Caulerpa cupressoides. limnol. Oceanogr.,* **29**, 374-379.

Xu,Y., Feng,L., Jeffrey,P.D., Shi,Y., and Morel,F.M. (2008). Structure and metal exchange in the cadmium carbonic anhydrase of marine diatoms. *Nature,* **452**, 56-61.

Zwolinski,B.I., Eyring,H., and Reese,C.E. (1949). Diffusion and membrane permeability. *J. Phys. Chem.,* **53**, 1426-1453.

In: Algae: Nutrition, Pollution Control and Energy Sources
Editor: Kristian N. Hagen, pp. 247-254
ISBN: 978-1-60692-008-4

Chapter 12

CHROMATOGRAPHIC DETERMINATION OF FREE D- AND L-AMINO ACIDS IN MARINE ALGAE

Eizo Nagahisa* and Takehiko Yokoyama
School of Fisheries Sciences, Kitasato University, Japan.

1. Introduction

It has been generally accepted that D-amino acids are uncommon in higher organisms and that L-amino acids are of primary importance biologically. However, in recent years, evidence that D-amino acids also play an important role in higher organisms has increased. Furthermore, as a result of bacterial metabolism and chemically induced racemization, free D-amino acids are common constituents of microbially fermented foods, beverages, and various other processed foods. Microbially and chemically induced racemization of amino acids in food proteins and the nutritional utilization and safety of D-amino acids have been the focus of increasing attention.

The discovery of naturally occurring D-amino acids in organisms, in both free and combined forms, and the elucidation of their physiological functions are dependent on the availability of reliable, sensitive and rapid analytical methods. Numerous HPLC methods, more than can be cited here, have been used for the detection of D-amino acids in organisms and foods. However, it is generally believed that the main factor in the appearance of D-amino acids in higher organisms is racemization of L-amino acids by endogenous racemase enzymes. In fact, alanine racemase has been found in several higher organisms, including plants (Gamburg and Rekoslavskaya,1992) and marine invertebrates (Matsushima *et al.*,1982, Fujita *et al.*,1997, Shibata *et al.*,2000, Yoshikawa *et al.*,2002), in addition to microorganisms. Recently, we found an alanine racemase in the marine diatom *Thalassiosira* sp. (Yokoyama *et al.*,2005).

In this report, we introduce a new method for the detection of D-amino acids and present the results of this analysis of D-amino acids in marine macro- and micro-algae.

Keywords: D-aspartate, D-alanine, marine algae, phytoplankton, diatom

---

* E-mail address: nagahisa@kitasato-u.ac.jp, FAX: +81-0192-44-3930. Correspondence concerning this article should be addressed to Eizo Nagahisa,

2. Materials and Methods

Materials

Instruments

High Performance Liquid Chromatograph (HPLC): with a fluorescence detector and an integrator (Jasco LC-800 series, Japan)

Column: Reversed phase column (TSK-ODS 80T, 4.6x250 mm, Tosoh, Japan)

Homogenizer: Polytron homogenizer (PT10135, Switzerland) and Ultrasonic homogenizer (Tomy UD200, Japan)

Centrifuge: High speed cooling centrifuge (KUBOTA 7930, Japan) and Table top centrifuge (Tomy LC-120, Japan)

Thermo-incubator: with illumination lamp (EYELA MTI-201B, Japan)

Reagents

The following reagent grade chemicals were purchased from Wako Pure Chem. Ind. (Japan): DL-amino acids, *o*-phthalaldehyde (OPA), *N*-acetyl-L-cysteine (NAC) and sodium borate. All other chemicals were analytical grade reagents.

Methods

Sources of Algae

Most of the macro-algae were collected at Okirai Bay (Iwate Prefecture, Japan). The freshwater algae, *Botrydiopsis alpina*, *Chlorella pyrenoidosa*, *Chlorella vulgaris* and *Scenedesmus obliquus*, were obtained from the Chlorella Kogyo Co. (Tokyo, Japan). Marine diatoms were isolated by the capillary method, and inoculated singly into each medium (Kotaki *et al.*,2000). Axenic strains were isolated by the method of Douglas and Bates (1992) and were cultured in germ-free conditions. *Asterionella* sp. and *Thalassiosira* sp. were cultured in T1/seawater medium (Ogata *et al.* 1987) at 15°C, and *Nitzschia navis-varingica* and *Pseudonitzschia pungens* were cultured in an f/2/seawater medium (Guillard, 1983) at 20°C under an illumination intensity of 60-70 μE, with a 16 h light:8 h dark cycle, using cool-white fluorescent lamps. All species of microalgae were harvested after about two weeks in culture, when the population had reached the stationary phase.

Sample Preparation

The macro-alga samples (whole tissue wet weight 20 g) were rinsed well with sterilized artificial seawater (Aquamarine, Yasu Chem. Co., Japan). Then the macro-alga was finely sliced and homogenized with 3 volumes (w/v) of 99% ethanol using the Polytron homogenizer. The homogenate was centrifuged at 10,000×g at 4°C for 20 min. The precipitate was again extracted with 3 volumes of 80% ethanol and this process was repeated 3 times. The supernatants were combined and evaporated to dryness under reduced pressure at 40-45°C. The residue was dissolved in a small amount of distilled water and rinsed in diethyl ether to remove pigments and fatty materials. This aqueous extract was evaporated to dryness and dissolved in a small amount of deionized water.

Micro-algal cells were collected from the culture medium (500 mL) by centrifugation (400×g, 5 min), and the resulting cell pellet (about 10^9 cells) was rinsed three times with 5 volumes (w/v) of fresh medium. The pellet was then homogenized with 3 volumes of 99% ethanol using an ultrasonic homogenizer, 3 times for 2 min at 4°C. The homogenate was centrifuged at 10,000×g at 4°C for 20 min. The precipitate was then extracted using the same procedures described above for the macro-alga. Both extracts from macro- and micro-alga were stored at –30°C until the HPLC analysis was conducted.

Derivatization of Amino Acids

Each sample extract or standard DL-amino acid solution (12.5 μmol/L in 0.1 N HCl) was passed through a Millipore filter (Millex-LG; 0.20 μm pore size). Twenty μL of an OPA/NAC solution (8 mg of OPA and 10 mg of NAC in 1mL of methanol) and then 70 μL of 0.08 M sodium borate was mixed with 10 μL of the filtrate in a 1 mL vial for HPLC.

HPLC Analysis

After the derivatization reaction was incubated for 2 min, 10 μL of derivative solution was injected onto the column and analyzed by HPLC.

Chromatography was carried out automatically by the method of Nimura *et al.* (1986), with the alterations described here. Samples were eluted with a mixture of solvent A (50 mM sodium acetate buffer at pH 5.35) and solvent B (80% methanol/solvent A) at 40°C; a flow rate of 1.0 mL/min. The excitation and emission wavelengths for fluorescence detection were 348 nm and 450 nm, respectively. The elution gradient was set as follows: 0-50 min, 0-40% solvent B in solvent A; 50-60 min, 40-60% solvent B in solvent A; and 60-85 min, 60-100% solvent B in solvent A.

Calculations

$$C_{HW} = \frac{\text{peak area of standard amino acid (AA)}}{\text{content of the standard AA (pmol) injected onto column}}$$

$$\text{AA content } (\mu\text{ mol/g}) = \frac{\text{peak area} \times \text{total derivative vol. } (100\,\mu\text{ L}) \times \text{total extract vol. (mL)} \times \text{dilution}}{C_{HW} \times \text{injection vol.} \times \text{extract vol.} (10\,\mu\text{ L}) \times \text{weight of sample alga (g)} \times 10^6}$$

Identification of D-Aspartate and D-Alanine Peaks

The peak fraction corresponding to D-aspartate isolated from the algal extracts by HPLC was collected and concentrated with an evaporator, and then analyzed again by HPLC. The peak corresponding to D-alanine was identified by treatment with D-amino acid oxidase. The reaction mixture consisted of 0.1 mL of algal extract, 100 μL of FAD, and 0.2 U of D-amino acid oxidase (from porcine kidney, Biozyme Lab., Blaenavon, UK) in 50 mM Tris-HCl buffer, pH 8.0, in a final volume of 1 mL. The reaction was started by adding the enzyme solution to the algal extract at 40°C. After incubating for 60 min, the reaction was stopped by heating for 3 min at 100°C, and the reaction mixture was then passed through a Millipore filter. The chromatographic conditions were the same as those described for Figure 1, with the exception that the elution gradient was: 0-10 min, 0-15% solvent B; 10-30 min, 15-25% solvent B; and 30-50 min, 25-30% solvent B in solvent A.

3. Results and Discussion

Chromatograms of Standard L- and D-Amino Acids and Sample Extract

Chromatograms of standard L- and D-amino acids are shown in Figure 1. The enantiomeric separation of some amino acids derivatives, such as cysteine, histidine, lysine, methionine, phenylalanine, proline and tryptophan was not possible by this analytic method.

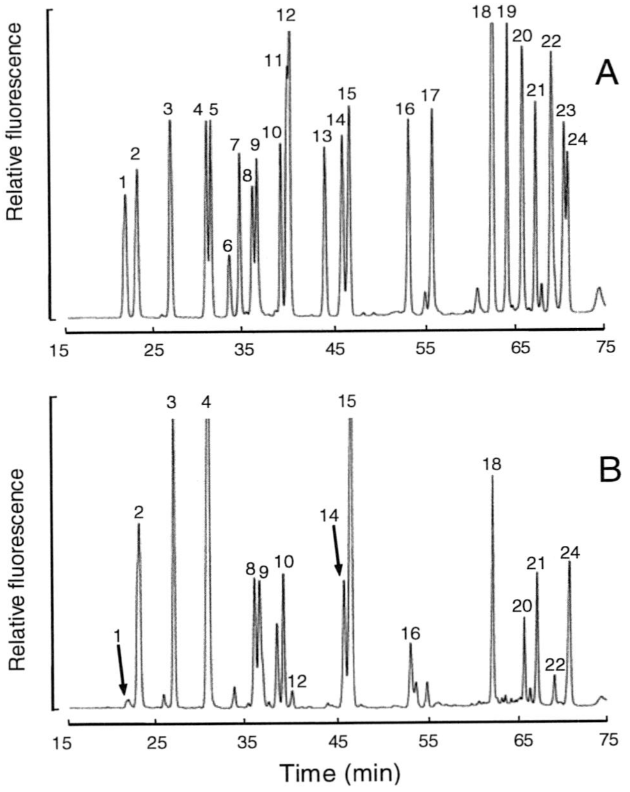

Figure 1. Typical HPLC Chromatograms of (A) the Converted D- and L-Enantiomers of Standard Amino Acids, and (B) an Extract of the Marine Diatom *Asterionella* sp. The number peaks correspond to: *1,* D-Asp; *2,* L-Asp; *3,* Ser; *4,* L-Glu; *5,* D-Glu; *6,* His; *7,* D-Thr; *8,* Gly; *9,* L-Thr; *10,* L-Arg; *11,* D-Arg; *12,* Tau; *13,* β-Ala; *14,* D-Ala; *15,* L-Ala; *16,* L-Tyr; *17,* D-Tyr; *18,* Met + L-Val; *19,* D-Val; *20,* Phe; *21,* L-Ile; *22,* D-Ile + Lys; *23,* D-Leu; *24,* L-Leu.

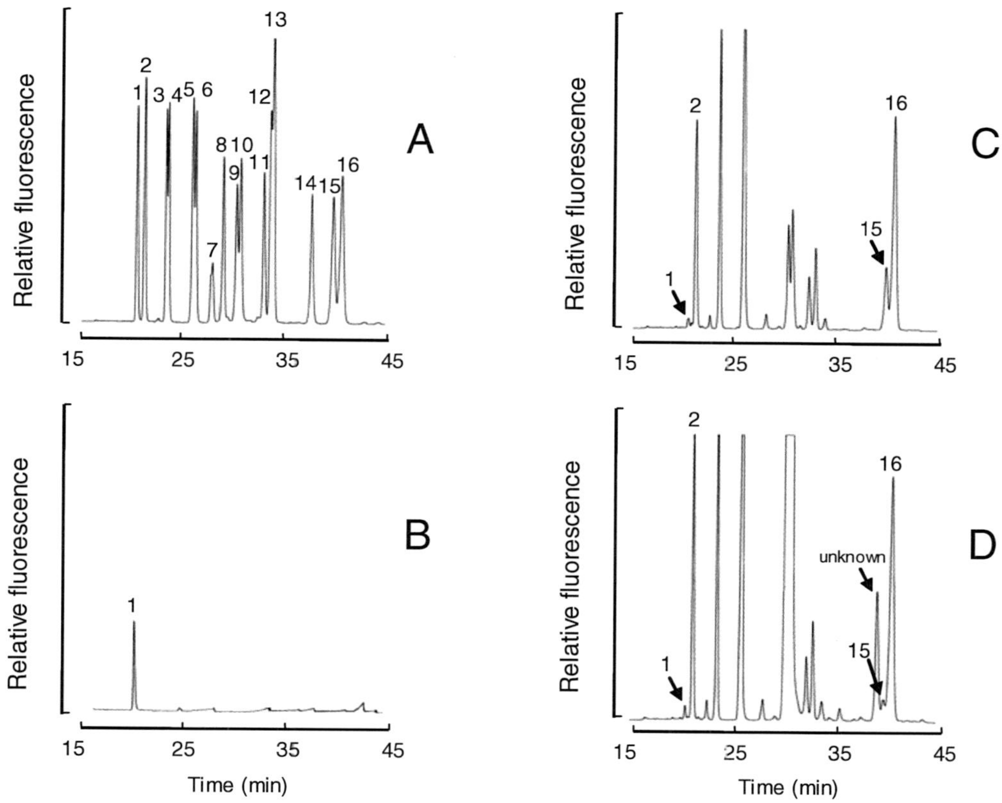

Figure 2. HPLC Chromatograms of (A) Standard DL-Amino Acids, (B) the Peak Corresponding to D-Aspartate Isolated from the Extract of *Asterionella* sp., and the Extract of *Asterionella* sp. before (C) and after (D) treatment with D-Amino Acid Oxidase. The number peaks correspond to *1*, D-Asp; *2*, L-Asp; *3*, D-Ser; *4*, L-Ser; *5*, L-Glu; *6*, D-Glu; *7*, His; *8*, D-Thr; *9*, Gly; *10*, L-Thr; *11*, L-Arg; *12*, D-Arg; *13*, Tau; *14*, β-Ala; *15*, D-Ala; *16*, L-Ala

The chromatographic pattern of the isolated, refractionated D-aspartate samples showed a single peak with the same retention time as the standard D-aspartate (Figure 2B). The chromatographic pattern of a sample extract treated with D-amino acid oxidase lacked the D-alanine peak, as shown in Figure 2D.

These results show that the peak 1 and peak 14 in Figure 1B is D-aspartate and D-alanine, respectively.

Applications

Results of the analysis of marine macro- and micro-algal extracts are shown in Table 1 and Table 2, respectively. Free D-aspartate and D-alanine were detected in both micro- and macro-algae, but other D-amino acids were not detected.

In macro-algae, D-aspartate was detected in small amounts, or not at all, in *Chlorophyta*, while the amino acid was strongly detected in most species of *Phaeophyta* and *Rhodophyta* (6 and 8 species, respectively). The content of D-aspartate was higher in *Phaeophyta* than in

Rhodophyta. In contrast, the D-alanine content in *Chlorophyta* and *Rhodophyta* was generally higher than that of D-aspartate. The D-alanine content in most species belonging to the genus *Phaeophyta* was approximately the same level as D-aspartate.

Table 1. Distribution of D and L-Amino Acids in Marine Macroalgae*.

Species	Asp (nmol/g wet wt.)		(%)	Ala (nmol/g wet wt.)		(%)
	D	L	D/(D+L)	D	L	D/(D+L)
Chlorophyta						
Monostroma nitidum				62	550	10
Ulva pertusa	Tr	120	—	38	590	6.1
Enteromorpha compressa	2.2	12	15	40	581	6.4
Enteromorpha linza	2.5	12	17			
Enteromorpha intestinalis	ND	7.6	—	38	1085	3.4
Chaetomorpha moniligera	ND	98.6	—	15	39	28
Codiu fragile	ND	43.2	—			
Phaeophyta						
Heterochordaria abietina				83	733	10
Scytosiphon lomentarius				40	216	16
Lamiraria japonica	5.9	37	14	ND	460	—
Costaria costata	66	37	64	13	83	14
Undaria pinnatifida	12	30	29	ND	11500	—
Hizikia fusiformis	98	89	52	96	588	14
Sargassum fulvellum	1.2	6.6	15	65	183	26
Sargassum yezoensis	49	57	46			
Sargassum nigrifolium				78	255	23
Coccophora langsdorfii				68	2300	2.9
Rhodophyta						
Grateloupia filicina	3.4	38	8.2			
Gloiopeltis furcata	5.5	75	6.8	44	625	6.6
Chondrus ocellatu	ND	2.6	—	ND	731	—
Caroopeltis affinis	3	36	7.7			
Chondrus yendoi	9.0	38	19			
Rhodymenia palmata	Tr	106	—			
Lomentaria catenata	5.5	33	14			
Nemalion vermiculare				ND	203	—
Neodilsea yendoana				ND	652	—
Ahnfeltia paradoxu				39	80	33
Rhodomela larix	66	201	25			

Tr: Below 1.0 nmol/g wet wt.; ND: not detect; Blank column: Not determine.*: These data were quoted from our previous papers (Nagahisa *et al*., 1992, 1995) with some alteration.

In micro-algae, D-aspartate was detected in all species of freshwater algae and marine diatoms examined, while D-alanine was detected only in the diatoms (Table 2). L-Isomers were detected in all specimens examined.

Table 2. Distribution of D- and L-Amino Acids in the Microalgae*.

Species	Asp (nmol/10^6 cells)		(%)	Ala (nmol/10^6 cells)		(%)
	D	L	D/(D+L)	D	L	D/(D+L)
Freshwater alga						
*Botrydiopsis alpina**	Tr	0.08	—	ND	0.19	—
*Chlorella pyrenoidosa**	0.01	0.12	7.69	ND	0.77	—
*Chlorella vulgaris**	0.03	0.16	15.79	ND	2.11	—
*Scenedesmus obliquus**	0.01	0.10	9.09	ND	0.74	—

Table 2. Continued

Species	Asp (nmol/10^6 cells)		(%)	Ala (nmol/10^6 cells)		(%)
	D	L	D/(D+L)	D	L	D/(D+L)
Marine diatom						
Asterionella sp.	0.09	4.77	1.85	0.40	5.19	7.16
Nitzschia navis-varingica	Tr	4.87	—	0.27	8.45	3.10
Pseudo-nitzschia pungens	0.10	2.05	4.65	0.24	3.61	6.23
Thalassiosira sp.	0.23	3.80	5.71	0.31	12.3	2.46

Tr, trace < 0.01; ND, not detected. ; These data were quoted from our previous paper (Yokoyama *et al.*, 2003)

*Obtained from Chlorella Kogyo Co. Ltd. (Tokyo Japan).

Each value for a marine diatom is the mean obtained from three or four samples taken from the respective cultures, and each value for a freshwater alga was obtained from one sample per culture. Each sample was collected after the alga had been cultured for 2 weeks (stationary phase).

References

Douglas D.J, Bates S.S. (1992) Production of domoic acid, a neurotoxic amino acid, by an axenic culture of the marine diatom *Nitzschia pungens f. ultiseries* Hasle. *Can. J. Fish. Aquat. Sci.*, **49**:85-90

Fujita E., Okuma E, Abe H. (1997) Occurrence of alanine racemase in crustaceans and the changes of the properties during seawater acclimation of crayfish. *Comp. Biochem. Physiol.* **116A**:83-87

Gamburg K.Z.and Rekoslavskaya N.I.(1992) Formation and functions of D-amino acids in plants. *Siberian Institute of Plant Physiology and Biochemistry,* **38**(6):1236-1246

Guillard, R.R.L.(1983) Culture of phytoplankton for feeding marine invertebrates. *Culture of Marine Invertebrates*, 108-132

Kotaki Y., Koike K., Yoshida M., Chu VT, Nguyen TMH, Nguyen CH, Fukuyo Y., Kodama M. (2000) Domoic acid production in *Nitzschia* sp.(Bacillariophyceae) isolated from a shrimp-culture pond in Do Son, Vientnam. *J.Phycol.* **36**:1057-1060

Matsushima O., Katayama H., Yamada K., Kado Y.(1984) Occurrence of free D-alanine and alanine racemase activity in bivalve molluscs with special reference to intracellular osmoreguration. *Mar. Biol. Lett,* **5**:217-225

Nagahisa E., Kan-no N., Sato M., Sato Y. (1992) Occurrence of free D-aspartic acid in marine macroalgae. *Biochem. Int.,* **28**:11-19

Nagahisa E., Kan-no N., Sato M., Sato Y. (1995) Occurrence of free D-alanine in marine macroalgae. *Biosci. Biotechnol. Biochem.,* **59**:2176-2177

Nimura N., Kinoshita T. (1986) *o*-Phthalaldehyde-*N*-acetyl-L-cysteine as a chiral derivatization reagent for liquid chromatographic optical resolution of amino acid enantiomers and its application to conventional amino acid analysis. *J. Chromatogr.,* **352**:169-177

Ogata T., Ishimaru T., Kodama M. (1987) Effect of water temperature and light intensity on growth rate and toxicity change in *Protogonyaulax tamarensis*. *Mar. Biol.* **95**:217-220

Shibata K., Shirasuna K., Motegi K., Kera Y., Abe H., Yamada R. (2000) Purification and properties of alanine racemase from crayfish *Procambarus clarkii*. *Comp. Biochem. Physiol.* **126B**:599-608

Yokoyama T., Kan-no N., Ogata T., Kotaki Y., Sato M., Nagahisa E. (2003) Presence of free D-amino acids in microalgae. *Biosci. Biotechnol. Biochem*. 67(2):388-392

Yokoyama T., Tanaka Y., Sato M., Kan-no N., Nakano T., Yamaguchi T., Nagahisa E. (2005) Alanine racemase activity in the microalga *Thalassiosira* sp. *Fisheries Sci.* **71**:924-930

Yoshikawa N., Dhome N., Takio K., Abe H. (2002) Purification, properties, and partial amino acid sequences of alanine racemase from the muscle of the blacktiger prawn *Penaeus monodon*. *Comp. Biochem. Physiol.,* **133B**:445-453

In: Algae: Nutrition, Pollution Control and Energy Sources
Editor: Kristian N. Hagen, pp. 255-263
ISBN: 978-1-60692-008-4

Chapter 13

REMOVING ALGAE WITH ELECTRO-COAGULATION/FLOTATION

Jia-Qian Jiang*[*1], Y.L. Xu[1], O.N. Mwabonje[1] and M. Chipps[2]

[1] CEHE, C5, Faculty of Engineering and Physical Science,
University of Surrey, Guildford, Surrey, GU2 7XH, UNITED KINGDOM
[2] Research and Development, Thames Water, Spencer House,
Manor Farm Road, Reading, RG2 0JN, UNITED KINGDOM

Abstract

This study demonstrates that by operating at a low current density, 2.5 A/m^2, electrocoagulation/flotation (ECF) performed superior to the chemical coagulation (CC) for the removal of algae. For the similar metal ion doses compared, the ECF can remove 11-12% more *Chlorophyll-a* than CC method. A colloidal titration study shows that the charge of flocs generated by the ECF was more positive than that by CC over the full doses studied and then the charge effect contributes to the superior performance of the ECF. In addition to this, a direct reduction of algal numbers on the surface of anodes is speculated and the flotation of aggregated flocs by hydrogen bubbles, resulting from the ECF process, should contribute to the overall algal removal.

Keywords: Algae-laden water; charge effect; chemical coagulation; colloidal titration; electrocoagulation/flotation.

Introduction

In water treatment, algae's blooming has caused many problems, including uncomfortable tastes and odours, clogging of filters, and the formation of disinfection by-products. In particular, some types of algae (e.g., blue-green algae) can be toxic to humans and other organisms. If an excessive growth of algae occurs, endogeneous toxins (e.g., microcystin and nodularin) emitted act as hepatoxins, while anatoxin and saxitoxin act as neurotoxins; these

[*] E-mail address: J.Jiang@surrey.ac.uk (Corresponding author)

may cause serious damage to both humans and animals alike (WHO, 2004). Therefore, the blooming of algae not only worsens water treatment performance and then deteriorates water quality, but also results in the toxic effect on human beings and animals.

Copper sulphate has been used to control nuisance algae in reservoirs, but is considered to be not a sustainable approach to control the algae blooming. Conventional coagulation/precipitation with dissolved air flotation, or sand filtration, or direct filtration, is still the main treatment process for algae control, but the process is not high efficient. Pre-oxidation with ozone or chlorine or chlorine dioxide has been demonstrated to enhance the algal removal with coagulation, but all these oxidants could result in the formation of disinfection by-products, e.g., THMs, chlorite, and bromate, which have been suspected to be hazardous to health, and therefore, the use of pre-oxidation for algal control might not be a suitable approach.

Water treatment with an electrocoagulation process has been studied since 1980s and been used in some cases in pilot- and full-scale (e.g., Jiang et al, 2002a, b). The principle of the electrocoagulation can be seen in equations 1-6, and the amount of metal ions (e.g., Al) dissolution can be calculated in terms of Faraday's Law (equation 7). In the electrocoagulation, aluminium or iron plate is used as anode to produce freshly formed Al/Fe hydroxides that co-precipitate the dissolved and particulate contaminants rapidly, and this can be used to remove algae. Also, the algae could migrate towards the anodes and this will lead to the direct reducing algae cell numbers on the surface of anodes. Moreover, the electrocoagulation process generates hydrogen bubbles, which effectively float the resulting aggregates (flocs) and should enhance the solid/liquid separation. Combining all these features together, it can be expected that electrocoagulation could be a potential technology for removing algae. However, the optimum electrocoagulation reactor configuration and operating conditions are to be determined through a preliminary batch scale study; this is the objective of this paper.

Proposed anode reactions:

$$Al \rightarrow Al^{3+} + 3e^- \qquad (1)$$

$$2H_2O + 2e^- \rightarrow H_2 \uparrow + 2OH^- \qquad (2)$$

Reducing algae cell numbers on the anodes (3)

Proposed cathode reactions:

$$2H_2O + 2e^- \rightarrow H_2 \uparrow + 2OH^- \qquad (4)$$

$$O_2 + 2H_2O + 4e^- \rightarrow 4OH^- \qquad (5)$$

Bulk (neutral) solution:

$$Al^{3+} + 3H_2O \rightarrow Al(OH)_3 + 3H^+ \qquad (6)$$

$$W/(g\ Al) = \frac{M_{Al} It}{nF} \Phi_e \qquad (7)$$

Materials and Methods

Construction of an Electrocoagulation Reactor

The reactor was built in a perspex electrolysis chamber with anodes and cathodes made of either aluminium plates (maximum Al 97%, Rudgwick Metals) or steel plates (maximum Fe 99%, Rudgwick Metals) (240 mm x 50 mm x 2.5mm). The reactor was configured by four electrodes and bipolar connection via the water. The effective electrode size was 175mm in depth and 46 mm in width, which gives the total anode surface area of 17.5x4.6x3 = 240 cm^2. The gap between two electrodes is 3mm. A schematic diagram of the electrocoagulation treatment system can be seen in figure 1, which consists of an electrocoagulation reactor, a DC power supply (AP500/7030, HiTek Power Ltd, UK), and a multimeter (IDM 93N, ISO-TECH). The connection of the multimeter between the reactor and power supply is to monitor the constant current applied for the experiment.

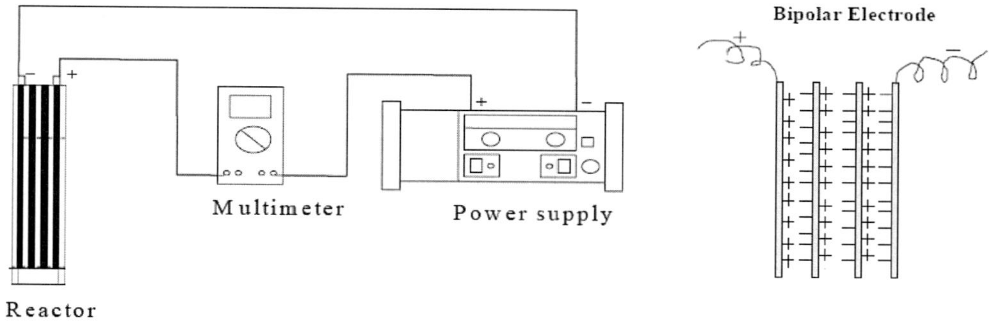

Figure 1. Configuration of electrocoagulation system.

Experimental Conditions

Table 1 summarises all test conditions used for the experiments. The operating current of electrocoagulation were between 10 and 160 mA, the current density was between 1.25 and 20 A/m^2, the electrolysis time was either 2 or 5 min.

Table 1. Operating conditions of electrocoagulation

Electrode configuration	Al plates x 4	Steel (Fe) plates x 4	Al x 2 + Fe x 2
Operating current (mA)	10; 20; 40; 80; 160		
Current density (A m^{-2})	1.25; 2.5; 5; 10; 20		
Electrolysis time (min)	2; 5		
Raw water pH range	6, 7, and 8		

Test waters were university pond water and a synthetic algal model water provided by the Thames Water. The quality parameters of raw waters were analysed before and after treatment. All tests were duplicated and the results were presented as average values with the deviation within 5%. The quality parameters of raw waters measured were pH, suspended solids (SS), turbidity, and conductivity which follows AWWA Standard Methods (1995). *Chlorophyll-a* was measured to determine the algal content in water samples by a wide adopted method (Holm-Hansen, 1978).

The charge of samples before and after treatment was measured using an established colloid titration method (Kawamura and Tanaka, 1966) and the procedures of which are described as follows.

10 mL of 0.001N MGC (a cationic colloid) and 6 drops of indicator (0.1% TB solution) were added into each flask containing the water sample to be determined. The content of the flasks were mixed with a swirling motion for about ten seconds. The excess amount of MGC was then titrated with 0.001N PVSK (a negative charged colloid), until the colour turns from toluidine blue to purple. The same procedure was used for the blank sample (distilled water). The charge of water samples can be calculated using the equation as shown below:

$$\text{Charge of a water sample (in the units of meq/L)} = \frac{(V_{PVSK/blank} - V_{PVSK/sample}) x [PVSK]}{V_{sample}}$$

where, $V_{PVSK/blank}$ = the volume of PVSK used for the titration of blanks (ml), $V_{PVSK/sample}$ = the volume of PVSK used for the titration of a given water sample, [PVSK] = the concentration of PVSK reagent which was 1 meq/L, and V_{sample} = the volume of a water samples which was 10 mL.

Results

Raw Water Qualities

The raw water qualities can be seen in table 2. Due to variations of temperature and climate conditions, the quality characteristics of the university pond water varied.

Table 2. Test water quality characteristics (July – Nov. 2005)

Source of test water	University pond water		Model water from the Thames Water
	Range	mean	
pH_0	7.7 - 8.7	8.1	9.9
Chlorophyll-a (μg/l)	110-290	200	54.8
Suspended solids (mg/l)	42 - 100	68.7	
Turbidity (NTU)	25.8 - 44.9	34.9	
Charge ((meq/l)x 10^3))	-4 - -6	-4.5	
Conductivity (μS/cm)	390 - 557	520.6	353

Algae in the university pond water can be seen in plate 1.

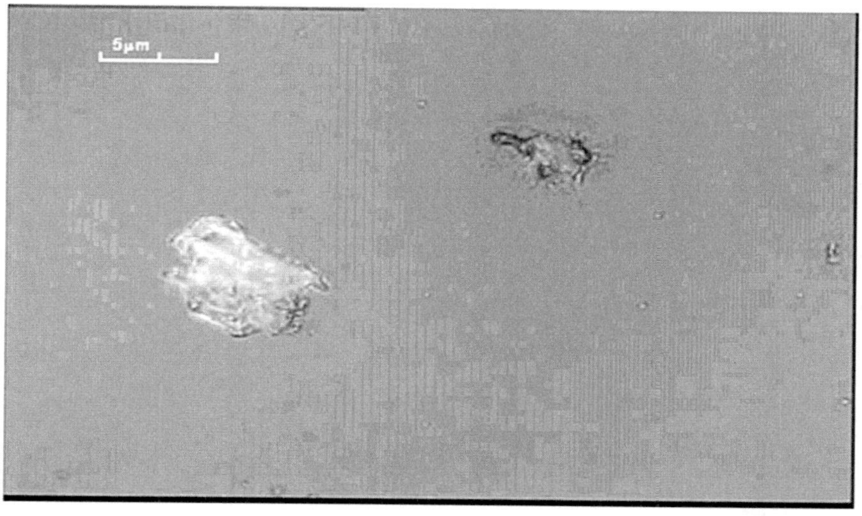

Plate 1. The size of algal species present in the university pond water.

The Effect of Current Density, pH and Electrode Configuration

The effect of current density, water pH and two electrode configurations (Al electrodes and Al+Fe electrodes) on the *Chlorophyll-a* reduction can be seen in figures 2 and 3. The test water used was provided by Thame Water (see table 2).

For both Al and Al+Fe electrode configurations, water pH 7 seems to be favourable to the *Chlorophyll-a* reduction. This is especially true for the Al+Fe electrode configuration, where at a low current density, 2.5 A/m^2, the Chlorophyll-*a* reduction was almost 100% (figure 3). For Al electrode configuration and pH7, the current density of 2.5 A/m^2 was also adequate to remove *Chlorophyll-a* at a high reduction efficiency (>90%, figure 2).

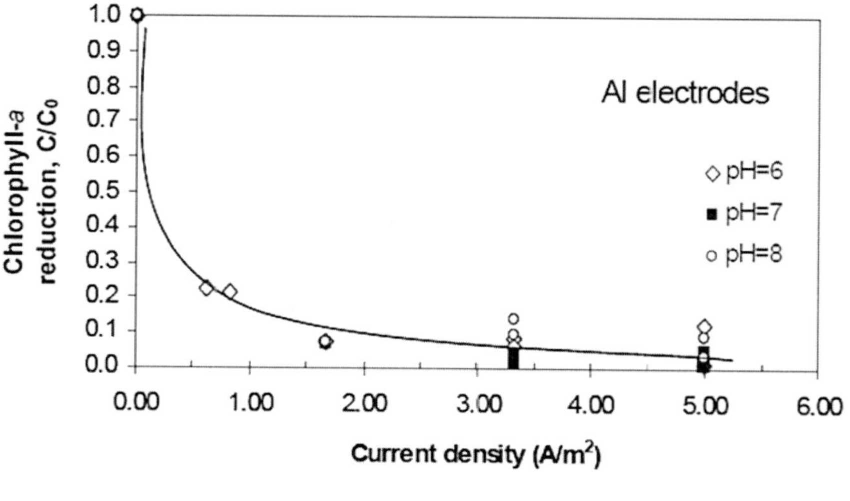

Figure 2. The effect of current density on the reduction of Chlorophyll-*a*, Al electrode configuration.

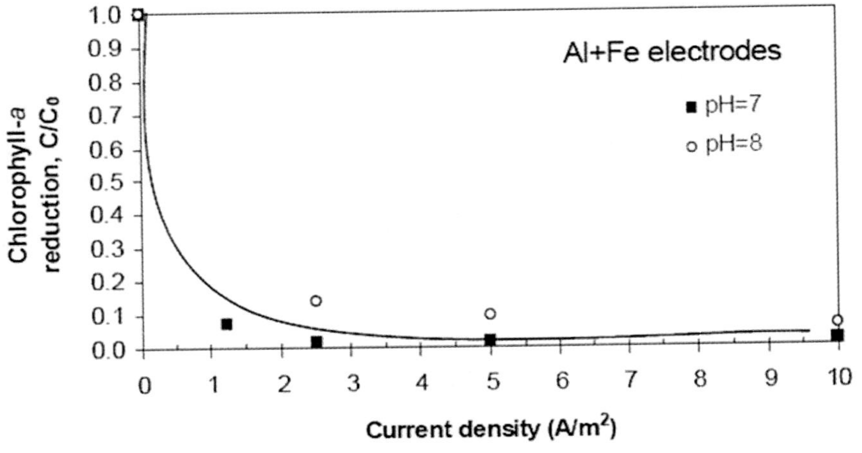

Figure 3. The effect of current density on the reduction of Chlorophyll-*a*, Al+Fe mixing electrode configuration.

Comparative Performance of Electrocoagulation/Flotation and Chemical Coagulation

In order to compare the algal removal performance of electrocoagulation/flotation (ECF) and chemical coagulation (CC), the university pond water was used as test water and the quality characteristics of which can be seen in table 2. Similar to previously, the ECF reactor was configurated with either Al electrodes or Al + Fe mixing electrodes, water pH was adjusted to pH7 and the current density used was 2.5 A/m^2. The coagulants used were aluminium sulphate (AS) and ferric sulphate (FS). A standard jar test procedure was used to conduct the coagulation tests which involved a fast mixing at 275 rpm for 1 min, a slow mixing at 35 rpm for 20 min and a 2-h sedimentation.

Table 3 shows that for the similar dose compared (expressed as mM Al or Al+Fe per litre of water), the ECF can remove 11-12% more *Chlorophyll*-a from the pond water than CC method. This has led to considerations of the cause of a superior performance of ECF in comparison with CC.

Table 3. Comparative performance of electrocoagulation/flotation (ECF) and chemical coagulation (CC)

	Current density (A/m^2)	Time (min)	Fe dose (mg/l)	Al dose (mg/l)	Total dose as mM (Al or Al+Fe)	pH_0	Chlorophyll–*a* (μg/l)	% removal
Raw water	0	0	0	0	0	9.87	54.86	--
ECF								
Al electrodes	2.5	2	0	7.6	0.28	7	0.39	99.3
Al and Fe electrodes	2.5	2	5.91	5.07	0.29	7	0.35	99.4
CC								
AS			0	8	0.30	7	6.34	88.4
FS			14	0	0.25	7	7.61	86.1

Discussion

A superior performance of ECF in the treatment of algal-laden water was observed in this study. For the similar dose compared, the ECF can remove 11-12% more Chlorophyll-*a* than the CC method. One assumption for this could be that as well as chemical dose (e.g., Al ions) and the corresponding reactions, the ECF could directly reduce the Chlorophyll-*a* concentration on the surface of anodes. However, we can only speculate this as there is no direct evidence for the assumption. Nevertheless, this study has tried to compare the charge of metal flocs produced by either ECF or CC methods and aimed at explaining the superior performance of the ECF by charge effects. A colloidal titration method was used to determine the charge of coagulated samples by ECF and CC. An example of the results is presented in figure 4, where, Al-ECF produced high positive charge than Al-CC over the dose range studied. This is a direct evidence showing that the high positive charge of the flocs generated from ECF contributes to a superior algal removal performance.

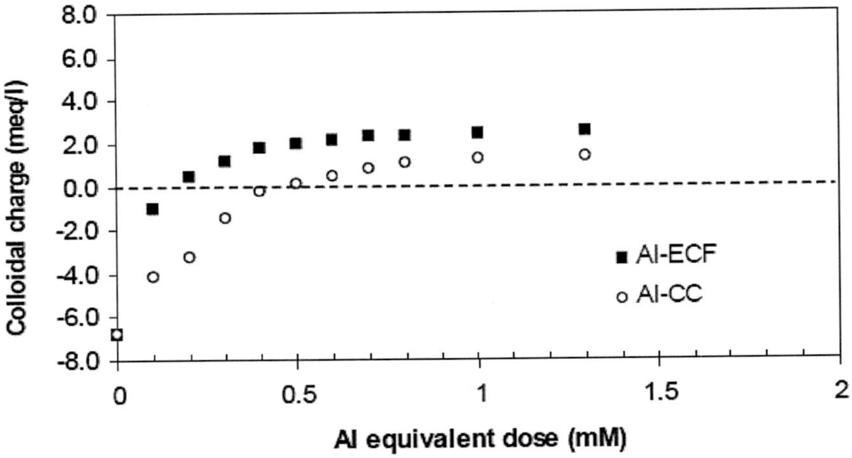

Figure 4. The colloidal charge of coagulated samples treated by electrocoagulation/flotation (Al-ECF) and chemical coagulation (Al-CC).

The charge effect on the algal removal has been observed by a previous research (Jiang et al., 1993). Due to possessing high positive charged species, the pre-polymerised metal coagulants have demonstrated superior performance in reducing algae numbers in comparison with conventional metal coagulants, such as AS. In a recent study of electrocoagulation (Jiang et al, 2006), the floc generated by the ECF has shown to possess a great fraction of Al_{13} species, this further explains why the ECF can remove more Chlorophyll-*a* than CC method.

Conclusions

The study demonstrates that the electrocoagulation/flotation (ECF) performed superior to the chemical coagulation (CC) in the respect of algal removal. For the similar metal ion doses compared, the ECF can remove 11-12% more *Chlorophyll-a* than the CC method. A colloidal titration study demonstrates that the charge of flocs generated by the ECF was more positive than that resulting from the CC over the full doses studied and then the charge effect contributes to the superior performance of the ECF. In addition to this, a direct reduction of algal numbers on the surface of anodes is speculated and the flotation of aggregated flocs by hydrogen bubbles generated in the ECF process should contribute to the overall algal removal.

Acknowledgement

The authors thank the financial support of this project from the Thames Water, UK. The views stated in this paper do not necessarily represent that of Thames Water.

References

Holm-Hansen O. (1978). *OIKOS,* **30**: 438 – 447.

Jiang J.Q., Graham N.J.D., Harward C. (1993). *Water Science And Technology*, **27** (11), 221-230.

Jiang J.Q., Graham N., André C.M., Kelsall G.H., Brandon N.P. (2002a). *Water Research*, **36**, 4068-4082.

Jiang J.Q., Graham, N., André C.M., Kelsall, G.H., Brandon N.P., Chipps M.J. (2002b). *Water Science and Technology: Water Supply*, **2**, 289–297.

Jiang J.Q., Xu Y., Quill K., Simon J., Shettle K. (2006). *Environmental Chemistry*, **3**(5), 350-354.

Kawamura S., Tanaka Y. (1966). *Water and Sewage Works*, 348-357.

Standard Methods for the Examination of Water and Wastewater, 19th ed., Amer. Public Health Assoc.; Amer. Wat. Works Assoc.; Wat. Environ. Feder., Washington DC, 1995.

Guidelines for Drinking Water Quality, Vol. 1: Recommendations, 3rd ed., World Health Organisation, Geneva, 2004, pp. 192-196.

In: Algae: Nutrition, Pollution Control and Energy Sources
Editor: Kristian N. Hagen, pp. 265-300
ISBN: 978-1-60692-008-4

Chapter 14

MICROALGAE IN NOVEL FOOD PRODUCTS

L. Gouveia*[,1], A.P. Batista[2], I. Sousa[3], A. Raymundo[2] and N.M. Bandarra[4]

[1] Instituto Nacional de Engenharia, Tecnologia e Inovação - INETI-DER - Unidade Biomassa, Estrada do Paço do Lumiar, 1649-038 Lisboa, Portugal
[2]Núcleo de Investigação de Engenharia Alimentar e Biotecnologia. Instituto Piaget - ISEIT de Almada. Quinta da Arreinela de Cima, 2800-305 Almada, Portugal
[3]DAIAT – Instituto Superior de Agronomia / Technical University of Lisbon Tapada da Ajuda, 1349-017 Lisboa, Portugal
[4]Departamento de Inovação Tecnológica e Valorização dos Produtos da Pesca Instituto de Investigação das Pescas e do Mar - IPIMAR. Av. Brasília, 1449-006, Lisboa, Portugal

Abstract

The implications of diet on health sustainability have assumed a major importance, supported by considerable epidemiological evidences, and is well recognized by the scientific community and general public, on developed countries. Microalgae are able to enhance the nutritional content of conventional food and feed preparation and hence to positively affect humans and animal health due to their original chemical composition, namely high protein content, with balanced amino acids pattern, carotenoids, fatty acids, vitamins, polysaccharides, sterols, phycobilins and other biologically active compounds, more efficiently than traditional crops. The aim of this chapter is to review the most important features of microalgae in animal and human nutrition, particularly in the development of novel design-foods rich in carotenoids and polyunsaturated fatty acids with antioxidant effect and other beneficial health properties.

---

* E-mail address: luisa.gouveia@ineti.pt

1. Introduction

Modern food industry leads to an increase of cheaper, healthier and more convenient products. The use of natural ingredients, like polyunsaturated fatty acids (PUFA) and antioxidant pigments, exhibiting high impact on functional properties is important to reduce chronic diseases incidence, which are strongly considered of capital importance in Europe, where aging population and welfare costs are fatal for public resources management. The impact of natural substances introduced in the diet via "usual" foods is proved to be efficient at long term and do not present the drawbacks of traditional therapeutic actions based on medicines of short term impact.

Microalgae are an enormous biological resource, representing one of the most promising sources for new products and applications (Pulz and Gross, 2004). They can be used to enhance the nutritional value of food and animal feed, due to their well balanced chemical composition. Moreover, they are cultivated as a source of highly valuable molecules such as polyunsaturated fatty acids, pigments, antioxidants, pharmaceuticals and other biologically active compounds. The application of microalgal biomass and/or metabolites is an interesting and innovative approach for the development of healthier food products.

Microalgal biotechnology is similar to conventional agriculture, but has received quite a lot of attention over the last decades, because they can reach substantially higher productivities than traditional crops and can be extended into areas and climates unsuitable for agricultural purposes (e.g. desert and seashore lands). Microalgae production is an important natural mechanism to reduce the excess of atmospheric CO_2 by biofixation and recycling of fixed C in products, ensuring a lower greenhouse effect, reducing the global environmental heating and climate changes. Microalgae cultivation also presents less or no seasonality, are important as feed to aquaculture and life-support systems, and can effectively remove nutrients (or pollutants) (e.g nitrogen and phosphorus) from water. Microalgal systems for sunlight driven environmental and production applications can clearly contribute to sustainable development and improved management of natural resources. Lately, microalgae have been seen with a great potential as a sustainable feedstock for biodiesel production, in substitution for oil from vegetable crops (Campbell, 1997), and also for hydrogen production (Dutta *et al.*, 2005).

This chapter reviews the main applications of microalgae in feed and food products focusing the authors' work on this subject, for the last years.

2. Microalgae

Microalgae use by indigenous populations has occurred for centuries. However, the cultivation of microalgae is only a few decades old (Borowitzka, 1999) and among the 30000 species that are believed to exist (Chaumont, 1993; Radmer and Parker, 1994), only a few thousands strains are kept in collections, a few hundred are investigated for chemical content and just a handful are cultivated in industrial quantities (Olaizola, 2003).

Some of the most biotechnologically relevant microalgae are the green algae (Chlorophycea) *Chlorella vulgaris*, *Haematococcus pluvialis*, *Dunaliella salina* and the

Cyanobacteria *Spirulina maxima* which are already widely commercialized and used, mainly as nutritional supplements for humans and as animal feed additives.

Chlorella vulgaris has been used as an alternative medicine in the Far East since ancient times and it is known as a traditional food in the Orient. It is widely produced and marketed as a food supplement in many countries, including China, Japan, Europe and the US, despite not possessing GRAS status. *Chlorella* is being considered as a potential source of a wide spectrum of nutrients (e.g. carotenoids, vitamins, minerals) being widely used in the healthy food market as well as for animal feed and aquaculture. *Chlorella* is important as a health promoting factor on many kinds of disorders such as gastric ulcers, wounds, constipation, anemia, hypertension, diabetes infant malnutrition and neurosis (Yamaguchi, 1997). It is also attributed a preventive action against atherosclerosis and hypercholesterolemia by glycolipids and phospholipids, and antitumor actions by glicoproteins, peptides and nucleotides (Yamaguchi, 1997). However the most important substance in *Chlorella* seems to be a beta-1,3-glucan, which is an active immunostimulator, a free-radical scavenger and a reducer of blood lipids (Spolaore *et al.*, 2006).

Haematococcus pluvialis has been identified as the organism which can accumulate the highest level of astaxanthin in nature (1.5-3.0% dry weight). This carotenoid pigment is a potent radical scavenger and singlet oxygen quencher, with increasing amount of evidence suggesting that surpasses the antioxidant benefits of β-carotene, vitamin C and vitamin E. *Haematococcus* is currently the prime natural source of this pigment for commercial exploitation, particularly in aquaculture salmon and trout farming (Lorenz and Cysewski, 2000). Another natural source, *Phaffia rhodozyma* (*Xanthophyllomyces dendrorhous*) yeast requires a large amount of feed for sufficient pigmentation (Dufossé *et al.*, 2005).

Dunaliella salina is an halotolerant microalga, naturally occurring in salted lakes, that is able to accumulate very large amounts of β-carotene, a valuable chemical mainly used as natural food colouring and provitamin A (retinol). The *D. salina* community in Pink Lake, Victoria (Australia) was estimated to contain up to 14% of this carotenoid in their dry weight (Aasen *et al.*, 1969), and in culture some *Dunaliella* strains may also contain up to 10% and more β-carotene, under nutrient-stressed, high salt and high light conditions (Ben-Amotz and Avron, 1980; Oren, 2005). Apart from β-carotene *Dunaliella* produces another valuable chemical, glycerol.

Arthrosphira (*Spirulina*) grows profusely in certain alkaline lakes in Mexico and Africa and has been used as food by local populations since ancient times (Yamaguchi, 1997). It is extensively produced around the world (3000 tons/year) and broadly used in food and feed supplements, due of its high protein content and its excellent nutritive value, such as high γ-linolenic acid (GLA; 18:3ω6) level (Ötles and Pire, 2001; Shimamatsu, 2004). In addition, this microalga has various possible health promoting effects: the alleviation of hyperlipidemia, suppression of hypertension, protection against renal failure, growth promotion of intestinal *Lactobacillus*, suppression of elevated serum glucose level (Spolaore *et al.*, 2006), anticarcinogenic effect and have hypocholesterolemic properties (Reinehr and Costa, 2006). *Spirulina* is also the main source of natural phycocyanin, used as a natural food and cosmetic colouring (blue colour extract) and as biochemical tracer in immunoassays, among other uses (Ötles and Pire, 2001; Kato, 1994; Shimamatsu, 2004).

Recently, attention has been drawn on the marine microalgae *Isochrysis galbana* and *Diacronema vlkianum* (Haptophyceae) due to their ability to produce long chain polyunsaturated fatty acids (LC-PUFA), mainly eicosapentaenoic acid (EPA, 20:5ω3) and also docosahexaenoic acid (DHA, 22:6ω3), that are accumulated as oil droplets in prominent lipid bodies in the cell (Liu and Lin, 2001). These microalgae have been used as feed species for commercial rearing of many aquatic animals, particularly larval and juvenile molluscs, crustacean and fish species (Fidalgo *et al.*, 1998). For example, in a relative ranking of microalgal diets for clam *Mercenaria mercenaria*, the microalga *I. galbana* was shown as the most suitable source of nutritional for rapid growth (Wikfors *et al.*, 1992), while *D. vlkianum* resulted in high growth rates and low mortality for the Pacific oyster *Crassostrea gigas* larvae (Ponis *et al.*, 2006). These microalgae are also potentially promising for the food industry as a valuable source of LC-PUFA's, in alternative to fish oils, supplying also sterols, tocopherols, colouring pigments and other nutraceuticals (Bandarra *et al.*, 2003; Donato *et al.*, 2003).

3. Bioactive Molecules

As with any higher plant, the chemical composition of algae is not an intrinsic constant factor but varies over a wide range. Environmental factors, such as temperature, illumination, pH-value, mineral contents, CO_2 supply, or population density, growth phase and algae physiology, can greatly modified chemical composition. Table 1 presents indicative values of a gross chemical composition of different algae and compared with the composition of selected conventional foodstuffs.

Microalgae can biosynthesize, metabolize, accumulate and secrete a great diversity of primary and secondary metabolites, many of which are valuable substances with potential applications in the food, pharmaceutical and cosmetics industries (Yamaguchi, 1997).

3.1. Pigments

One of the most obvious and arresting characteristic of the algae is their colour. In general, each phylum has its own particular combination of pigments and an individual colour. Aside chlorophylls, as the primary photosynthetic pigment, microalgae also form various accessory or secondary pigments, such as phycobiliproteins and a wide range of carotenoids. These natural pigments are able to improve the efficiency of light energy utilization of the algae and protect them against solar radiation and related effects. Their function as antioxidants in the plant shows interesting parallels with their potential role as antioxidants in foods and humans (Van den Berg *et al.*, 2000). Therefore, microalgae are recognized as an excellent source of natural colorants and nutraceuticals and it is expected they will surpass synthetics as well as other natural sources due to their sustainability of production and renewable nature (Dufossé *et al.*, 2005).

Table 1. General composition of different human sources and microalgae (% dry matter) (adapted from Becker, 1994; Spolaore *et al.*, 2006 and Natrah *et al.*, 2007)

Commodity	Protein	Carbohydrate	Lipid
Baker's yeast	39	38	1
Meat	43	1	34
Egg	47	4	41
Milk	26	38	28
Rice	8	77	2
Soya	37	30	20
Anabaena cylindrical	43-56	25-30	4-7
Chaetoceros calcitrans	36	27	15
Chlamydomonas rheinhardii	48	17	21
Chlorella pyrenoidosa	57	26	2
Chlorella vulgaris	51-58	12-17	14-22
*Chlorella vulgaris**	38	33	5
Chlorella vulgaris (carotenogenic)*	12	25	28
*Diacronema vlkianum**	38	25	18
Dunaliella salina	57	32	6
Euglena gracilis	39-61	14-18	14-20
Haematococcus pluvialis (carotenogenic)*	10	40	41
Isochrysis galbana	48	27	15
Porphyridium cruentum	28-39	40-57	9-14
Scenedesmus obiquus	50-56	10-17	12-14
Scenedesmus dimorphus	8-18	21-52	16-40
Spirogyra sp.	6-20	33-64	11-21
Spirulina maxima	60-71	13-16	6-7
*Spirulina maxima**	45	21	4
Spirulina platensis	46-63	8-14	4-9
Synechococcus sp.	63	15	11
Tetraselmis maculate	52	15	3

*Values experimentally determined by the authors (Batista *et al.*, 2007a).

3.1.1. Chlorophylls

All algae contain one or more type of chlorophyll: chlorophyll-*a* is the primary photosynthetic pigment in all algae (Figure 1) and is the only chlorophyll in cyanobacteria (blue-green algae) and rhodophyta. Like all higher plants, chlorophyta and euglenophyta contain chlorophyll-*b* as well; chlorophylls -*c*, -*d* and –*e* can be found in several marine algae and fresh-water diatoms. Chlorophylls amounts are usually about 0.5-1.5% of dry weight (Becker, 1994).

Apart from their use as food and pharmaceutical colorants, chlorophyll derivatives can exhibit health promoting activities. These compounds have been traditionally used in medicine due to its wound healing and anti-inflammatory properties as well as control of calcium oxalate crystals and internal deodorization (Ferruzi and Blakeslee, 2007). Recent epidemiological studies from *The Netherlands Cohort Study* (Balder *et al.*, 2006) has provided evidence linking chlorophyll consumption to a decreased risk of colorectal cancer.

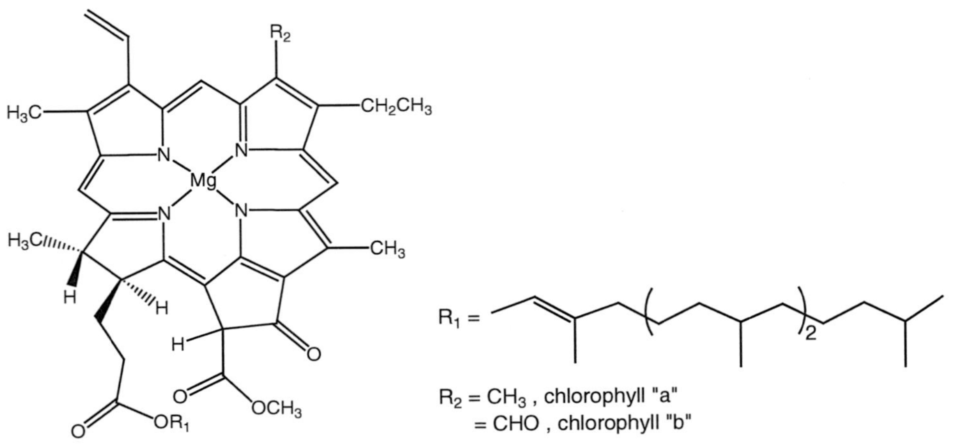

Figure 1. Chemical structures of chlorophyll a and b.

3.1.2. Carotenoids

Carotenoids are naturally occurring pigments that are responsible for the different colours of fruits, vegetables and other plants (Ben-Amotz and Fishler, 1998). Carotenoids are usually yellow to red, isoprenoid polyene pigments derived from lycopene (Figure 2). They are synthesized *de novo* by photosynthetic organisms and some other microorganisms (Borowitzka, 1988). In animals the carotenoids ingested in the diet are accumulated and/or metabolized by the organism, being present in meat, eggs, fish skin (trout, salmon), in the carapace of Crustacea (shrimp, lobster, Antartic krill, crawfish), and in the subcutaneous fat, the skin, the egg yolks, the liver, the integuments, and in the feathers of birds (poultry) (Breithaupt, 2007).

In the algae the carotenoids seem to function primarily as photoprotective agents and as accessory light harvesting pigment, thereby protecting the photosynthetic apparatus against photo damage (Ben-Amotz *et al.*, 1987). They also play a role in phototropism and phototaxis (Borowitzka, 1988). Some microalgae can undergo a carotenogenesis process, in response to various environmental and cultural stresses (e.g. light, temperature, salts, nutrients), where the alga stops growth and changes dramatically its carotenoid metabolism, accumulating secondary carotenoids as an adaptation to severe environments (Bhosale, 2004).

The consumption of a diet rich in carotenoids has been epidemiologically correlated with a lower risk for several diseases particularly those in which free radicals are thought to play a role in initiation, such as arteriosclerosis, cataracts, age-related macular degeneration, multiple sclerosis and cancer (Stahl and Sies, 2005; Tapiero *et al.*, 2004). However, unexpected results from intervention studies (ATBC, 1994; Omenn *et al.*, 1996) with β-carotene suggest that the threshold between the beneficial and adverse effects of some carotenoids is low and provides a strong stimulus to further understanding the functional effects of specific carotenoids (Van den Berg *et al.*, 2000).

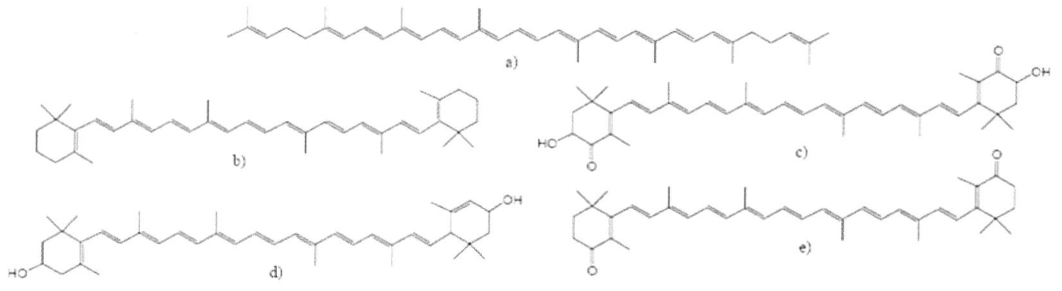

Figure 2. Chemical structures of some carotenoids. a) lycopene, b) β-carotene, c) astaxhanthin, d) lutein, e) canthaxanthin.

More than 600 known carotenoids were reported in nature and about 50 have provitamin-A activity, which includes α-carotene, β-carotene and β-cryptoxanthin (Faure *et al.*, 1999). However, only very few carotenoids are used commercially: β-carotene and astaxanthin and, of lesser importance, lutein, zeaxanthin, lycopene and bixin which are used in animal feeds, pharmaceuticals, cosmetics and food colourings.

The main carotenoids produced by microalgae are β-carotene from *Dunaliella salina* and astaxanthin from *Haematococcus pluvialis*.

β-carotene serves as an essential nutrient and has high demand in the market as a natural food colouring agent, as an additive to cosmetics and also as an health food (Raja *et al.*, 2007). β-carotene is routinely used in soft-drinks, cheeses and butter or margarines. Is well regarded as being safe and indeed positive health effects are also ascribed to this carotenoids due to a pro-vitamin A activity (Baker and Gunther, 2004).

The benefits of astaxanthin are said to be numerous, and include enhancing eye health, improving muscle strength and endurance and protecting the skin from premature ageing, inflammation and UVA damage, is a strong coloring agent and has many functions in animals such as growth, vision, reproduction, immune function, and regeneration (Blomhoff *et al.* 1992; Tsuchiya *et al.* 1992; Beckett and Petrovich, 1999). Some reports support the assumption that daily ingestion of astaxanthin may protect body tissues from oxidative damage as this might be a practical and beneficial strategy in health management. It has also been suggested that astaxanthin has a free radical fighting capacity worth 500 times that of vitamin E (Dufossé *et al.*, 2005).

3.1.3. Phycobiliproteins

Besides chlorophyll and carotenoid lipophilic pigments, Cyanobacteria (blue-green algae), Rhodophyta (red algae) and Cryptomonads algae contain phycobiliproteins, deep colored water-soluble fluorescent pigments, which are major components of a complex assemblage of photosynthetic light-harvesting antenna pigments - the phycobilisomes (Glazer, 1994). Phycobiliproteins are formed by a protein backbone covalently linked to tetrapyrrole chromophoric prosthetic groups, named phycobilins (Figure 3). The main natural resources of phycobiliproteins are the cyanobacterium *Spirulina* (*Arthrospira*) for phycocyanin (blue) and the rhodophyte *Porphyridium* for phycoerythrin (red).

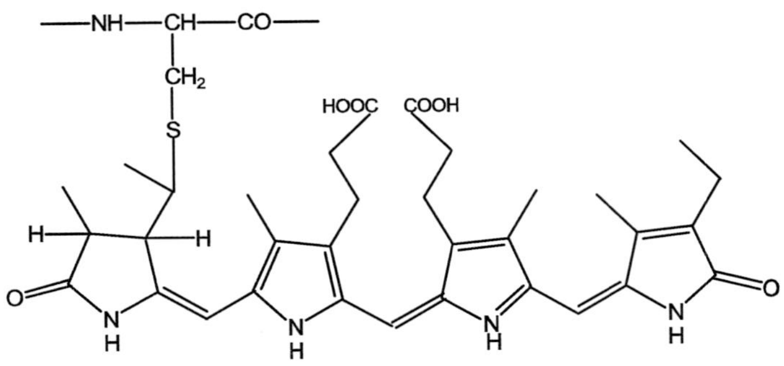

Figure 3. Chemical structure of a phycocyanobilin attached by thioether linkage to the apoprotein.

This group of pigments possesses a large spectrum of applications, which is evidenced by the recent work of Sekar and Chandramohan (2007) that screened 297 patents on phycobiliproteins from global patent databases. They are extensively used for fluorescence applications, as highly sensitive fluorescence markers in clinical diagnosis and for labeling antibodies used in multicolour immunofluorescence or fluorescence-activated cell-sorter analysis (Becker, 1994).

Phycocyanin is currently used in Japan and China as a natural colouring, in food products like chewing gums, candies, dairy products, jellies, ice creams, soft drinks (*e.g.* Pepsi® blue) and also in cosmetics such as lipsticks, eyeliners and eye shadows (Sekar and Chandramohan, 2007). In a recent study, phycocyanin was considered a more versatile blue colorant than gardenia and indigo, providing a bright blue color in jelly gum and coated soft candy, despite its lower stability towards heat and light (Jespersen *et al.*, 2005). A rising number of investigations revealed several pharmacological properties attributed to phycocyanin including, antioxidant, anti-inflammatory, neuroprotective and hepatoprotective effects (Romay *et al.* 2003; Benedetti *et al.*, 2004; Bhat and Madyastha, 2000).

3.2. Fatty Acids

Some microalgae synthesize fatty acids with particular interest (Figure 4), namely GLA (*Arthrospira*), arachidonic acid (AA, 20:4ω6) (*Porphyridium*), EPA (*Nannochloropsis*, *Phaeodactylum*, *Nitzschia*, *Isochrysis*, *Diacronema*) and DHA (*Crypthecodinium*, *Schizochytrim*) (Bandarra *et al.*, 2003; Donato *et al.*, 2003; Chini Zittelli *et al.*, 1999; Molina Grima *et al.*, 2003; Spolaore *et al.*, 2006). These LC-PUFA (more than 18 carbons) can not be synthesized by higher plants and animals, only by microalgae which supply whole food chains (Pulz and Gross, 2004). Is estimated that only healthy human adults are able to elongate 18:3ω3 to EPA in an extend lower than 5% and convert EPA to DHA in a rate inferior to 0.05%, being inhibit in childhood and elderly life (Burdge and Calder, 2005; Wang *et al.*, 2006). This statement confirms the importance of the inclusion of these long chain fatty acids in daily diet.

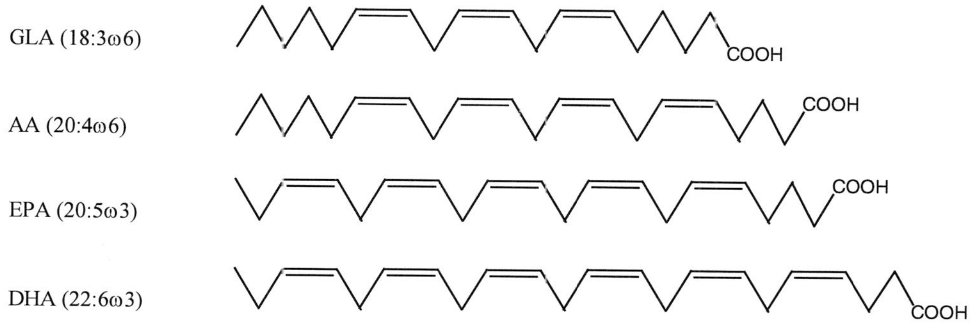

Figure 4. Chemical structure of polyunsaturated fatty acids of high pharmaceutical and nutritional value.

Fish and fish oils are the main sources of LC-PUFA, still global fish stocks are declining due to general fishing methods and over-fishing and the derived oils are sometimes contaminated with a range of pollutants, heavy metals, toxins and typical fishy smell, unpleasant taste and poor oxidative stability (Certik and Shimizu, 1999; Luiten *et al.*, 2003). The production of LC-PUFA from microalgae biotechnology is an alternative approach, and currently microalgal DHA from *Crypthecodinium* and *Ulkenia* is commercially available by the Martek (USA) and Nutrinova (Germany) companies (respectively), for application in infant formulas, nutritional supplements and functional foods (Pulz and Gross, 2004; Spolaore *et al.*, 2006).

PUFA's ω-3, especially DHA, are essential in infant nutrition, being important building blocks in brain development, retinal development and ongoing visual, cognitive, as well as important fatty acids in human breast milk (Ghys *et al.*, 2002; Wroble *et al.*, 2002; Arteburn *et al.*, 2007; Crawford, 2000). LC-PUFA ω-3 consumption has been associated with the regulation of eicosanoid production (prostaglandins, prostacyclins, tromboxanes and leucotrienes) which are biologically active substances that influence various functions in cells and tissues (e.g. inflammatory processes) being important in the prophylaxis and therapy of chronic and degenerative diseases including reduction of blood cholesterol, protection against cardiovascular, coronary heart diseases, atherosclerosis, diabetes, hypertension, rheumatoid arthritis, rheumatism, skin diseases, digestive and metabolic diseases as well as cancer (Simopoulos, 2002; Bønaa *et al.*, 1990; Sidhu, 2003; Thies *et al.*, 2003). Other important role is attributed to gene expression regulation, as well as cholesterol and fasting triacylglycerol (TAG) decreases (Calder, 2004).

The evidence of a dietary deficiency in long-chain omega3 fatty acids is firmly linked to increased morbidity and mortality from coronary heart disease.

3.3. Tocopherols and Sterols

Tocopherols have a widespread occurrence in nature being present in both photosynthetic (e.g. leaves) and non-photosynthetic (e.g. seedlings) tissues of higher plants and algae. However *Euglena* microalga has the highest tocopherols content among the several genera of yeast, molds and algae tested (Kusmic *et al.*, 1999).

Studies covering a wide range of phytoplankton have suggested that the growth rates of bivalves are related to the kind and amount of sterols present in the diet phytoplankton (Wikfors *et al.*, 1991). On the other hand, it has been found that many polyhydroxysterols from marine organisms have anticancer, cytotoxic and other biological activity (Cui *et al.*, 2000; Tang *et al.*, 2002; Han *et al.*, 2003; Volkman, 2003).

3.4. Proteins

The high protein content of various microalgae species is one of the main reasons to consider them as an unconventional source of protein (Soletto *et al.*, 2005), well illustrated by the great interest in microalgae as single cell protein (SCP) during the 1950s. In addition, the amino acid pattern of almost all algae compares favorably with that of other food proteins. Since the cells are capable of synthesize all amino acids, they can provide the essential ones to humans and animals (Guil-Guerrero *et al.*, 2004). As other bioactive compounds synthesized by microalgae, amino acids composition, especially the free amino acids, varies greatly between species as well as with growth conditions and growth phase (Borowitzka, 1988). Protein or amino acids may therefore be by-products of an algal process for the production of other fine chemicals, or with appropriate genetic enhancement, microalgae could produce desirable amino acids in sufficiently high concentrations (Borowitzka, 1988).

3.5. Polysaccharides

Polysaccharides are widely used in the food industry primarily as gelling and/or thickening agents. Many commercially used polysaccharides like agar, alginates and carrageenans are extracted from macroalgae (e.g. *Laminaria, Gracilaria, Macrocystis*) (Borowitzka, 1988). Nevertheless, most microalgae produce polysaccharides and some of them could have industrial and commercial applications, considering the fast growth rates and the possibility to control the environmental conditions regulating its growth. The most promising microalga for commercially purposes is the unicelular red alga *Porphyridium cruentum*, which produces a sulphated galactan exopolysaccharide that can replace carrageenans in many applications. Another example is *Chlamydomonas mexicana*, which releases up to 25% of its total organic production as extracellular polysaccharides and which as found application as a soil conditioner in the USA (Borowitzka, 1988). Certain highly sulphated algal polysaccharides also present pharmacological properties acting on the stimulation of the human immune system (Pulz and Gross, 2004).

3.6. Vitamins and Minerals

Microalgae biomass represents a valuable source of nearly all essential vitamins (*e.g.* A, B_1, B_2, B_6, B_{12}, C, E, nicotinate, biotin folic acid and pantothenic acid) and a balanced mineral content (e.g. Na, K, Ca, Mg, Fe, Zn and trace minerals) (Becker, 2004). The high levels of vitamin B_{12} and Iron in some microalgae, like *Spirulina*, makes them particularly suitable as nutritional supplements for vegetarian individuals. The vitamin content of an alga depends on the genotype, the stage in the growth cycle, the nutritional status of the alga, the light intensity (photosynthetic rate). The vitamin content is therefore amenable to manipulation by varying the culture conditions as well as by strain selection or genetic engineering. However, vitamins cell content fluctuates with environmental factors, the harvesting treatment and the biomass drying methods (Brown *et al.*, 1999; Borowitzka, 1988).

3.7. Antioxidants

Microalgae are photoautotrophic organisms that are exposed to high oxygen and radical stresses, and consequently have developed several efficient protective systems against reactive oxygen species and free radicals (Pulz and Gross, 2004). Hence, there is increasing interest in using microalgae as natural antioxidants source for cosmetics (e.g. sun-protecting) and functional food/nutraceuticals.

Natrah *et al.* (2007) reported a stronger antioxidant activity exhibited by methanolic microalgal crude extracts (from e.g. *Isochrysis galbana, Chlorella vulgaris, Nannochloropsis oculata, Tetraselmis tetrathele, Chaetoceros calcitrans*) when compared with α-tocopherol, but lower than the synthetic antioxidant BHT. However BHT and BHA synthetic antioxidants, are questionable in terms of their safe use, since they are believed to be carcinogenic and tumorigenic if given in high doses (Schildermann *et al.*, 1995; Aruoma, 2003).

3.8. Pharmaceuticals and Other Biologically Active Compounds

The microalgae represent a very large, relatively unexploited reservoir of novel compounds, many of which are likely to show biological activity, presenting unique and interest structures and functions (Yamaguchi, 1997). In the last decades marine microorganisms, particularly Cyanobacteria, have been screened for new pharmaceuticals and antibiotics. Published data until 1996 revealed 208 cyanobacterial compounds with biological activity while in 2001 the number of compounds screened was raised to 424, including lipoproteins (40%), alkaloids, amides and others (Burja *et al.*, 2001). The reported biological activities comprise cytotoxic, antitumor, antibiotic, antimicrobial (antibacterial, antifungal, antiprotozoa), antiviral (e.g. anti-HIV) activities as well as biomodulatory effects like immunosuppressive and anti-inflammatory (Burja *et al.*, 2001; Singh *et al.*, 2003). The cytotoxic activity, important for anticancer drugs development, is likely related to defense strategies in the highly competitive marine environment, since usually only those organisms lacking an immune system are prolific producers of secondary metabolites such as toxins (Burja *et al.*, 2001).

4. Microalgae in Animal Nutrition

Several microalgae (e.g. *Chlorella, Tetraselmis, Spirulina, Nannochloropsis, Nitzchia, Navicula, Chaetoceros, Scenedesmus, Haematococcus, Crypthecodinium*), macroalgae (e.g. *Laminaria, Gracilaria, Ulva, Padina, Pavonica*) and fungi (*Mortierella, Saccharomyces, Phaffia, Vibrio marinus*) can be used in both terrestrial and aquatic animal feed (Harel and Clayton, 2004).

Feeds can be formulated by using vegetable protein sources, vegetable oil sources, fishmeal, mineral and vitamin premixes in order to reach appropriate nutritional properties for each animal group and promote health and welfare benefits (Harel and Clayton, 2004). Using even very small amounts of microalgal biomass can positively affect the physiology of animals by improved immune response, resulting in growth promotion, disease resistance, antiviral and antibacterial action, improved gut function, probiotic colonization stimulation, as well as by improved feed conversion, reproductive performance and weight control (Harel and Clayton, 2004). The external appearance of the animals may also be improved, resulting in healthy skin and a lustrous coat, for both farming animals (poultry, cows, breeding bulls) and pets (cats, dogs, rabbits, ornamental fishes and birds) (Certik and Shimizu, 1999).

Since feed corresponds to the most important exogenous factor influencing animal health and also the major expense in animal production, the use of alternative high quality protein supplements replacing conventional protein sources is encouraged. Considering that animal feed stands at the beginning of the food chain, increasing public and legislative interest is evident, especially considering intensive breeding conditions and the recent trend to avoid *"chemicals"* like antibiotics (Breithaupt, 2007). The large number of nutritional and toxicological evaluations already conducted has demonstrated the suitability of algae biomass as a valuable feed supplement (Becker, 1994). In fact, 30% of the current world algal production is sold for animal feed applications (Becker, 2004).

4.1. Poultry

The replacement of conventional protein in broilers rations was done by several feeding trials and authors, using various microalgae species, namely *Chlorella*, *Euglena*, *Oocystis*, *Scenedesmus*, *Spirulina*, with incorporation % depending on algae specie (usually up to 10%) (Becker, 1994). In laying hens no differences were found in egg production rate and egg quality (size, weight, shell thickness, solid content of the egg, albumin index, etc) and feed conversion efficiency, between control and birds receiving 12% sewage-growth *Chlorella* (Becker, 1988). Algae may serve as almost the sole source of protein in layers ration (Becker, 1988) and the yolk can have a distinct intense orange colour in layers feed the algal diet (Becker, 2004).

For pigmentation purposes of broilers and/or egg yolks the diet must contain a carotenoids source. Traditionally, dehydrated alfalfa meal and yellow corn were used (Marusich *et al.*, 1960; Becker, 2004). However, today, feed mills use low-cost raw material to provide high energy diets and control the pigment content by appropriate supplementation. Petals of Aztec marigold (*Tagetes erecta*), rich in lutein, have been reported to be very effective as yolk pigmenting agent as well as synthetic canthaxanthin (Madiedo and Sunde, 1964). For laying hens feeds, canthaxanthin should not exceed 8 mg/kg since at extremely high dosages minute crystals may be formed in the retina by a reversible deposition process (Breithaupt, 2007).

In the last decades, microorganisms such as microalgae, have been tested for pigmentation purposes in poultry. *Dunalliella bardawil* can be a source of vitamin A and a yolk enhancing agent when administrated to laying hens (Avron *et al.*, 1952). Gouveia *et al.* (1996a) reported the effect of carotenoids present in *Chlorella vulgaris* microalga biomass upon pigmentation of egg yolk comparable with commercially synthetic pigments used.

Haematococcus microalga can also be used as a natural feed colourant of broiler chickens (Kenneth, 1989; Waldenstedt *et al.*, 2003).

Studies with chickens fed red microalga *Porphyridium* sp. biomass (at 5% and 10% diet incorporation), showed a reduced blood cholesterol level and a modified fatty acid composition in egg yolk, in spite of no differences in body weight, egg number, and egg weight (Ginzberg *et al.*, 2000). Chickens fed with algal biomass consumed 10% less food for both groups, and their serum cholesterol levels were significantly lower (by 11% and 28% for the groups fed with 5% and 10% supplement, respectively) as compared with the respective values of the control group. Egg yolk of chickens fed with algae tended to have reduced cholesterol levels (by 10%) and increased linoleic acid and arachidonic acid levels (by 29% and 24%, respectively). In addition, the color of the egg yolk was darker as a result of the higher carotenoid levels (2.4 fold higher) for chickens that fed with 5% supplement (Ginzberg *et al.*, 2000).

Algae are, in general, officially approved in several countries as chicken feed and do not require new testing or approval. However, it has to be decided from case to case how restrictive the different algae species are regarded as feed supplements (Becker, 1994). In the European Union the Regulation (EC) No. 1831/2003 determines the use of additives in animal nutrition and sets out rules for the authorization, marketing and labeling of feed additives.

4.2. Pigs

Aside from poultry, pigs appear to be a potential group for which algae could be used as feed supplement. *Chlorella* and *Scenedesmus* were used for substituting soybean meal and cotton seed meal in concentrations up to 10%, without differences in feed conversion efficiency (Hintz *et al.*, 1966; Hintz and Heitmann, 1967). Microalgal biomass is a feed ingredient of good nutritional quality and suited very well for rearing pigs. It can replace conventional proteins like soybean meal or fishmeal and no difficulties in acceptability of algae were reported for these animals (Becker, 1994).

Spirulina has also been tested as additive in short-term and long-term experiences (Fevrier and Seve, 1975) and all parameters studied remained identical and no differences in reproductive capacity were observed. The authors recommended 25% of microalgal biomass incorporation, while Yap *et al.* (1982) assumed 33% incorporation, without negative symptoms.

4.3. Ruminants

It should be expected that ruminants represent the group of animals most suitable for feeding with algae, since these animals are able to digest even unprocessed algal material (e.g. cell walls). However a limited number of trials have been done due the large amount of algae required to perform appropriate feeding experiments with these animal species. Sheep's, lambs and cattle's shows an inability to digest efficiently the carbohydrate fraction of the algae (*Chlorella*, *Scenedesmus obliquus* and *Scenedesmus quadricauda*) (Hintz *et al.*, 1966; Davis *et al.*, 1975). Better digestibility was obtained with *Spirulina* constituting 20% of a complete sheep diet. Calves revealed a minor difference between control and untreated fresh *Scenedesmus* alga feeding animals (Calderon *et al.*, 1976).

4.4. Aquaculture

Microalgae feeds are currently used mainly for the culture of larvae and juvenile shell- and finfish, as well as for raising the zooplankton required for feeding of juvenile animals (Benemann, 1992; Chen, 2003). They are required for larval nutrition during a brief period, either for direct consumption in the case of molluscs and peneid shrimp or indirectly as food for the live prey, mainly rotifers, copepods and *Artemia* nauplii, which in turn are used for crustaceans and fish larvae feeding (Brown *et al.*, 1997; Duerr *et al.*, 1998; Muller-Feuga, 2000; Xu *et al.*, 2007).

In 1999, the production of microalgae for aquaculture reached 1000 t (62% for mollusks, 21% for shrimps and 16% for fish) for a global world aquaculture production of 43×10^6 t of plants and animals (Muller-Feuga, 2000). The most frequently used species in aquaculture are *Chlorella, Tetraselmis, Isochrysis, Pavlova, Phaeodactylum, Chaetoceros, Nannochloropsis, Skeletonema* and *Thalassiosira* (Yamaguchi, 1997; Borowitzka, 1997; Apt and Behrens, 1999; Muller-Feuga, 2000).

Microalgae contain essential nutrients which determine the quality, survival, growth and resistance to disease of cultured species. These illustrate the importance of the control of microalgal biochemical composition for the success of aquaculture feed chains, opening new perspectives for the study of fish larval nutrition and the development of microalgae-based feeds for aquaculture (Fábregas *et al.*, 2001). To support a better balanced nutrition for animal growth, it is often advised to use mixed microalgae cultures, in order to have a good protein profile, adequate vitamin content and high polyunsaturated fatty acids, mainly EPA, AA and DHA, recognized as essential for survival and growth during the early stages of life of many marine animals (Volkman *et al.*, 1989). One of the beneficial effects attributed to adding algae is an increase in ingestion rates of food by marine fish larvae which enhance growth and survival as well as the quality of the fry (Naas *et al.* 1992). In addition, the presence of algae in rearing tanks of European sea bass larvae has been shown to increase digestive enzyme secretion (Cahu and Zambonino-Infante 1998).

Aquatic species, such as salmonids (salmon and trout), shrimp, lobster, seabream, goldfish and koi carp under intensive rearing conditions need a supplementation of carotenoids pigments in their diet, to attain their characteristic muscle colour. In addition to pigmenting effects, carotenoids, namely astaxanthin and canthaxanthin, exert benefits on animal health and welfare, promote larval development and provide growth and performance stimulatory effects in farmed fish and shrimp (Baker and Gunther, 2004). These effects were proved by Torrissen (1984) during the early star-feeding of Atlantic salmon reared in fresh water, by Christiansen *et al.* (1995) in Atlantic salmon parr and by Torrissen and Christiansen (1995) that proposed a minimum of 10 mg astaxanthin or canthaxanthin per kg of diet for all fish and crayfish; also for non-salmonid fish, authors have reported growth benefits with carotenoids supplementation, for instance in carp and tilapia (Segner *et al.*, 1989) and in crustacean, such as prawn *Panaeus japonicus* (Chien and Jeng, 1992; Nègre-Sadargues *et al.*, 1993).

A positive metabolic role of carotenoids in the nutrition of larval fish and survival of young fry was also discussed by Reitain *et al.* (1997), Shahidi *et al.* (1998), Planas and Cunha (1999) and Lazo *et al.* (2000). However, the inclusion of 45 mg carotenoids in the diet (Rema and Gouveia, 2005) besides this effectiveness on skin pigmentation, was not sufficient to induce any differences in growth and survival of larvae and juvenile goldfish, independently of the source (natural or synthetic).

Nevertheless, given carotenoids high costs, efforts have been deployed to evaluate the potential of some natural pigments obtained from the red yeast *Phaffia rhodozyma* (Bon *et al.*, 1997), the marine bacteria *Agrobacterium aurantiacum* (Yokoyama and Miki, 1995), the green algae *Haematococcus pluvialis* (Harker *et al.*, 1996; Yuang and Chen, 2000), *Chlorella zofingiensis* (Bar *et al.*, 1995) and *Chlorella vulgaris* (Gouveia *et al.*, 1996b) as dietary carotenoid sources. Numerous reports show that carotenogenic microalgae appear as suitable source of carotenoids in fish feeds.

Haematococcus pluvialis was assayed in rainbow trout for colouring purposes (Sommer *et al.*, 1991, 1992, Choubert and Heinrich, 1993) in spite of less flesh pigmentation than by synthetic astaxanthin, due the esterified form of astaxanthin and a low availability of the pigment inside the alga spore. However, Gomes *et al.* (2002) proved their efficiency on skin pigmentation of gilthead seabream and Gouveia *et al.* (2003, 2005) in ornamental goldfish and koi carp.

Chlorella vulgaris biomass proved to be efficient, comparable with synthetic astaxanthin and canthaxanthin, for pigmentation purposes, in rainbow trout (Gouveia *et al.*, 1996c, 1997, 1998), gilthead seabream (Gouveia *et al.*, 2002), ornamental goldfish and koi carp (Gouveia *et al.* 2003, 2005) and shrimps (Passos *et al.*, in preparation).

Spirulina (rich in β-carotene) is usually used in aquaculture feeds up to 5-20% as a fish and shrimp feed (Benemann, 1992) and to enhances the red and yellow patterns in carp while leaving a brilliant white colour (Gouveia *et al.*, 2003, 2005; Spolaore *et al.*, 2006) and in ornamental goldfish (Gouveia *et al.*, 2003, 2005).

Haslea ostrearia, a diatom, induces a blue-green colour on the gills and labial palps of oysters, which increase market's value by 40% (Spolaore *et al.*, 2006).

5. Microalgae in Human Nutrition

In early 1950's microalgae were considered to be a good supplement and/or fortification in diets for malnourished children and adults, as a single cell protein but nowadays microalgae for human nutrition is marketed in different forms of tablets, capsules and liquids (Spolaore *et al.*, 2006).

Some nutritional studies were done with humans and the authors suggest that the algae daily consumption should be restricted to about 20 g, with no harmful side effects occurence, even after a prolonged period of intake (Becker, 1988). Gross *et al.* (1978) performed a study feeding algae (*Scenedesmus obliquus*) to children (5 g/daily) and adults (10 g/daily), incorporated into their normal diet, during four-week test period. Hematological data, urine, serum protein, uric acid concentration and weight changes were measured, and no changes in the analyzed parameters were found, except a slight increase in weight, especially important for children. The same authors also carried out a study with a slightly (group I) and seriously (group II) malnourished infant during three weeks. The four-years-old children of group I (10 g algae/daily) showed a significant increase in weight (27 g/day) compared with the other children of the same group who received a normal diet, and no adverse symptoms were recorded. The second group was nourished with a diet enriched with 0.87 g algae/kg body weight, substituting only 8% of the total protein and the daily increase in weight was about sevenfold (in spite of a low protein contribution) and all anthropogenic parameters were normal. The authors concluded that the significant improvement in the state of the health was attributed not only to the algal protein but also to therapeutic factors.

However, adults are very resistant in the acceptation of novel foods with microalgae incorporation, which was demonstrated by Feldheim (1972) and Gross and Gross (1978) because it often affects conservative ethnic factors, including religious and socio-economic aspects (Becker, 1994) being much easier with children's who are more willing to accept uncommon preparations. This was demonstrated in Mexico, where a beverage formed by 50% of a suspension of *Spirulina* ("green milk") was given, without problems, as bottle feed to babies (Jacket, 1974).

6. New Trends in Microalgae Food Applications

All over the world commercial production of microalgae for human nutrition is already a reality. Numerous combinations of microalgae or mixtures with other health foods can be found in the market in the form of tablets, powders, capsules, pastilles and liquids, as nutritional supplements (Table 2). They can also be incorporated into food products (*e.g.* pastas, biscuits, bread, snack foods, candies, yoghurts, soft drinks), providing the health-promoting effects that are associated with microalgal biomass, probably related to a general immune-modulating effect (Belay, 1993). In spite of some reluctance for novel foods in the past, nowadays there is an increasing consumer demand for more natural food products, presenting health benefits. Functional foods supplemented with microalgae biomass are sensorily much more convenient and variable, thus combining health benefits with attractiveness to consumers (Pulz and Gross, 2004). In some countries (Germany, France, Japan, USA, China, Thailand), food production and distribution companies have already started serious activities to market functional foods with microalgae and cyanobacteria (Pulz and Gross, 2004). Food safety regulations for human consumption are the main constraint for the biotechnological exploitation of microalgal resources, but successful cases such as the approval (9 December 2002) of the marine diatom *Odontella aurita* by Innovalg (France) as a novel food, following EC Regulation 258/97, broadens perspectives.

In the last years, our research group in Portugal aims to develop a range of novel attractive healthy foods, prepared from microalgae biomass, rich in carotenoids and polyunsaturated fatty acids with antioxidant effect and other beneficial properties. At the same time toxicological studies involving all the microalgae to be incorporated are also been conducted. Traditional foods, like mayonnaises, gelled desserts, biscuits, pasta and breakfast cereals, largely consumed on daily basis on different European diets, can be used as vehicles to those nutraceuticals. This strategy avoids the hassled of changing food habits; considering that Europeans are getting older and have strong cultural motivations, being highly resistant to food innovations. The impact of natural substances introduced in the diet via "*usual*" foods is proved to be efficient at long term and do not present the drawbacks of traditional therapeutic actions based on medicines of short term impact.

Table 2. Major microalgae commercialized for human nutrition (Adapted from Pulz and Gross, 2004; Spolaore *et al.*, 2006 and Hallmann, 2007)

Microalga	Major Producers	Products	World production (t/year)
Spirulina (*Arthrosphira*)	Hainan Simai Pharmacy Co. (China) Earthrise Nutritionals (California, USA) Cyanotech Corp. (Hawaii, USA) Myanmar Spirulina factory (Myanmar)	powders, extracts tablets, powders, extracts tablets, powders, beverages, extracts tablets, chips, pasta and liquid extract	3000
Chlorella	Taiwan Chlorella Manufacturing Co. (Taiwan) Klötze (Germany)	tablets, powders, nectar, noodles powders	2000
Dunaliella salina	Cognis Nutrition and Health (Australia)	powders β-carotene	1200
Aphanizomenon flos-aquae	Blue Green Foods (USA) Vision (USA)	capsules, crystals powder, capsules, crystals	500

The viability of incorporating microalgal biomass in food systems is conditioned by the applied processing type and intensity (e.g. thermal, mechanical), by the nature of the food matrix (e.g. emulsion, gel, aerated dough systems) and to the interactions with other food components (e.g. proteins, polysaccharides, lipids, sugars, salts). Besides colouring and nutritional purposes, introducing microalgal ingredients in food systems, can also impart significant changes in its microstructure and rheological properties (Batista *et al.*, 2006a). These aspects are particularly focused in our research.

6.1. Oil-in-Water Emulsions

The development of coloured oil-in-water emulsions using natural sources, especially from microalgal origin, is an interesting field to investigate. The attainment of appealing and stable colourations is an important innovation for these types of products. Due to the antioxidant properties that most natural pigments present it is also possible to improve the resistance to oil oxidation, which is particularly advantageous in high fat products like emulsions.

6.1.1. Emulsions Coloured with Natural Pigments

The addition of natural pigments, typically present in microalgae, to oil-in-water (o/w) emulsions was studied by Batista *et al.* (2006a, 2006b). The emulsions were prepared with 3% (w/w) pea protein isolate and 65% (w/w) vegetable oil, according to previous studies that successfully replaced egg yolk protein by leguminous proteins in o/w emulsions (Raymundo *et al.*, 2002). Commercial lutein oil dispersion (FloraGlo®, Kemin, USA) and phycocyanin extracted from *Sprirulina* (*Arthrospira*) *maxima* laboratory cultures (INETI, Portugal) (Reis *et al.*, 1998) were used, at concentrations ranging from 0.25% to 1.25% (w/w). Emulsions containing both pigments, in different proportions (total pigment concentration of 0.5% w/w) were also prepared. Regarding carotenoids lipophilic character, lutein was added to the emulsions dispersed oil phase while phycocyanin, being an hydrophilic proteinaceous pigment, was added to the continuous aqueous phase, prior to the emulsification process.

Lutein (yellow) and Phycocyanin (blue) imparted appealing and innovative colourations to food emulsions, as can be observed in Figure 5. However, the addition of these pigments had significant implications on the emulsions structural and rheological properties. The effects were markedly different for the two pigments used. Their distribution between the continuous (aqueous) and dispersed (oil) phase and its interactions with the emulsifier molecules at the interface seems to be of major importance (Batista *et al.*, 2006b).

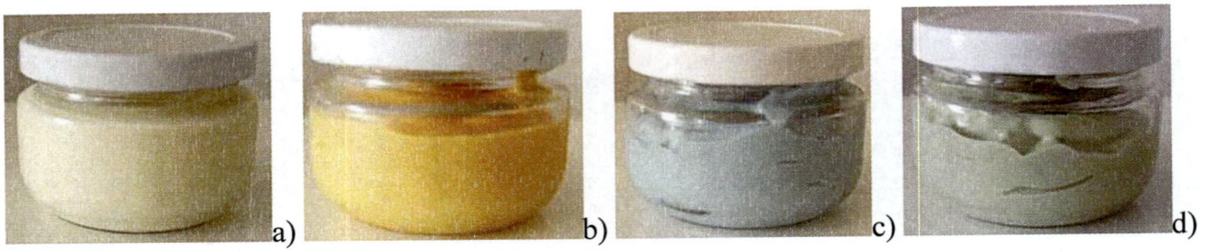

Figure 5. Oil-in-water (o/w) pea protein-stabilized emulsions, a) without pigment addition (control), b) with 0.50% (w/w) lutein, c) with 0.50% (w/w) phycocyanin, d) with both pigments in equal proportion 50L:50P (0.50% total piment).

The addition of lutein had a negative impact on the emulsion microstructure and rheological characteristics (Figure 6), although there were no significant differences between samples with different lutein concentrations. Adding lutein to the oil fraction could have modified the nature of the emulsions' dispersed phase, namely the strength of the attractive interactions between molecules and the effectiveness of their packing in the condensed phase (McClemments, 1999). Recent studies (Granger *et al.*, 2003; Rampon *et al.*, 2004) have suggested that not only the surfactant molecules, *i.e.* emulsifiers and proteins, but also the fat used in the emulsions formulation participates in the development of the interface characteristics and rheological properties. Lutein molecules are mainly lipophylic molecules but present polar hydroxyl groups in both ends of the conjugated polyisoprenoid chain, so it is possible to interact with hydrophylic domains of the pea protein emulsifier, creating weaker and disordered layers. Santipanichwong and Suphantarika (2007) also reported emulsion destabilization by the addition of lutein in reduced-fat mayonnaises with spent brewers' yeast as fat replacer.

On the other hand, phycocyanin addition resulted in a significant improvement of the emulsions rheological properties (Figure 6) which increased linearly with phycocyanin concentration (Batista *et al.*, 2006a). The presence of phycocyanin protein molecules may have contributed to a marked increase in the viscosity of the aqueous continuous phase, thus retarding the oil droplet association movements and consequently enhancing emulsion stability. It is also possible that phycocyanin protein molecules interact in the interfacial protein adsorbed layer at the surface of oil droplets, reinforcing in this case the pea protein emulsifier film and imparting stability to emulsions. In fact, previous studies (Chronakis *et al.*, 2000) have demonstrated that a protein isolate from blue-green algae (*Spirulina platensis* strain Pacifica), containing phycocyanin, was capable of reducing the interfacial tension at the aqueous/air interface at relatively lower bulk concentrations compared to common food proteins.

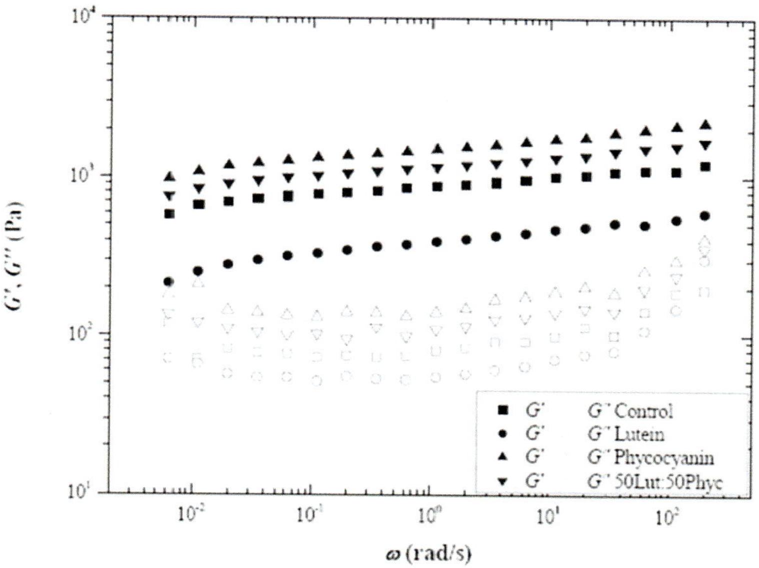

Figure 6. Mechanical spectra of o/w emulsions without pigment addition (control), with 0. 50% lutein, 0.50% phycocyanin, and with both pigments in equal proportion 50L:50P (0.50% total pigment).

When using combinations of both pigments, an increase of the rheological and parameters with phycocyanin proportion was apparent, and a synergetic effect was observed when using small amounts (< 50% proportion) of lutein.

6.1.2. Emulsions Coloured with Microalgal Biomass

The use of the microalgae *Haematococcus pluvialis* (carotenogenic) and *Chlorella vulgaris* (before and after carotenogenesis) to colour oil-in-water pea protein-stabilized emulsions was also investigated by the authors (Gouveia *et al.*, 2006), obtaining a wide range of attractive and stable colours (Figure 7). These microalgae were cultivated in the Biomass Unit of the Department of Renewable Energies from INETI (Portugal).

Figure 7. Oil-in-water (o/w) pea protein-stabilized emulsion with 0.25%, 0.50% and 0.75% (w/w) (from left to right) of *Haematococcus pluvialis* (top) and *Chlorella vulgaris* biomass (carotenogenic) (bottom).

The colour stability of the emulsions was evaluated, through the evolution of the L*a*b* parameters (CIELAB system) along six weeks. The primary and secondary oxidation products of the emulsions were also determined, and an enhanced resistance to oxidation was evidenced by emulsions containing microalgae (Gouveia *et al.*, 2006). The incorporation of *Haematococcus pluvialis* provided higher oxidation stability over time, in comparison with *Chlorella vulgaris.* It should be considered that during carotenogenesis *Haematococcus pluvialis* accumulates mainly astaxanthin while canthaxanthin is the dominant carotenoid in *Chlorella vulgaris*. The higher oxidation stability of astaxanthin as already been reported, and is related to the fact that antioxidant effectiveness of carotenoids increases as the number of the conjugated double bounds of carotenoids increased (Yen and Chen, 1995). However, microalgal biomass may be considered as multi-component antioxidant systems, which are generally more effective due to synergistic or additive interactions between the different antioxidant components.

The addition of microalgal components improved the emulsion texture parameters which should be related with a higher stability level. It can also be observed that 0.75% (w/w) biomass seems to be an optimal concentration level, since the three emulsions presented similar firmness values (2.3-2.5 N) (Figure 8). At higher concentrations the emulsions became excessively firm, which could be related to an increase on the viscosity of the aqueous phase.

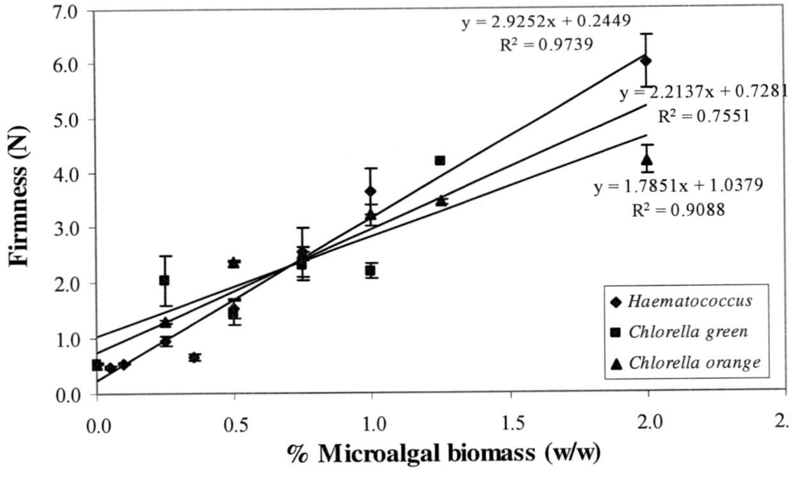

Figure 8. Firmness values of oil-in-water pea protein-stabilized food emulsions coloured with different concentrations of *Haematococcus pluvialis*, *Chlorella vulgaris* green and *Chlorella vulgaris* orange biomass.

The capacity of the *Chlorella vulgaris* biomass as a fat mimetic, and its emulsifier ability, has also been studied (Raymundo *et al.*, 2005). Pea protein emulsions with *Chlorella vulgaris* addition (both green and orange - carotenogenic) were prepared at different protein (2-5% w/w) and oil (50-65% w/w) contents, characterized in terms of rheological behaviour. It was observed that emulsions with 55% oil and 2% microalga were more structured than the emulsions with 65% oil and no microalgal biomass addition (Figure 9). This behaviour can be explained by the increase of the viscosity of the continuous phase of the emulsion, by the microalgal material. This result supports the potential use of microalgae material to act as a fat mimetic, besides the possible advantages as colouring and antioxidant agent. The development of the emulsion structure did not occur when microalgal biomass fully replaced the vegetable protein as an emulsifier, and phase separation was instantaneous.

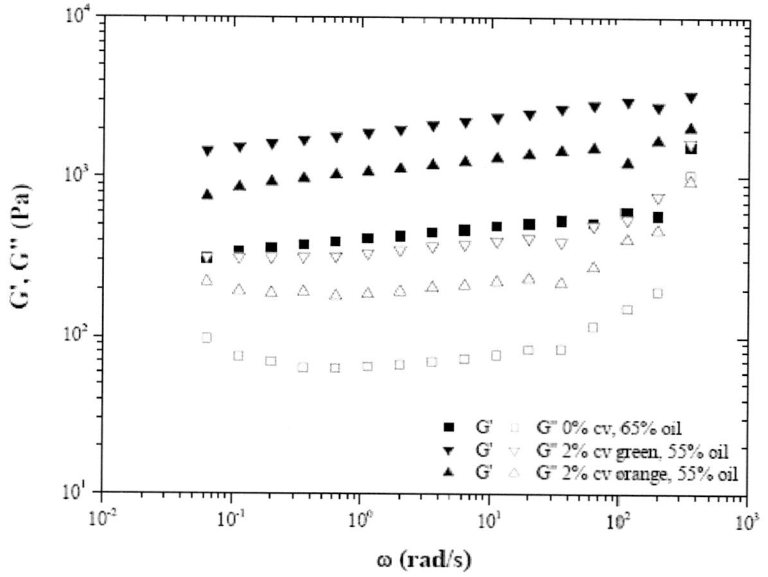

Figure 9. Mechanical spectra of pea protein-stabilized o/w emulsions with and without 2% w/w Chlorella green and orange biomass, at different oil contents.

6.2. Biscuits

Short dough cookies and biscuits are widely consumed food products, appreciated for their taste, versatility, convenience, conservation, texture and appearance. The use of natural ingredients, exhibiting functional properties and providing specific health benefits beyond traditional nutrients, is a very attractive way to design new food products, with an important market niche presently exhibiting pronounced growth.

6.2.1. Biscuits Coloured with Chlorella vulgaris Biomass

A study was undertaken to determine the effects of adding *Chlorella vulgaris* biomass as a colouring ingredient in traditional butter biscuits (Gouveia *et al.*, 2007a). The cookies were manufactured at a pilot scale, according to an optimized formulations from previous studies (Piteira *et al.*, 2004), and stored for three months at room temperature, protected from light and air.

Chlorella vulgaris biscuits presented an accentuated green tonality (Figure 10), which increased with the amount of added biomass. In general, colour parameters (CIELAB system) remained very stable along the storage period. However, it seems not necessarily to use biomass concentrations above 1% (w/w), since the green tonality (-a*) differences are no longer significant ($p<0.05$), and higher algal concentrations are related with some colour variations along time (Figure 11a).

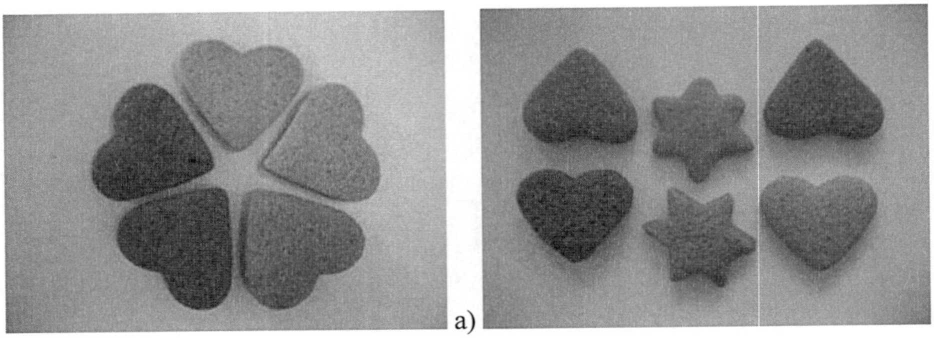

Figure 10. Biscuits with *Chlorella vulgaris* biomass, a) at various concentration levels (0.0-3.0%), b) and in comparison with Haematococcus pluvialis (pink) and *Chlorella vulgaris* (orange) carotenogenic biomass addition.

The texture profile of the biscuits was also evaluated, and a significant increase of their firmness was evidenced with an increase of added microalgal biomass (Figure 11b). These results evidence the positive effect of the alga in the biscuit structure, reinforcing the short dough system. Biscuits are considered solid emulsions of sucrose, lipids and non-gelatinized starch (Hoseney *et al.*, 1988), being this morphology is responsible for the biscuits structure and texture. The main factor affecting these properties is the moisture content and water mobility, which is highly affected by the interaction with hydroxyl groups present in the matrix (Hoseney *et al.*, 1988). The replacement of a small amount of flour by microalgae biomass, resulted in the inclusion of a complex biomaterial, rich in different proteins and polysaccharides. These molecules have an important role on the water absorption process, which promotes the increase of biscuits firmness, resulting in more compact structures.

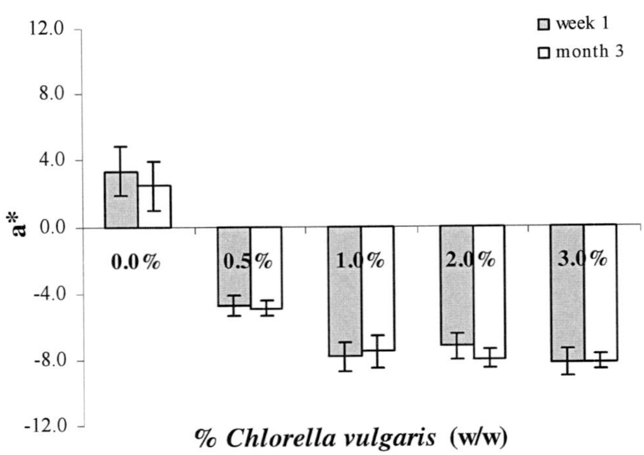

a)

Figure 11. Continued on next page.

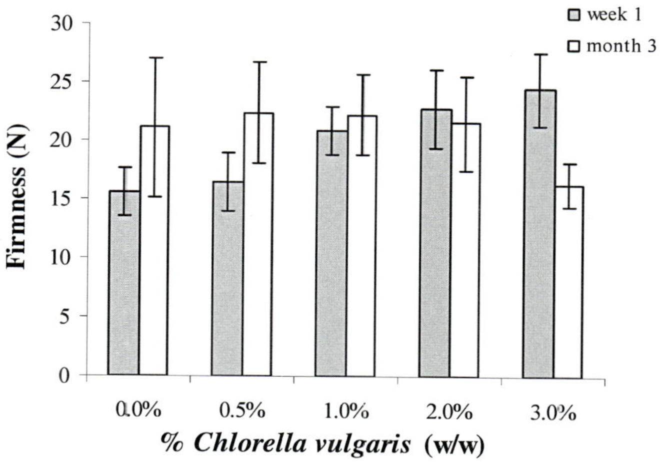

b)

Figure 11. Green chromaticity a* (a) and firmness values (b) of biscuits with different concentrations of *Chlorella vulgaris* biomass, after one week and three months storage.

6.2.2. Biscuits with Isochrysis galbana Biomass Rich in Polyunsaturated Fatty Acids

A similar study was performed, this time using *Isochrysis galbana* biomass (Gouveia *et al.*, 2007b), that was cultivated in the Department of Aquaculture of IPIMAR (Portugal). An enhancement of the biscuits texture properties and high stability of colour and texture along three months storage was observed, as previously reported for *Chlorella* biscuits (Gouveia *et al.*, 2007a). The biscuits presented quite different tonalities, turning from green to a brownish and duller colour when increasing the biomass concentration from 1.0% to 3.0% (Figure 12).

Figure 12. Biscuits with different incorporation levels (0.0%, 1.0%, 3.0%) of biomass.

The main interest in using *Isochrysis galbana* biomass is due to its high levels of long chain omega-3 polyunsaturated fatty acids, especially EPA and DHA (Bandarra *et al.*, 2003). The biscuits fatty acids profile is clearly related to butter (Özkanli and Kaya, 2007), with predominance of saturated (~60%) and monounsaturated fatty acids (~30%), mainly palmitic acid (30-40%) and oleic acid (18:1ω9) (20-25%), respectively. Polyunsaturated fatty acids corresponded to 6-9% (4-5% linoleic acid; 18:2ω6), the highest levels being for 3% *Ig* biscuits (55% linoleic acid, 15% EPA, 6% α-linolenic acid and 3% DHA).

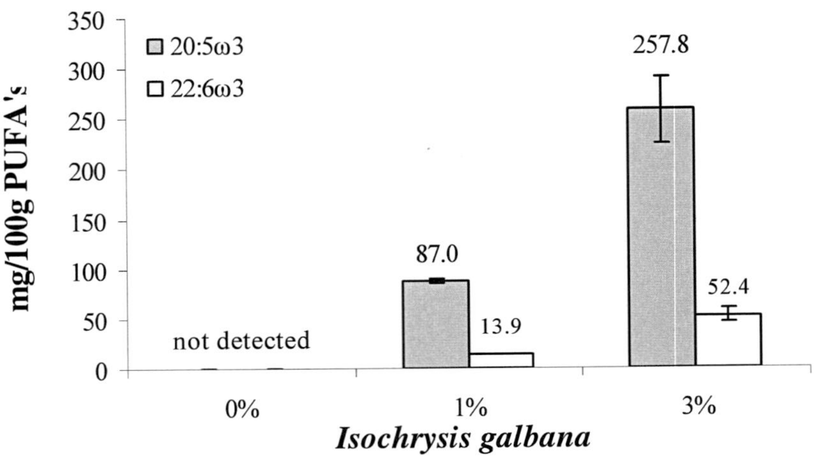

Figure 13. Omega-3 polyunsaturated fatty acids, of biscuits with 0%, 1% and 3% *Isochrysis galbana* biomass incorporation.

In spite of the drastic thermal processing (high temperatures) during biscuits manufacturing, the addition of microalgal biomass leads to the presence of ω3 fatty acids (absent in control biscuits) which remain stable along storage time (Figure 13). The thermal resistance of fatty acids should be due to its presence in an encapsulated form, inside the microalga cell. Ig biscuits presented LC-PUFA-ω3 levels (EPA+ DHA) of 100 mg/100g and 30 mg/100g biscuit, for 1% and 3% microalgal biomass incorporation, respectively (Figure 13). These values reflect an important source of PUFA-ω3 with a moderate biscuit consumption, as the recommendations for dietary intake in healthy adults is 500 mg/day (ISSFAL, 2004).

6.3. Food Gels

Most recently, our group is studying the incorporation microalgal biomass in food gelled products, based on protein and polysaccharide mixed biopolymer systems. Gelled vegetable desserts, alternative to dairy desserts, with pea protein isolate (4% w/w), κ-carrageenan (0.15%) and starch (2.5%), optimized in previous studies (Nunes *et al.*, 2003, 2006) were used as model systems. The gels were prepared with different microalgae - *Spirulina (Arthrosphira) maxima*, *Chlorella vulgaris* (green and orange, after carotenogenesis), *Haematococcus pluvialis* (red, carotenogenic) and *Diacronema vlkianum* – were evaluated in terms of colour and texture and compared with gels prepared with commercial pigments – phycocyanin, astaxanthin, β-carotene, canthaxanthin and lutein. The microalgae gels imparted less intense tonalities (Figure 14) and texture modifications, compared to the other colour pigments addition.

The pea protein/κ-carrageenan/starch mixed gel systems with 0.75% (w/w) microalgal biomass addition were characterized in terms of rheological behaviour, including monitoring of the viscoelastic functions (G', G") during gelification (cooling process) and maturation kinetics (Batista *et al.*, 2007).

(a) (b)

Figure 14. Gelled vegetable desserts with incorporation of microalgae biomass (a) and commercial pigments (b).

The incorporation of these biomaterials seemed to be beneficial, especially for *Haematococcus pluvialis* which promoted a structural reinforcement expressed by improved rheological properties (Figure 15). This may be related to its significantly higher fat content (40.7%) in relation to other microalgae. The influence of fat content on gelling behaviour has been studied in milk gelled systems (Houzé *et al.*, 2005; Vélez-Ruiz *et al.*, 2005) being concluded that using high fat milk rather than skim milk results in stronger gels, which is usually attributed to fat droplets acting as active filler particles embedded in the protein matrix.

However, the addition of *Spirulina* promoted a drastic reduction on the gels rheological parameters (Figure 15), which should be related with a thermodynamic incompatibility between the microalgal protein and other components of the mixed gelled system. In fact, Chronakis (2001) reported that proteins isolated from *Spirulina* are quite intricate biomaterials, likely to be protein and/or protein-pigment (phycocyanin) complexes rather than individual protein molecules. Therefore, *Spirulina* denaturation and gel formation is a complex phenomenon *per se*, which can probably interfere with the gelling process of the biopolymers present in the mixed gel system.

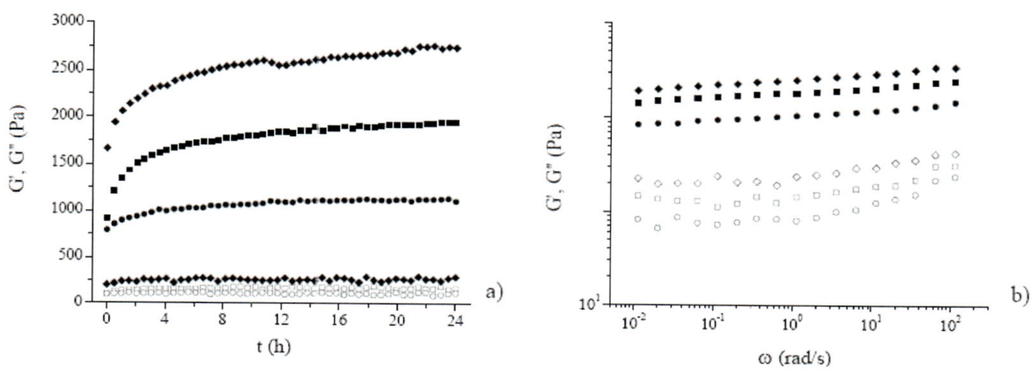

Figure 15. Maturation kinetic curves (a) and mechanical spectra (b), at 5°C, of pea protein/κ-carrageenan/starch gels (■) with 0.75% (w/w) *Haematococcus pluvialis* (♦) and *Spirulina maxima* (●) biomass. G' (filled symbol), G" (open symbol).

Further research is required in order to better understand the gelation mechanism of these microalgae and the specific interactions with each biopolymer present in the complex mixed gel system, as well as the influence of processing conditions (e.g. temperature, heating and cooling rates).

Conclusion

The combination of the exceptional nutritional value of microalgae with colouring and therapeutical properties, associated with an increase demand of natural products, make microalgae worth exploring for utilization in the future in feed, food, cosmetic and pharmaceutical industries, with recognized advantages over the traditional ingredients.

In the actual scenario with multiple therapeutics, many believe that simple dietary interventions or nutritional supplements may be more natural, acceptable and feasible method of providing health benefits.

The choise of the right food to eat in an early stage of life associated with a healthy lifestyle can have important benefits in future life. A healthy diet based on microalgae novel food products can have important benefits for all age groups.

The great results obtained by the authors in the preparation of common food products with microalgae incorporation providing attractive and healthier food with enormous potential as a functional food ingredient.

In the near future, the authors' intent to continue the development of healthier food products, preparing other widespread food products, such as pasta, salt crackers, extruded products with the incorporation of microalgal biomass, as vehicles of functional ingredients, namely pigments, antioxidants and PUFA.

Acknowledgements

This work is part of a research *project "Pigments, antioxidants and PUFA's in microalgae based food products – functional implications"* (PTDC/AGR-ALI/65926/2006) sponsored by the Portuguese Foundation for the Science and Technology (*"Fundação para a Ciência e a Tecnologia"* - FCT). A.P. Batista acknowledges the PhD research grant from FCT (SFRH/21388/BD/2005).

References

Aasen, A.J., Eimhjellen, K.E., and Liaaen-Jensen, S. (1969). An extreme source of β-carotene. *Acta Chemica Scandinavica*, **23**, 2544-2545.

Apt, K.E., and Behrens, P.W. (1999). Commercial developments in microalgal biotechnology. *Journal of Phycology*, **35**, 215-226.

Arteburn, L.A., Oken, H.A., Hoffman, J.P., Bailey-Hall, E., Cheng, G., Rom, D., Hamersley, J., and McCarthy, D. Bioequivalence of docosahexaenoic acid from different algal oils in capsules and in DHA-fortified food. *Lipids*, In press (DOI: 10.1007/s11745-007-3098-5).

Aruoma, O.I. (2003). Methodological considerations for characterizing potential antioxidant actions of bioactive components in plant foods. *Mutation Research*, **523**, 9-20.

ATBC Cancer Prevention Study Group (1994). The effect of vitamin E and beta carotene on the incidence of lung cancer and other cancers in male smokers. *The New England Journal of Medicine*, **330**, 1029-1035.

Avron, M., Ben-Amotz, A., and Edelstein, S. (1952). Use of *Dunalliella bardawil* as a source of vitamin A and a yolk-color enhancing agent. Commonwealth of Australia Patent Act 71864/87.

Bandarra, N.M., Pereira, P.A., Batista, I., and Vilela, M.H. (2003). Fatty acids, sterols and α – tocopherol in *Isochrysis galbana*. *Journal of Food Lipids*, **18**, 25-34.

Baker, R., and Gunther, C. (2004). The role of carotenoids in consumer choice and the likely benefits from their inclusion into products for human consumption. *Trends in Food Science and Technology*, **15**, 484-488.

Balder, HF, Vogel, J., Jansen, M.C., Weijenberg, M.P., van den Brandt, P.A., Westenbrink, S., van der Meer, R., and Goldbohm, R.A. (2006). Heme and chlorophyll intake and risk of colorectal cancer in the Netherlands cohort study. *Cancer Epidemiology Biomarkers and Prevention*, **15**,717-725.

Bar, E., Rise, M., Vishkautsan, M., and Arad, S. (1995). Pigment and structural changes in *Chlorella zofingiensis* upon light and nitrogen stress. *Journal of Plant Physiology*, **146**, 527-534.

Batista, A.P., Raymundo, A., Sousa, I., and Empis, J. (2006a). Rheological characterization of coloured oil-in-water food emulsions with lutein and phycocyanin added to the oil and aqueous phases. *Food Hydrocolloids*, **20**, 44-52.

Batista, A.P., Raymundo, A., Sousa, I., Empis, J., and Franco, J.M. (2006b). Colored food emulsions – implications of pigment addition on the rheological behaviour and microstructure. *Food Biophysics*, **1,** 216-227.

Batista, A.P., Gouveia, L., Nunes, M.C., Franco, J.M., and Raymundo, A. (2007). Microalgae biomass as a novel functional ingredient in mixed gel systems. In Gums *and Stabilisers for the Food Industry* – 14th Edition. Eds. P.A. Williams, G.O. Phillips. Royal Society of Chemistry. *In press.*

Becker, E.W. (1988). Micro-algae for human and animal consumption. In M.M. Borowitzka, and L.J. Borowitzka (Eds), *Micro-algal biotechnology* (pp. 222-256). Cambridge University Press.

Becker, E.W. (1994). *Microalgae: biotechnology and microbiology*. Cambridge University Press.

Becker, E.W. (2004). Microalgae in human and animal nutrition. In A. Richmond (Ed), *Handbook of microalgal culture* (pp. 312-351). Oxford: Blackwell.

Beckett, B.R., and Petkovich, M. (1999). Evolutionary conservation in retinoid signalling and metabolism. *American Zoology*, *39*, 783-795.

Belay, A. (1993). Current knowledge on potential health benefits of *Spirulina platensis*. *Journal of Applied Phycology*, **5**, 235-240.

Ben-Amotz, A., and Avron, M. (1980). Glicerol, β-carotene and dry algal meal production by commercial cultivation of *Dunaliella*. In G. Shelef, and C.J. Soeder (Eds), *Algae Biomass* (pp. 603-610). Amsterdam: Elsevier/North Holland Biomedical Press.

Ben-Amotz, A., Gressel, J., and Avron, M. (1987). Massive accumulation of phytoene induced by norflurazon in *Dunaliella bardawil* (Chlorophyceae) prevents recovery from photoinhibition. *Journal of Phycology*, **23**, 176-181.

Ben-Amotz, A., and Fishler, R. (1998). Analysis of carotenoids with emphasis on 9-*cis*-β-carotene in vegetables and fruits commonly consumed in Israel. *Food Chemistry*, **62**, 515-520.

Benedetti, S., Benvenuti, F., Pagliarani, S., Francogli, S., Scoglio, S., and Canestrari, F. (2004). Antioxidant properties of a novel phycocyanin extract from the blue-green alga *Aphanizomenon flos-aquae. Life Sciences*, **55**, 2353-2362.

Benemann, J.R. (1992). Microalgae aquaculture feeds. *Journal of Applied Phycology*, **4**, 233-245.

Bhat, V.B., and Madyastha, K.M. (2000). C-Phycocyanin: a potent peroxyl radical scavenger in vivo and in vitro. *Biochemical and Biophysical Research Communications*, **275**, 20-25.

Bhosale, P. (2004). Environmental and cultural stimulants in the production of carotenoids from microorganisms. *Applied Microbiology and Biotechnology*, **63**, 351-361.

Blomhoff, R., Green, M.H., and Norum, K.R. (1992). Vitamin A: physiological and biochemical processing. *Annual Reviews in Nutrition*, **12**, 37-57.

Bon, J.A., Leathers, T.D., and Jayaswal, R.K. (1997). Isolation of astaxanthin-overproducing mutants of *Phaffia rhodozyma*. *Biotechnology Letters*, **19**, 109-112.

Bønaa, K.H., Bjerve, K.S., Straume, B., Gram, I.T., and Thelle, D. (1990) Effect of eicosapentaenoic and docosahexaenoic acids on blood pressure in hypertension: a population-based intervention trial from the Tromsø study. *New England Journal of Medicine*, **322**, 795-801.

Borowitzka, M.A. (1988). Vitamins and fine chemicals from micro-algae. In M.A. Borowitzka, and L.J. Borowitzka (Eds), *Micro-algal biotechnology* (pp. 153-196). Cambridge, UK: Cambridge University Press.

Borowitzka, M.A. (1997). Microalgae for aquaculture: opportunities and constraints. *Journal of applied phycology*, **9**, 393-401.

Borowitzka, M.A. (1999). Commercial production of microalgae: ponds, tanks, tubes and fermenters. *Journal of Biotechnology*, **70**, 313-321.

Breithaupt, D.E. (2007). Modern application of xanthophylls in animal feeding - a review. *Trends in Food Science and Technology*, **18**, 501-506.

Brown, M.R., Jeffrey, S.W., Volkman, J.K., and Dunstan, G.A. (1997). Nutritional properties of microalgae for mariculture. *Aquaculture*, **151**, 315-331.

Brown, M.R., Mular, M., Miller, I., Farmer, C., and Trenerry, C. (1999). The vitamin content of microalgae used in aquaculture. *Journal of Applied Phycology*, **11**, 247-255.

Burdge, G.C., and Calder, P.C. (2005). alfa-Linolenic acid metabolism in adult humans:the effects of gender and age on conversion to longer-chain polyunsaturated fatty acids. *European Journal of Lipid Science and Technology*, **107**, 426–439.

Burja, A.M., Banaigs, B., Abou-Mansour, E., Burgess, J.G., Wright, P.C. (2001). Marine cyanobacteria - a prolific source of natural products. *Tetrahedron*, **57**, 9347-9377.

Cahu, C.L., and Zambonino-Infante, J.L. (1998). Algal addition in sea bass (*Dicentrarchus labrax*) larvae rearing: effect on digestive enzymes. *Aquaculture*, **161**, 479-489.

Calder, P.C. (2004). Review - n–3 Fatty acids and cardiovascular disease: evidence explained and mechanisms explored. *Clinical Science*, **107**, 1–11.

Calderon, C.J.F., Merino, Z.H., and Barragán, M.D. (1976). Valor alimentico del alga espirulina (*Spirulina geitleri*) para ruminants. *Tecnica Pecuaria en Mexico*, **31**, 42-46.

Campbell, C.J. (1997). *The Coming Oil Crisis*. Essex, England: Multi-Science Publishing Company and Petroconsultants.

Certik, M., and Shimizu, S. (1999). Biosíntesis and regulation of microbial polyunsaturated fatty acid production. *Journal of Biosciences and Bioengineering*, **87**, 1-14.

Chaumont, D. (1993). Biotechnology of algal biomass production: a review of systems for outdoor mass culture. *Journal of Applied Phycology*, **5**, 593-604.

Chen, Y.-C. (2003). Immobilized *Isochrysis galbana* (Haptophyta) for long-term storage and applications for feed and water quality control in clam (*Meretrix lusoria*) cultures. *Journal of Applied Phycology*, **15**, 439-444.

Chien, Y.H., and Jeng, S.C. (1992). Pigmentation of kuruma prawn, *Panaeus japonicus* Bate, by various pigment sources and levels and feeding regimes. *Aquaculture*, **102**, 333-346.

Chini Zittelli, G., Lavista, F., Bastianini, A., Rodolfi, L., Vincenzini, M., and Tredici, M.R. (1999). Production of eicosapentaenoic acid by *Nannochloropsis* sp. cultures in outdoor tubular photobioreactors. *Journal of Biotechnology*, **70**, 299-312.

Choubert, G., and Heinrich, O. (1993) Carotenoid-pigments of the green-alga *Haematococcus pluvialis* – assay on rainbow trout, *Oncorhynchus mykiss*, pigmentation in comparison with synthetic astaxanthin and canthaxanthin. *Aquaculture*, **112**, 217-226.

Christiansen, R., Glette, J., Lie, O., Torrissen, O.J., and Waagbo, R. (1995). Antioxidant status and immunity in Atlantic salmon, *Salmo salar*, fed semi-purified diets with and without astaxanthin supplementation. *Journal of Fish Diseases*, **18**, 317-328.

Chronakis, I.S., Galatanu, A.N., Nylander, T., and Lindman, B. (2000). The behaviour of protein preparations from blue-green algae (*Spirulina platensis* strain Pacifica) at the air/water interface. *Colloids and Surfaces A: Physicochemical and Engineering Aspects*, **173**, 181-192.

Chronakis, I.S. (2001). Gelation of edible blue-green algae protein isolate (*Spirulina platensis* strain Pacifica): thermal transitions, rheological properties, and molecular forces involved. *Journal of Agricultural and Food Chemistry*, **49**, 888-898.

Clark, D.C., Wilde, P.J., Wilson, D.R., and Wustneck, R. (1992). The interaction of sucrose esters with β-lactoglobulin and α-casein from bovine milk. *Food Hydrocolloids*, **6**, 173-186.

Crawford, M. (2000). Placental delivery of arachidonic acid and docosahexaenoic acids: implication of the lipid nutrition of preterm infants. *American Journal of Clinical Nutrition*, **71**, 275S-284S.

Cui, J.G., Wang, Y., Lin, G.W., and Zeng, L.M. (2000). Polyhydroxylated sterols with biological activities in marine organism. *Natural Products Research Developments, 12*, 94-100.

Davis, I.F., Sharkey, M.J., and Williams, D. (1975). Utilization of sewage algae in association with paper in diets of sheep. *Agriculture and Environment*, **2**, 333-338.

Donato, M., Vilela, M.H., and Bandarra, N.M. (2003). Fatty acids, sterols, α-tocopherol and total carotenoids composition of *Diacronema vlkianum*. *Journal of Food Lipids*, **10**, 267-276.

Duerr, E.O., Molnar, A., and Sato, V. (1998). Cultured microalgae as aquaculture feeds. *Journal of Marine Biotechnology*, **75**, 65-70.

Dufossé, L., Galaup, P., Yaron, A., Arad, S.M., Blanc, P., Murthy, K.N.C., and Ravishankar, G.A. (2005). Microorganisms and microalgae as sources of pigments for food use: a scientific oddity or an industrial reality?. *Trends in Food Science and Technology*, **16**, 389-406.

Dutta, D., De, D., Chaudhuri, S., and Bhattacharya, S.K. (2005). Hydrogen production by Cyanobacteria. *Microbial Cell Factories*, **4**, 36-46.

Fábregas, J., Otero, A., Dominguez, A., and Patino, M. (2001). Growth rate of the microalga *Tetraselmis suecica* changes the biochemical composition of *Artemia* species. *Marine Biotechnology*, **3**, 256-263.

Faure, H., Fayol, V., Galabert, C., Grolier, P., Moel, G.L., Steghens, J., Kappel, A.V., Nabet, F. (1999). Carotenoids: 1. Metabolism and physiology. *Annales de Biologie Clinique*, **57**, 169-183.

Feldheim, W. (1972). Studies on use of microalgae in human nutrition. 1. Nutrition tests with algae containing diets in Thailand. *International Journal for Vitamin and Nutrition Research*, **42**, 600-606. Cited by Becker, E.W. (1988).

Ferruzi, M.G., and Blakeslee, J. (2007). Digestion, absorption, and cancer preventive activity of dietary chlorophyll derivatives. *Nutrition Research*, **27**, 1-12.

Fevrier, C., and Seve, B. (1975). Incorporation of *Spirulina maxima* in pig diets. *Annales de la Nutrition et de l'Alimentation*, **29**, 625-650.

Fidalgo, J.P., Cid, A.T.E., Sukenik, A., and Herrero, C. (1998). Effects of nitrogen source and growth phase on proximate biochemical composition, lipid classes and fatty acid profile of the marine microalgae *Isochrysis galbana. Aquaculture*, **166**, 105-116.

Ghys, A., Bakkere, E., Hornstra, G., Van der Hout, M. (2002). Red blood cell and plasma phospholipid arachidonic and docosahexaenoic acid levels at birth and cognitive development at 4 years of age. *Early Human Development*, **69**, 83-90.

Ginzberg, A., Cohen, M., Sod-Moriah, U., Shany, S., Rosenshtrauch, A., and Arad, S. (2000). Chickens fed with biomass of the red microalga *Porphyridium* sp. have reduced blood cholesterol level and modified fatty acid composition in egg yolk. *Journal of Applied Phycology*, **12**, 325-330.

Glazer, A.N. (1994). Phycobiliproteins - a family of valuable, widely used fluorophores. *Journal of Applied Phycology*, **6,** 105-112.

Gomes, E., Dias, J., Silva, P., Valente, L., Empis, J., Gouveia, L., Bowen, J., and Young, A. (2002). Utilization of natural and synthetic sources of carotenoids in the skin pigmentation of gilthead seabream (*Sparus aurata*). *European Food Research and Technology*, **214**, 287–293.

Gouveia, L., Veloso, V., Reis, A., Fernandes, H.L., Empis, J., and Novais, J.M. (1996a). *Chlorella vulgaris* used to colour egg yolk. *Journal of the Science of Food and Agriculture*, **70**, 167–172.

Gouveia, L., Veloso, V., Reis, A., Fernandes, H.L., Empis, J., and Novais, J.M. (1996b). Evolution of pigment composition in *Chlorella vulgaris. Bioresource Technology*, **57**, 157-163.

Gouveia, L., Gomes, E., and Empis, J. (1996c). Potential use of a microalgal (*Chlorella vulgaris*) in the pigmentation of rainbow trout (*Oncorhynchus mykiss*) muscle. *Zeitschrift fur Lebensmittel Untersuchung und Forschung*, **202**, 75-79.

Gouveia, L., Gomes, E., and Empis, J. (1997) Use of *Chlorella vulgaris* in rainbow trout, *Oncorhynchus mykiss*, diets to enhance muscle pigmentation. *Journal of Applied Aquaculture*, **7,** 61–70.

Gouveia, L., Choubert, G., Gomes, E., Rema, P., and Empis, J. (1998). Use of *Chlorella vulgaris* as a carotenoid source for salmonids: Effect of dietary lipid content on pigmentation, digestibility and muscular retention. *Aquaculture International*, **6**, 269–279.

Gouveia, L., Choubert, G., Gomes, E., Pereira, N., Santinha, J., and Empis, J. (2002). Pigmentation of gilthead seabream, *Sparus aurata* (Lin 1875), using *Chlorella vulgaris* (Chlorophyta, Volvocales) microalga. *Aquaculture. Research*, **33**, 987-993.

Gouveia, L., Rema, P., Pereira, O., and Empis, J. (2003). Colouring ornamental fish (*Cyprinus carpio* and *Carassius auratus*) with microalgal biomass. *Aquaculture Nutrition*, **9,** 123–129.

Gouveia, L., and Rema, P. (2005). Effect of microalgal biomass concentration and temperature on ornamental goldfish (*Carassius auratus*) skin pigmentation. *Aquaculture Nutrition*, **11**, 19-23.

Gouveia, L., Batista, A.P., Raymundo, A., Sousa, I., and Empis, J. (2006). *Chlorella vulgaris* and *Haematococcus pluvialis* biomass as colouring and antioxidant in food emulsions. *European Food Research and Technology*, **222**, 362-367.

Gouveia, L., Batista, A.P., Miranda, A., Empis, J., and Raymundo, A. (2007a). *Chlorella vulgaris* biomass used as colouring source in traditional butter cookies. *Innovative Food Science and Emerging Technologies*, **8,** 433-436.

Gouveia, L., Coutinho, C., Mendonça, E., Batista, A.P., Sousa, I., Bandarra, N.M., and Raymundo, A. (2007b). Sweet biscuits with Isochrysis galbana microalga biomass as a functional ingredient. *Journal of the Science of Food and Agriculture, In press.*

Granger, C., Barey, P., Combe, N., Veschambre, P., and Cansell, M. (2003). Influence of the fat characteristics on the physicochemical behaviour of oil-in-water emulsions based on milk proteins-glycerol esters mixtures. *Colloids and Surfaces B: Biointerfaces*, **32**, 353-363.

Gross, R., Gross, U., Ramirez, A., Cuadra, K., Collazos, C., and Feldheim, W. (1978). Nutritional tests with green *Scenedesmus* with health and malnourished children. *Archiv fur Hydrobiologie, Beihefte Ergebnisse der Limnologie*, **11**, 161-173.

Gross, U., and Gross, R. (1978). Acceptance and product selection of food fortified with the microalga *Scenedesmus. Archiv fur Hydrobiologie, Beihefte Ergebnisse der Limnologie*, **11**, 174-183.

Guill-Guerrero, J.L., Navarro-Juárez, R., López-Martínez, J.C., Campra-Madrid, P., Rebolloso-Fuentes, M.M. (2004). Functional properties of the biomass of the three microalgal species. *Journal of Food Engineering*, **65**, 511-517.

Hallmann, A. (2007). Algal transgenics and biotechnology. *Transgenic Plant Journal*, **1,** 81-98.

Han, L., Cui, J.G., and Huang, C.S. (2003). Bioactive polyhydroxy sterols and their sapogenins from marine organisms. *Acta Chimica Sinica*, **23**, 305-311.

Harel, M., and Clayton, D. (2004). *Feed formulation for terrestrial and aquatic animals*. US Patent 20070082008 (WO/2004/080196).

Harker, M., Tsavalos, A.J., and Young, A.J. (1996). Autotrophic growth and carotenoid production of *Haematococcus pluvialis* in a 30 liter air-lift photobioreactor. *Journal of Fermentation and Bioengineering*, **82**, 113-118.

Hintz, H.F., and Heitmann, H. (1967). Sewage-grown algae as a protein supplement for swine. *Animal Production*, **9**, 135-141.

Hintz, H.F., Heitmann, H., Weird, W.C., Torell, D.T., and Meyer, J.H. (1966). Nutritive value of algae grown on sewage. *Journal of Animal Science*, **25**, 675-681.

Hoseney, R.C., Wade, P., and Finley, J.W. (1988). Soft wheat products. In Y. Pomeranz (Ed), *Wheat Chemistry and Technology*, volume II, 3rd edition (pp. 407-456). American Association of Cereal Chemists Monograph Series.

Houzé, G., Cases, E., Colas, B., and Cayot, P. (2005). Viscoelastic properties of acid milk gel as affected by fat content at low level. *Interational Dairy Journal*, **15**, 1006-1016.

International Society for the Study of Fatty Acids and Lipids (ISSFAL) (2004). *Recommendations for dietary intake of polyunsaturated fatty acids in healthy adults.*

Jacket, J. (1974). Utilisations biologiques des *Spirulines. Bulletin de l'Académie veterinaire de France*, **47**, 133-143.

Jespersen, L., Strømdahl, L.D., Olsen, K., and Skibsted, L.H. (2005). Heat and light stability of three natural blue colorants for use in confectionery and beverages. *European Food Research and Technology*, **220**, 261–266

Kato, T. (1994). Blue pigment from *Spirulina. New Food Industry*, **29**, 17-21.

Kenneth, G.S. (1989). Pigmentation supplements for animal feed compositions. United States Patent, n° 4.871.551.

Kusmic, C., Barsacchi, R., Barsanti, L., Gualteri, P., and Passarelli, V. (1999). *Euglena gracilis* as a source of the antioxidant vitamin E. Effects of culture conditions in the wild strain and in the natural mutant WZSL. *Journal of Applied Phycology*, **10**, 555-559.

Lazo, P., Dinis, M.T., Holt, J., Faulk, C., and Arnold, C. (2000). Co-feeding microparticulate diets with algae: toward eliminating the need of zooplankton at first feeding in larval red drum (*Sciaenops ocellatus*). *Aquaculture*, **188**, 339-351.

Liu, C.P., and Lin, L.P. (2001). Ultrastructural study and lipid formation of *Isochrysis* sp. CCMP1324. *Botanical Bulletin Academia Sinica*, **42**, 207-214.

Lorenz, R.T., and Cysewski, G.R. (2000). Commercial potential for *Haematococcus* microalgae as a natural source of astaxanthin. *Trends in Biotechnology*, **18**, 160-167.

Luiten, E.E.M., Akkerman, I., Koulman, A., Kamersmans, P., Reith, H., Barbosa, M.J., Sipkema, D., and Wijffels, R.H. (2003). Realizing the promises of marine biotechnology. *Biomolecular Engineering*, **20**, 429-439.

Madiedo, G., and Sunde, M. (1964). The effect of algae, dried lake weed, alfalfa and ethoxyquin on yolk colour. *Poultry Science*, **43**, 1056-1061.

Marusich, W.L., De Ritter, E., and Bauerfeind, J.C. (1960). Evaluation of carotenoid pigments for colouring egg yolks. *Poultry Science*, **40**, 1338-1345.

McClements, D.J. (1999). *Food Emulsions: Principles, Practice and Techniques*. CRC Press.

Molina Grima, E., Belarbi, E.H., Acién Fernández, F.G. Robles Medina, A., and Chisti, Y. (2003). Recovery of microalgal biomass and metabolites: process options and economics. *Biotechnology Advances*, **20**, 491-515.

Muller-Feuga, A. (2000). The role of microalgae in aquaculture: situation and trends. *Journal of Applied Phycology*, **12**, 527-534.

Naas, K.E., Naess, T., and Harboe, T. (1992). Enhanced 1st feeding of Halibut Larvae (*Hippoglossus hippoglossus* L) in green water. *Aquaculture*, **105**, 143-156.

Natrah, F., Yosoff, F.M. Shariff, M., Abas, F., and Mariana, N.S. Screening of Malaysian indigenous microalgae for antioxidant properties and nutritional value. *Journal of Applied Phycology*, In press (DOI 10.1007/s10811-007-9192-5).

Nègre-Sadargues, G., Castillo, R., Petit, H., Sancé, S., Martinez, R.G., Milicua, J.C., Choubert, G., and Trilles, J.P. (1993). Utilization of synthetic carotenoids by the prawn *Panaeus japonicus* reared under laboratory conditions. *Aquaculture*, **110**, 151-159.

Nunes, M.C., Batista, P., Raymundo, A., Alves, M.M., and Sousa, I. (2003). Vegetable proteins and milk puddings. *Colloids and Surfaces B: Biointerfaces*, **31**, 21-29.

Nunes, M.C., Raymundo, A., and Sousa, I. (2006). Rheological behaviour and microstructure of pea protein / κ–carrageenan / starch gels with different setting conditions. *Food Hydrocolloids*, **20**, 106-113.

Olaizola, M. (2003). Commercial development of microalgal biotechnology: from the test tube to the marketplace. *Biomolecular Engineering*, **20**, 459-466.

Omenn, G.S., Goodman, G.E., Thornquist, M.D., Balmes, J., Cullen, M.R., Glass, A., Keogh, J.P., Meyskens, F.L., Valanis, B., Williams, J.H., Barnhart, S., Cherniack, M.G., Brodkin, C.A., and Hammar, S. (1996). Risk factors for lung cancer and for intervention effects in CARET, the beta-carotene and retinol efficacy trial. *Journal of the National Cancer Institute*, **88**, 1550-1559.

Oren, A. (2005). A hundred years of *Dunaliella* research: 1905-2005. *Saline Systems*, *1*, 2.

Ötles, S., and Pire, R. (2001). Fatty acid composition of *Chlorella* and *Spirulina* microalgae species. *Journal of AOAC International*, **84**, 1708-1714.

Özkanli, O., and Kaya, A. (2007). Storage stability of butter oils from sheep's non-pasteurized and pasteurized milk. *Food Chemistry*, **100**, 1026-1031.

Passos, R., Lagreze, F., Moriel, D., Bonfim, T., Gouveia, L., Maraschin, M. and Beirão, L. (2007). Pigmentation of pacific white shrimp (*Litopenaeus vannamei*, Boone, 1931) with carotenoids from natural sources (in preparation).

Piteira, F., Nunes, M.C., Raymundo, A., and Sousa, I. (2004). Effect of principal ingredients on quality of cookies with dietary fibre. In P.A. Williams, and G.O. Phillips (Eds), *Gums and Stabilisers for the Food Industry* **12** (pp. 475-483). Cambridge, UK: Royal Society of Chemistry.

Planas, M., and Cunha, I. (1999). Larviculture of marine fish: problems and perspectives. *Aquaculture*, **177**, 171-190.

Ponis, E., Probert, I., Véron, B., Mathieu, M., and Robert, R. (2006). New microalgae for the Pacific oyster *Crassostrea gigas* larvae. *Aquaculture*, **253**, 618-627.

Pulz, O., and Gross, W. (2004). Valuable products from biotechnology of microalgae. *Applied Microbiology and Biotechnology*, **65**, 635-648.

Radmer, R.J., and Parker, B.C. (1994). Commercial applications of algae: opportunities and constraints. *Journal of Applied Phycology*, **6**, 93-98.

Rampon, V., Brossard, C., Mouhous-Riou, N., Bousseau, B., Llamas, G., and Genot, C. (2004). The nature of the apolar phase influences the structure of the protein emulsifier in oil-in-water emulsions stabilized by bovine serum albumin. A front-surface fluorescence study. *Advances in Colloid and Interface Science*, **108-109**, 87-94.

Raja, R., Hemaiswarya, S., and Rengasamy, R. (2007). Exploitation of *Dunaliella* for β-carotene production. *Applied Microbiology and Biotechnology*, **74**, 517-523.

Raymundo, A., Franco, J.M., Empis, J., and Sousa, I. (2002). Optimization of the composition of low-fat oil-in-water emulsions stabilized by white lupin protein. *Journal of the American Oil Chemists Society*, **79**, 783-790.

Raymundo, A., Gouveia, L., Batista, A.P., Empis, J., and Sousa, I. (2005). Fat mimetic capacity of *Chlorella vulgaris* biomass in oil-in-water food emulsions stabilised by pea protein. *Food Research International*, **38**, 961-965.

Regulation (EC) No 1831/2003 of the European Parliament and the Council of 22 September 2003 on additives for use in animal nutrition. *Official Journal of the European Union*, L268/29,18.10.2003.

Regulation (EC) No 258/1997 of the European Parliament and the Council of 27 January 1997 on novel foods and food ingredients. *Official Journal of the European Union*, L043,14.02.1997.

Reis, A., Mendes, A., Lobo-Fernandes, H., Empis, J.A., and Maggiolly-Novais, J. (1998). Production, extraction and purification of phycobiliproteins from *Nostoc* sp. *Bioresource Technology*, **66**, 181-187.

Reitain, K., Rainuzzo, J.R., Oie, G., and Olsen, Y. (1997). A review of the nutritional effects of algae in marine fish larvae. *Aquaculture*, **155**, 207-221.

Rema, P., and Gouveia, L. (2005). Effect of various sources of carotenoids on survival and growth of goldfish (*Carassius auratus*) larvae and juvenile. *Journal of Animal and Veterinary Advances*, **4,** 654-658.

Reinehr, C.O., and Costa, J.A. (2006). Repeated batch cultivation of the microalga *Spirulina platensis*. *World Journal of Microbiology and Biotechnology*, **22**, 937-943.

Romay, C.H., Gonzalez, R., Ledon, N., Remirez, D., and Rimbau, V. (2003). Phycocyanin: a biliprotein with antioxidant, anti-inflammatory and neuroprotective effects. *Current Protein and Peptide Science*, **4**, 207-216.

Santipanichwong, R. and Suphantarika, M. (2007). Carotenoids as colorants in reduced-fat mayonnaise containing spent brewer's yeast b-glucan as a fat replacer. *Food Hydrocolloids*, **21**, 565-574.

Schilderman, P.A.E.L., ten Vaarwerk, F.J., Lutgerink, J.T., Van der Wurff, A., ten Hoor, F., and Kleinjans, J.C.S. (1995). Induction of oxidative DNA damage and early lesions in rat gastro-intestinal epithelium in relation to prostaglandin H synthese-mediated metabolism of butylated hydroxyanisole. *Food and Chemical Toxicology,* **33**, 99-109.

Segner, H., Arend, P., Von Poeppinghausen, K., and Schmidt, H. (1989). The effect of feeding astaxanthin to *Oreochromis* nioloticus and *Colisa* labiosa on the histology of the liver. *Aquaculture*, **79**, 381-390.

Sekar, S., and Chandramohan, M. (2007). Phycobiliproteins as a commodity: trends in applied research, patents and commercialization. *Journal of Applied Phycology, In Press* (doi: 10.1007/s10811-007-9188-1).

Shahidi, F., Metusalach, J., and Brown, J.A. (1998). Carotenoids pigments in seafoods and aquaculture. *Critical Reviews in Food Science and Nutrition*, **38**, 1-67.

Shimamatsu, H. (2004). Mass production of *Spirulina*, an edible microalga. *Hydrobiologia*, **512**, 39-44.

Sidhu, K.S. (2003). Health benefits and potential risks related to consumption of fish or fish oil. *Regulatory Toxicology and Pharmacology*, **38**, 336-344.

Simopoulos, A.P. (2002). The importance of the ratio of omega-6/omega-3 essential fatty acids. *Biomedicine and Pharmacotherapy*, **56**, 365-379.

Singh, S., Kate, B.N., and Banerjee, U.C. (2005). Bioactive compounds from Cyanobacteria and Microalgae: na overview. *Critical Reviews in Biotechnology*, **25**, 73-95.

Soletto, D., Binaghi, L., Lodi, A., Carvalho, J.C.M., and Converti, A. (2005). Batch and fed-batch cultivations of *Spirulina platensis* using ammonium sulphate and urea as nitrogen sources. *Aquaculture*, **243**, 217-224.

Sommer, T.R., Potts, W.T., and Morrissy, N.M. (1991). Utilization of microalgal astaxanthin by rainbow trout. *Aquaculture*, **94**, 79-88.

Sommer, T.R., Da Souza, F.M.L., and Morrisy, N.M. (1992). Pigmentation of adult rainbow trout, *Oncorhynchus mykiss*, using the green alga *Haematococcus pluvialis*. *Aquaculture*, **106**, 63-74.

Spolaore, P., Joannis-Cassan, C., Duran, E., and Isambert, A. (2006). Commercial applications of Microalgae- review. *Journal of Bioscience and Bioengineering*, **101**, 87-96.

Stahl, W.; Sies, H. (2005). Bioactivity and protective effects of natural carotenoids. *Biochimica et Biophysica Acta*, **1740**, 101-107.

Tang, H.F., Yi, Y.H., and Yao, X.S. (2002). Progress in the study of marine steroids. *Journal of Marine Drugs* (Chinese), **3**, 38-47.

Tapiero, H.; Townsend, D.M.; Tew, K.D. (2004). The role of carotenoids in the prevention of human pathologies. *Biomedicine and Pharmacotherapy*, **58**, 100-110.

Thies, F., Garry, J.M., Yagoob, P., Rerkasem, K., Williams, J., Shearman, C.P., Gallagher, P.J., Calder, P.C., and Grimble, R.F. (2003). Association of n-3 polyunsaturated fatty acids with stability of atherosclerotic plaques: a randomized controlled trial. *Lancet*, **361**, 477-485.

Torrissen, O.J. (1984). Pigmentation of salmonids-effect of carotenoids in eggs and star-feeding diet on survival and growth rate. *Aquaculture*, **43**, 185-193.

Torrissen, O.J., and Christiansen, R. (1995). Requirements of carotenoids in fish diets. *Journal of Ichthylogy*, **11**, 225-230.

Tsuchiya, M., Scita, G., Freisleben, H.L., Kagan, V.E., and Packer, L. (1992). Antioxidant radical-scavenging activity of carotenoids and etinoids compared to β-tocopherol. *Methods of Enzymology*, **213**, 460 – 472.

Van den Berg, H, Faulks, R., Granado, H.F., Hirschberg, J., Olmedilla, B., Sandmann, G., Southon, S., and Stahl, W. (2000). The potential for the improvement of carotenoid levels in foods and the likely systemic effects. *Journal of the Science of Food and Agriculture*, **80**, 880-912.

Vélez-Ruiz, J.F., González-Tomás, L., and Costell, E. (2005). Rheology of dairy custard model systems: influence of milk-fat and hydrocolloid type. *European Food Research and Technology*, **221**, 342-347.

Volkman, J.K., Jeffery, S.W., Nichols, P.D., Rogers, G.I., and Garland, C.D. (1989). Fatty acid and lipid composition of 10 species of microalgae used in mariculture. *Journal of Experimental Marine Biology and Ecology*, **128**, 219-240.

Volkman, J.K. (2003). Sterols in microrganisms. *Applied Microbiology and Biotechnology*, **60**, 495-506.

Waldenstedt, L., Inborr, J., Hansson, I., and Elwinger, K. (2003). Effects of astaxanthin-rich algal meal (*Haematococcus pluvialis*) on growth performance, caecal campylobacter and clostridial counts and tissue astaxanthin concentration of broiler chickens. *Animal Feed Science and Technology*, **18**, 119-132.

Wang, C., Harris, W.S., Chung, M., Lichtenstein, A.H., Balk, E.M., Kupelnick, B., Jordan, H.S., and Lau, J. (2006). n-3 Fatty acids from fish or fish-oil supplements, but not a-linolenic acid, benefit cardiovascular disease outcomes in primary and secondary prevention studies: a systematic review. *American Journal of Clinical Nutrition*, **84**, 5–17.

Wikfors, G.H., Gladu, P.K., and Patterson, G.W. (1991). In search of the ideal diet for oysters: recent progress with emphasis on sterols. *Journal of Shellfish Research*, **10**, 292-301.

Wikfors, G.H., Ferris, G.E., and Smith, B.C. (1992). The relationship between gross biochemical composition of cultured algal foods and growth of the hard clam, *Mercenaria mercenaria* (L.). *Aquaculture*, **108**, 135-154.

Wroble, M., Mash, C., Williams, L., and McCall, R.B. (2002). Should long chain polyunsaturated fatty acids be added to infant formula to promote development? *Applied Developments in Phychology*, **23**, 99-112.

Xu, Z., Yan, X., Pei, L, Luo, Q., and Xu, J. Changes in fatty acids and sterols during batch growth of *Pavlova viridis* in photobioreactor. *Journal of Applied Phycology*, In press (DOI: 10.1007/s10811-007-9230-3).

Yamaguchi, K. (1997). Recent advances in microalgal bioscience in Japan, with special reference to utilization of biomass and metabolites: a review. *Journal of Applied Phycology*, **8,** 487-502.

Yap, T.N., Wu, J.F., Pond, W.G., and Krook, L. (1982). Feasibility of feeding *Spirulina maxima*, *Arthrospira platensis* or *Chlorella* sp. to pigs weaned to a dry diet at 4 to 8 days of age. *Nutritional Reports International,* **35**, 543-552.

Yen, W.J., and Chen, B.H. (1995). Isolation of xanthophylls from Taiwanese orange peels and their effects on the oxidation stability of soybean oil. *Food Chemistry*, **53**, 417-425.

Yokoyama, A., and Miki, W. (1995). Composition and presumed bio-synthetic pathway of carotenoids in the astaxanthin-producing bacterium *Agrobacterium aurantiacum. FEMS Microbiology Letters*, **128**, 139-144.

Yuang, J.P., Chen, F. (2000). Purification of trans-astaxanthin from a high-yielding astaxanthin ester-producing strain of the microalga *Haematococcus pluvialis*. *Food Chemistry*, **68**, 443-448.

INDEX

A

B

C

D

E

F

G

H

I

J

K

L

M

N

Q

R

S

T

U

V

W

X

Y

Z